Universitext

Universitext

Universitext is a series of textbooks that presents material from a wide variety of mathematical disciplines at master's level and beyond. The books, often well class-tested by their author, may have an informal, personal even experimental approach to their subject matter. Some of the most successful and established books in the series have evolved through several editions, always following the evolution of teaching curricula, to very polished texts.

Thus as research topics trickle down into graduate-level teaching, first textbooks written for new, cutting-edge courses may make their way into *Universitext*.

For further volumes:
www.springer.com/series/223

Luis Barreira

Ergodic Theory, Hyperbolic Dynamics and Dimension Theory

 Springer

Luis Barreira
Departamento de Matemática
Instituto Superior Técnico
Lisboa
Portugal

ISBN 978-3-642-28089-4   ISBN 978-3-642-28090-0 (eBook)
DOI 10.1007/978-3-642-28090-0
Springer Heidelberg New York Dordrecht London

Library of Congress Control Number: 2012937287

Mathematics Subject Classification (2010): 37AXX, 37DXX, 37C45

Printed on acid-free paper

Springer is part of Springer Science+Business Media (www.springer.com)

To Claudia

Preface

This book is an introduction to the interplay of three main areas of research: *ergodic theory*, *hyperbolic dynamics*, and *dimension theory* of dynamical systems. This includes an introduction to the thermodynamic formalism, which is an important tool in dimension theory.

My main aim was to provide in a single volume a rigorous self-contained introduction to dimension theory of hyperbolic dynamics, including sufficiently high-level introductions to ergodic theory and the thermodynamic formalism. This caused that several topics of ergodic theory and hyperbolic dynamics had to be excluded, essentially to keep the size of the book under control. However, it should be emphasized that the same happened to several topics of dimension theory while making an effort to reach a good compromise between the various areas. On the other hand, it was necessary to include topics of ergodic theory and hyperbolic dynamics that are often absent in introductory texts, such as the construction of Markov partitions for repellers and an introduction to the thermodynamic formalism.

The book is directed primarily to graduate students interested in dynamical systems, as well as researchers in other areas who wish to learn ergodic theory or dimension theory of hyperbolic dynamics at an intermediate level and in a sufficiently detailed manner. In particular, the text can be used as a basis for a graduate course on ergodic theory and the thermodynamic formalism (using Chaps. 2–5) and for a graduate course on dimension theory of hyperbolic dynamics (using Chaps. 6–9, eventually referring to Chaps. 2–5 for any prerequisites).

The book can also be used for independent study: it is self-contained, and with the exception of some basic well-known statements and results from other areas, all statements are included with detailed proofs. Moreover, each chapter can essentially be read independently. I only assume some familiarity with basic material from measure theory and integration theory, which anyway is recalled in Appendix A.

The material is divided into four parts:

* Part I is dedicated to the foundations of ergodic theory and its interplay with symbolic dynamics and topological dynamics. In Chap. 2, we introduce the basic

notions and results of ergodic theory, including Poincaré's recurrence theorem and Birkhoff's ergodic theorem. We also give a large number of examples. In Chap. 3, we discuss additional topics of ergodic theory, including the existence of invariant measures for a continuous transformation of a compact metric space.

* Part II is an introduction to entropy theory and the thermodynamic formalism. In Chap. 4, we introduce the notions of metric entropy and topological entropy. We also establish the Shannon–McMillan–Breiman theorem and the variational principle for the topological entropy. Chapter 5 is an introduction to the thermodynamic formalism. In particular, we establish the variational principle for the topological pressure.

* Part III is dedicated to hyperbolic dynamics. In Chap. 6, we start by discussing some basic properties of hyperbolic sets. In addition, we discuss in detail some properties of the Smale horseshoe and of the hyperbolic automorphisms of $\mathbb{T}^2$. We also construct Markov partitions for any repeller. In Chap. 7, after establishing the existence of stable and unstable manifolds, we use the shadowing property to construct Markov partitions for any locally maximal hyperbolic set.

* Finally, Part IV is an introduction to dimension theory of hyperbolic dynamics. In Chap. 8 we give an introduction to the basic notions of dimension theory. In particular, we introduce the notions of Hausdorff dimension and box dimension. We also show how pointwise dimension can be used to estimate the dimension of a measure. In Chap. 9, we study the dimension of repellers and hyperbolic sets for conformal transformations using Markov partitions.

In addition, the book contains more than 150 exercises of variable level of difficulty.

There are no words that can adequately express my gratitude to Claudia Valls for her help, patience, encouragement, and inspiration, without which it would be impossible for this book to exist.

Lisbon, Portugal   *Luís Barreira*
December 2011

Contents

Chapter 1
Introduction

This chapter is a first introduction to the rich interplay of *ergodic theory*, *hyperbolic dynamics*, and *dimension theory*—the three main areas considered in the book. The main aim is to describe briefly the topics under consideration and illustrate their importance for the theory of dynamical systems. The exposition is nontechnical but also rigorous.

1.1 Dynamical Systems and Hyperbolicity

One of the paradigms of the theory of dynamical systems is that the local instability of trajectories influences the global behavior of the system and opens the way to the existence of stochastic behavior. Mathematically, the instability of trajectories corresponds to some amount of hyperbolicity.

Let $f: M \to M$ be a diffeomorphism and let $\Lambda \subset M$ be a compact f-invariant set. We say that Λ is a *hyperbolic set* for f if for every point $x \in \Lambda$ there exists a decomposition of the tangent space

$$T_x M = E^s(x) \oplus E^u(x)$$

satisfying

$$d_x f E^s(x) = E^s(f(x)) \quad \text{and} \quad d_x f E^u(x) = E^u(f(x)),$$

and there exist constants $\lambda \in (0, 1)$ and $c > 0$ such that

$$\|d_x f^n | E^s(x)\| \le c\lambda^n \quad \text{and} \quad \|d_x f^{-n} | E^u(x)\| \le c\lambda^n$$

for every $x \in \Lambda$ and $n \in \mathbb{N}$.

L. Barreira, *Ergodic Theory, Hyperbolic Dynamics and Dimension Theory*, Universitext,
DOI 10.1007/978-3-642-28090-0_1, © Springer-Verlag Berlin Heidelberg 2012

A hyperbolic set possesses a very rich structure and in particular families of stable and unstable manifolds. Given $x \in M$ and $\varepsilon > 0$, we consider the sets

$$V^s(x) = \{y \in B(x, \varepsilon) : d(f^n(y), f^n(x)) < \varepsilon \text{ for every } n > 0\}$$

and

$$V^u(x) = \{y \in B(x, \varepsilon) : d(f^n(y), f^n(x)) < \varepsilon \text{ for every } n < 0\},$$

where d is the distance on M and $B(x, \varepsilon) \subset M$ is an open ball of radius ε.

Theorem 1.1 (Hadamard–Perron). *If Λ is a hyperbolic set for a C^1 diffeomorphism, then there exists $\varepsilon > 0$ such that for each $x \in \Lambda$, the sets $V^s(x)$ and $V^u(x)$ are manifolds containing x such that*

$$T_x V^s(x) = E^s(x) \quad and \quad T_x V^u(x) = E^u(x).$$

The manifolds $V^s(x)$ and $V^u(x)$ are called, respectively, *stable* and *unstable manifolds*. Under the assumptions of Theorem 1.1, one can also show that the sizes of $V^s(x)$ and $V^u(x)$ are uniformly bounded away from zero, that is, there exists $\gamma = \gamma(\varepsilon) > 0$ such that

$$V^s(x) \supset B^s(x, \gamma) \quad and \quad V^u(x) \supset B^u(x, \gamma)$$

for every $x \in \Lambda$, where $B^s(x, \gamma)$ and $B^u(x, \gamma)$ are the open balls of radius γ with respect to the distances induced by d, respectively, on $V^s(x)$ and $V^u(x)$. The continuous dependence of the spaces $E^s(x)$ and $E^u(x)$ in $x \in \Lambda$ guarantees that there exists $\delta = \delta(\varepsilon) > 0$ such that if $d(x, y) < \delta$ for some points $x, y \in \Lambda$, then the intersection $V^s(x) \cap V^u(y)$ consists exactly of one point.

When the whole manifold M is a hyperbolic set for f, we say that f is an *Anosov diffeomorphism*. This class of diffeomorphisms was introduced and studied by Anosov in [4]. The notion of hyperbolic set was introduced by Smale in his seminal paper [99]. In a certain sense, Anosov diffeomorphisms and more generally diffeomorphisms with a hyperbolic set possess the strongest possible hyperbolicity.

1.2 Ergodic Theory and Nontrivial Recurrence

Now we introduce the notion of invariant measure, which is another fundamental departure point for the study of stochastic behavior. Namely, the existence of a finite invariant measure ensures that there is a *nontrivial recurrence*.

Let $T : X \to X$ be a measurable transformation. We say that a measure μ in X is *T-invariant* if

$$\mu(T^{-1}A) = \mu(A)$$

for every measurable set $A \subset X$. The study of measure-preserving transformations is the main theme of ergodic theory.

In order to describe rigorously the concept of nontrivial recurrence, we recall one of the basic but also fundamental results of ergodic theory—Poincaré's recurrence theorem. It states that any dynamics preserving a finite measure exhibits a nontrivial recurrence in any set A of positive measure, in the sense that the orbit of almost every point in A returns infinitely often to A.

Theorem 1.2 (Poincaré's recurrence theorem). *Let $T: X \to X$ be a measurable transformation and let μ be a T-invariant finite measure in X. If $A \subset X$ is a measurable set of positive measure, then*

$$\mathrm{card}\{n \in \mathbb{N} : T^n(x) \in A\} = \infty$$

for μ-almost every $x \in A$.

A modified version of Theorem 1.2 was first established by Poincaré in his seminal memoir on the three-body problem [78].

The simultaneous existence of *hyperbolicity* and *nontrivial recurrence* causes a very rich orbit structure (see [11, 43]). Roughly speaking, the nontrivial recurrence implies that there exist orbits returning arbitrarily close to the initial point. On the other hand, the existence of stable and unstable manifolds at these points together with their transversality often causes the existence of transverse homoclinic points, thus leading to (topological) Smale horseshoes and an enormous complexity. In other words, the ergodic theory of hyperbolic dynamics, bringing together hyperbolicity and nontrivial recurrence, is a natural playground for the study of stochastic behavior. Unfortunately, real-life situations are often much more complicated. In particular, the notion of hyperbolicity in Sect. 1.1 can be too stringent for the dynamics, thus leading to the consideration of weaker notions such as nonuniform hyperbolicity (see [10, 11]).

1.3   Hyperbolicity and Dimension Theory

We describe briefly in this section the relation between hyperbolicity and dimension.

We first introduce some basic notions of dimension theory. Let X be a separable metric space. Given $Z \subset X$ and $\alpha \in \mathbb{R}$, we define

$$m(Z, \alpha) = \lim_{\varepsilon \to 0} \inf_{\mathcal{U}} \sum_{U \in \mathcal{U}} (\mathrm{diam}\, U)^\alpha,$$

where the infimum is taken over all finite or countable covers of Z by open sets of diameter at most ε. The *Hausdorff dimension* of Z is defined by

$$\dim_H Z = \inf\{\alpha : m(Z, \alpha) = 0\}.$$

The *lower* and *upper box dimensions* of Z are defined, respectively, by

$$\underline{\dim}_B Z = \liminf_{\varepsilon \to 0} \frac{\log N(Z, \varepsilon)}{-\log \varepsilon} \quad \text{and} \quad \overline{\dim}_B Z = \limsup_{\varepsilon \to 0} \frac{\log N(Z, \varepsilon)}{-\log \varepsilon},$$

where $N(Z, \varepsilon)$ denotes the number of balls of radius ε needed to cover Z. It is easy to verify that

$$\dim_H Z \le \underline{\dim}_B Z \le \overline{\dim}_B Z. \tag{1.1}$$

In general, these inequalities may be strict, and the coincidence of the three dimensions is a relatively rare phenomenon (see [8, 29, 73]).

Now let μ be a finite measure in X. The *Hausdorff dimension* and the *lower* and *upper box dimensions* of μ are defined, respectively, by

$$\dim_H \mu = \lim_{\delta \to 0} \inf\{\dim_H Z : \mu(Z) \ge \mu(X) - \delta\},$$

$$\underline{\dim}_B \mu = \lim_{\delta \to 0} \inf\{\underline{\dim}_B Z : \mu(Z) \ge \mu(X) - \delta\},$$

$$\overline{\dim}_B \mu = \lim_{\delta \to 0} \inf\{\overline{\dim}_B Z : \mu(Z) \ge \mu(X) - \delta\}.$$

In general, these quantities need not coincide, respectively, with the Hausdorff dimension and the lower and upper box dimensions of the support of the measure, and thus, they contain additional information about the way in which μ is distributed on its support. It follows from (1.1) that

$$\dim_H \mu \le \underline{\dim}_B \mu \le \overline{\dim}_B \mu.$$

As in (1.1), in general, these inequalities may be strict. The following criterion for equality was established by Young in [105]: if μ is a finite measure in X and

$$\lim_{r \to 0} \frac{\log \mu(B(x, r))}{\log r} = d \tag{1.2}$$

for μ-almost every $x \in X$, then

$$\dim_H \mu = \underline{\dim}_B \mu = \overline{\dim}_B \mu = d.$$

The limit in (1.2), when it exists, is called the *pointwise dimension* of μ at x.

There is a vast theory relating hyperbolicity and dimension. In particular, the following result is due to Barreira, Pesin and Schmeling [12].

Theorem 1.3. *Let $f: M \to M$ be a $C^{1+\alpha}$ diffeomorphism, for some $\alpha > 0$, and let μ be a finite f-invariant measure whose support is a hyperbolic set. Then the limit*

$$\lim_{r \to 0} \frac{\log \mu(B(x, r))}{\log r} \tag{1.3}$$

exists for μ-almost every $x \in M$.

Theorem 1.3 also builds on seminal work of Ledrappier and Young in [54]. It plays a role in dimension theory that is similar to the central role played by the Shannon–McMillan–Breiman theorem (Theorem 4.4) in entropy theory. Under the assumptions of Theorem 1.3, when μ is ergodic, it follows from the above criterion of Young that

$$\dim_H \mu = \underline{\dim}_B \mu = \overline{\dim}_B \mu.$$

In fact, the almost everywhere existence of the limit in (1.3) guarantees the coincidence not only of these three dimensions but also of many other characteristics of dimension type (see [12, 73, 105]).

1.4 Geometric Constructions and Limit Sets

There are important differences between the dimension theory of invariant sets and the dimension theory of invariant measures. In particular, while virtually all dimensional characteristics of invariant measures on hyperbolic sets coincide, the study of the dimension of hyperbolic sets revealed that different dimensional characteristics frequently depend on properties of geometrical and number-theoretical nature. This justifies the interest in simpler models.

We start with the description of a geometric construction in $\mathbb{R}$. Consider constants $\lambda_1, \dots, \lambda_p \in (0, 1)$ and disjoint closed intervals $\Delta_1, \dots, \Delta_p \subset \mathbb{R}$ of lengths $\lambda_1, \dots, \lambda_p$. For each $k = 1, \dots, p$, choose again p disjoint closed intervals $\Delta_{k1}, \dots, \Delta_{kp} \subset \Delta_k$ of lengths $\lambda_k \lambda_1, \dots, \lambda_k \lambda_p$. Iterating this procedure, for each $n \in \mathbb{N}$, we obtain p^n disjoint closed intervals $\Delta_{i_1 \cdots i_n}$ of lengths $\prod_{k=1}^n \lambda_{i_k}$. The limit set of the construction is defined by

$$F = \bigcap_{n=1}^{\infty} \bigcup_{i_1 \cdots i_n} \Delta_{i_1 \cdots i_n}. \tag{1.4}$$

In [62], Moran showed that $\dim_H F = s$, where s is the unique root of the equation

$$\sum_{k=1}^{p} \lambda_k^s = 1. \tag{1.5}$$

It is remarkable that the Hausdorff dimension of the set F does not depend on the location of the intervals $\Delta_{i_1 \cdots i_n}$ but only on their lengths. Pesin and Weiss [75] extended the result of Moran to arbitrary symbolic dynamics in $\mathbb{R}^m$, using the thermodynamic formalism (see Sect. 1.5).

We also consider geometric constructions described in terms of a general symbolic dynamics. Given $p \in \mathbb{N}$, consider the family of sequences $X_p = \{1, \dots, p\}^{\mathbb{N}}$ and equip this space with the distance

$$d(\omega, \omega') = \sum_{k=1}^{\infty} e^{-k} |\omega_k - \omega_k'|. \tag{1.6}$$

We also consider the shift map $\sigma: X_p \to X_p$ such that $(\sigma\omega)_n = \omega_{n+1}$ for each $n \in \mathbb{N}$. A *geometric construction* in $\mathbb{R}^m$ is defined by:

1. A compact set $Q \subset X_p$ such that $\sigma^{-1}Q \supset Q$ for some $p \in \mathbb{N}$.
2. A decreasing sequence of compact sets $\Delta_{\omega_1\cdots\omega_n} \subset \mathbb{R}^m$ for each $\omega \in Q$ such that $\operatorname{diam}\Delta_{\omega_1\cdots\omega_n} \to 0$ where $n \to \infty$.

We also assume that

$$\operatorname{int}\Delta_{i_1\cdots i_n} \cap \operatorname{int}\Delta_{j_1\cdots j_n} \neq \varnothing$$

whenever $(i_1\cdots i_n) \neq (j_1\cdots j_n)$. The *limit set* F of the geometric construction is defined by (1.4) with the union taken over all vectors $(i_1, \ldots, i_n)$ such that $i_k = \omega_k$ for each $k = 1, \ldots, n$ and some $\omega \in Q$.

Now we consider the particular case when all sets $\Delta_{i_1\cdots i_n}$ are balls. Write $r_{i_1\cdots i_n} = \operatorname{diam}\Delta_{i_1\cdots i_n}$. The following result was established by Barreira [7].

Theorem 1.4 (Dimension of the limit set). *For a geometric construction modeled by $Q \subset X_p$ such that all sets $\Delta_{i_1\cdots i_n}$ are balls, if there exists a constant $\delta > 0$ such that*

$$r_{i_1\cdots i_{n+1}} \geq \delta r_{i_1\cdots i_n} \quad \text{and} \quad r_{i_1\cdots i_{n+m}} \leq r_{i_1\cdots i_n} r_{i_{n+1}\cdots i_m}$$

for every $(i_1 i_2 \cdots) \in Q$ and $n, m \in \mathbb{N}$, then

$$\dim_H F = \underline{\dim}_B F = \overline{\dim}_B F = s,$$

where s is the unique root of the equation

$$\lim_{n\to\infty} \frac{1}{n} \log \sum_{i_1\cdots i_n} r_{i_1\cdots i_n}^s = 0.$$

This result contains as particular cases the results of Moran and of Pesin and Weiss (see also Sect. 1.5), for which

$$r_{i_1\cdots i_n} = \prod_{k=1}^{n} \lambda_{i_k}$$

for some numbers $\lambda_1, \ldots, \lambda_p$. The value of the dimension is also independent of the location of the sets $\Delta_{i_1\cdots i_n}$. The proof of Theorem 1.4 is based on a nonadditive version of the thermodynamic formalism (see [9]).

1.5 Classical Thermodynamic Formalism

This section is a brief introduction to the classical thermodynamic formalism, considering only the case of symbolic dynamics.

Let $Q \subset X_p$ be a compact set such that $\sigma^{-1}Q \supset Q$. Given a continuous function $\varphi: Q \to \mathbb{R}$, the *topological pressure* of φ (with respect to σ) is defined by

$$P(\varphi) = \lim_{n \to \infty} \frac{1}{n} \log \sum_{i_1 \cdots i_n} \exp \sup \left(\sum_{k=0}^{n-1} \varphi \circ \sigma^k \right), \tag{1.7}$$

where the supremum is taken over all sequences $(j_1 j_2 \cdots) \in Q$ such that $(j_1 \cdots j_n) = (i_1 \cdots i_n)$. The *topological entropy* of $\sigma | Q$ is defined by

$$h(\sigma | Q) = P(0).$$

Topological pressure is the most basic notion of the thermodynamic formalism. It was introduced by Ruelle in [83] for expansive transformations and by Walters in [102] in the general case.

Now we present an equivalent description of the topological pressure. Let μ be a σ-invariant probability measure in Q and let ξ be a finite or countable partition of Q into measurable sets. We write

$$H_\mu(\xi) = - \sum_{C \in \xi} \mu(C) \log \mu(C),$$

with the convention that $0 \log 0 = 0$. The *Kolmogorov–Sinai entropy* of $\sigma | Q$ with respect to μ is defined by

$$h_\mu(\sigma | Q) = \sup_\xi \lim_{n \to \infty} \frac{1}{n} H_\mu \left(\bigvee_{k=0}^{n-1} \sigma^{-k} \xi \right),$$

where $\bigvee_{k=0}^{n-1} \sigma^{-k} \xi$ is the partition of Q formed by the sets

$$C_{i_1 \cdots i_n} = \bigcap_{k=0}^{n-1} \sigma^{-k} C_{i_{k+1}}$$

with $C_{i_1}, \ldots, C_{i_n} \in \xi$. The topological pressure satisfies the *variational principle*

$$P(\varphi) = \sup_\mu \left\{ h_\mu(\sigma | Q) + \int_Q \varphi \, d\mu \right\}, \tag{1.8}$$

where the supremum is taken over all σ-invariant probability measures μ in Q. A σ-invariant probability measure in Q is called an *equilibrium measure* for φ (with respect to $\sigma | Q$) if the supremum in (1.8) is attained at this measure, that is, if

$$P(\varphi) = h_\mu(\sigma | Q) + \int_Q \varphi \, d\mu.$$

There is a very close relation between dimension theory and the thermodynamic formalism. To illustrate this relation, we consider numbers $\lambda_1, \ldots, \lambda_p$ and the function $\varphi \colon Q \to \mathbb{R}$ defined by

$$\varphi(i_1 i_2 \cdots) = \log \lambda_{i_1}. \tag{1.9}$$

We have

$$P(s\varphi) = \lim_{n\to\infty} \frac{1}{n} \log \sum_{i_1\cdots i_n} \exp\left(s\sum_{k=1}^{n} \log \lambda_{i_k}\right)$$

$$= \lim_{n\to\infty} \frac{1}{n} \log \sum_{i_1\cdots i_n} \prod_{k=1}^{n} \lambda_{i_k}{}^{s}$$

$$= \lim_{n\to\infty} \frac{1}{n} \log \left(\sum_{i=1}^{p} \lambda_i{}^{s}\right)^{n}$$

$$= \log \sum_{i=1}^{p} \lambda_i^{s}.$$

Therefore, (1.5) is equivalent to

$$P(s\varphi) = 0. \tag{1.10}$$

Equation (1.10) was introduced by Bowen in [21]. It has a rather universal character: virtually all known equations used to compute or to estimate the dimension of an invariant set are particular cases of (1.10) or of an appropriate generalization (see [9]). For example, the result of Pesin and Weiss in [75] mentioned in Sect. 1.4 can be formulated as follows:

Theorem 1.5 (Dimension of the limit set). *For a geometric construction modeled by $Q \subset X_p$ such that the sets $\Delta_{i_1\cdots i_n}$ are balls of diameter $\prod_{k=1}^{n} \lambda_{i_k}$, we have*

$$\dim_H F = \underline{\dim}_B F = \overline{\dim}_B F = s,$$

where s is the unique root of the equation $P(s\varphi) = 0$ with φ as in (1.9).

1.6 Dimension Theory in Hyperbolic Dynamics

As observed above, one of the main motivations for the study of geometric constructions is the study of the dimension of hyperbolic sets. This approach can be effected using Markov partitions.

We first consider expanding maps. These are a noninvertible version of diffeomorphisms with a hyperbolic set. Let $g\colon M \to M$ be a differentiable map of a smooth manifold. We also consider a g-invariant compact set $J \subset M$. We say that J

is a *repeller* of g and that g is an *expanding map* on J if there exist constants $c > 0$ and $\beta > 1$ such that

$$\|d_x g^n v\| \geq c\beta^n \|v\|$$

for every $n \in \mathbb{N}$, $x \in J$ and $v \in T_x M$.

Now let J be a repeller of the map g. A finite cover of J by nonempty closed sets $R_1, \ldots, R_p$ is called a *Markov partition* of J if:

1. $\overline{\text{int } R_i} = R_i$ for each i.
2. $\text{int } R_i \cap \text{int } R_j = \varnothing$ whenever $i \neq j$.
3. $g(R_i) \supset R_j$ whenever $g(\text{int } R_i) \cap \text{int } R_j \neq \varnothing$.

The interior of each set R_i is computed with respect to the topology induced on J. Any repeller has Markov partitions of arbitrarily small diameter (see [85]).

Now we use Markov partitions to model repellers by geometric constructions. Let J be a repeller of a map g and let $R_1, \ldots, R_p$ be the elements of a Markov partition of J. We consider the $p \times p$ matrix $A = (a_{ij})$ with entries

$$a_{ij} = \begin{cases} 1 & \text{if } g(\text{int } R_i) \cap \text{int } R_j \neq \varnothing, \\ 0 & \text{if } g(\text{int } R_i) \cap \text{int } R_j = \varnothing. \end{cases}$$

Consider the space of sequences $X_p = \{1, \ldots, p\}^{\mathbb{N}}$ and the shift map $\sigma: X_p \to X_p$ (see Sect. 1.4). We call *topological Markov chain* with *transition matrix* A to the restriction of the shift map σ to the set

$$X_A = \{(i_1 i_2 \cdots) \in X_p : a_{i_n i_{n+1}} = 1 \text{ for every } n \in \mathbb{N}\}.$$

One can define a coding map $\chi: X_A \to J$ by

$$\chi(i_1 i_2 \cdots) = \bigcap_{k=0}^{\infty} g^{-k} R_{i_{k+1}}.$$

The map χ is surjective, satisfies

$$\chi \circ \sigma = g \circ \chi, \tag{1.11}$$

and is Hölder continuous (with respect to the distance in (1.6)). Even though in general χ is not invertible, identity (1.11) allows one to see χ as a dictionary transferring the symbolic dynamics $\sigma | X_A$ (and often the results at the level of symbolic dynamics) to the dynamics of g on J. In particular, the repeller can be seen as the limit set of a geometric construction (see Sect. 1.4), defined by the sets

$$\Delta_{i_1 \cdots i_n} = \bigcap_{k=0}^{n-1} g^{-k} R_{i_{k+1}}.$$

The map g is said to be *conformal* on J if $d_x g$ is a multiple of an isometry for every $x \in J$. When J is a repeller of a conformal map of class $C^{1+\alpha}$, one can show that there is a constant $C > 0$ such that

$$C^{-1} \prod_{k=0}^{n-1} \exp \varphi(g^k(x)) \leq \operatorname{diam} \Delta_{i_1 \cdots i_n} \leq C \prod_{k=0}^{n-1} \exp \varphi(g^k(x))$$

for every $x \in \Delta_{i_1 \cdots i_n}$, where the function $\varphi \colon J \to \mathbb{R}$ is defined by

$$\varphi(x) = -\log \|d_x g\|.$$

The topological pressure defined by (1.7) with $Q = X_A$ can be used to compute the dimension of the repeller.

Theorem 1.6 (Dimension of conformal repellers). *If J is a repeller of a $C^{1+\alpha}$ transformation g, for some $\alpha > 0$, such that g is conformal on J, then*

$$\dim_H J = \underline{\dim}_B J = \overline{\dim}_B J = s,$$

where s is the unique root of the equation $P(s\varphi) = 0$.

Ruelle showed in [85] that $\dim_H J = s$. The coincidence between the Hausdorff and box dimensions is due to Falconer [31]. It was also shown by Ruelle in [85] that if μ is the unique equilibrium measure of $-s\varphi$, then

$$\dim_H J = \dim_H \mu. \tag{1.12}$$

In fact, he showed that μ is equivalent to the s-dimensional Hausdorff measure on J.

Now we move to the study of the dimension of hyperbolic sets. Let Λ be a hyperbolic set for a diffeomorphism $f \colon M \to M$. We consider the functions $\varphi_s \colon \Lambda \to \mathbb{R}$ and $\varphi_u \colon \Lambda \to \mathbb{R}$ defined by

$$\varphi_s(x) = \log \|d_x f | E^s(x)\| \quad \text{and} \quad \varphi_u(x) = -\log \|d_x f | E^u(x)\|.$$

The set Λ is said to be *locally maximal* if there exists an open neighborhood U of Λ such that

$$\Lambda = \bigcap_{n \in \mathbb{Z}} f^n(U).$$

The following result is a version of Theorem 1.6 for hyperbolic sets:

Theorem 1.7 (Dimension of hyperbolic sets on surfaces). *If Λ is a locally maximal hyperbolic set for a C^1 surface diffeomorphism, and $\dim E^s(x) = \dim E^u(x) = 1$ for every $x \in \Lambda$, then*

$$\dim_H \Lambda = \underline{\dim}_B \Lambda = \overline{\dim}_B \Lambda = t_s + t_u,$$

where t_s and t_u are the unique real numbers such that

$$P(t_s\varphi_s) = P(t_u\varphi_u) = 0.$$

It follows from work of McCluskey and Manning [59] that $\dim_H \Lambda = t_s + t_u$. The coincidence between the Hausdorff and box dimensions is due to Takens [101] for C^2 diffeomorphisms and to Palis and Viana [68] in the general case. The result in Theorem 1.7 can be readily extended to the more general case of conformal maps. We say that $f: M \to M$ is *conformal* on a hyperbolic set Λ if $d_x f | E^s(x)$ and $d_x f | E^u(x)$ are multiples of isometries for every $x \in \Lambda$ (e.g., if M is a surface and $\dim E^s(x) = \dim E^u(x) = 1$ for every $x \in \Lambda$, then f is conformal on Λ).

One can also ask whether there is an appropriate generalization of property (1.12) in the present context, that is, whether there exists an invariant measure μ supported on Λ such that $\dim_H \Lambda = \dim_H \mu$. The answer to this question is almost always negative. More precisely, McCluskey and Manning [59] showed that such a measure exists if and only if there is a continuous function $\psi: \Lambda \to \mathbb{R}$ such that

$$t_s\varphi_s - t_u\varphi = \psi \circ f - f$$

on Λ. By Livschitz's theorem (see, e.g., [44]), this happens if and only if

$$\|d_x f | E^s(x)\|^{t_s} \|d_x f | E^u(x)\|^{t_u} = 1$$

for every $x \in \Lambda$ and $n \in \mathbb{N}$ such that $f^n(x) = x$.

The study of the dimension of repellers and hyperbolic sets for nonconformal maps is much less developed than the corresponding study for conformal maps. The main difficulty is the possibility of existence of distinct Lyapunov exponents associated to directions that may change from point to point. There exist however some partial results, for certain classes of repellers and hyperbolic sets, starting essentially with the seminal work of Douady and Oesterlé [28]. Falconer [32] computed the Hausdorff dimension of a class of nonconformal repellers (see also [30]), while Hu [41] computed the box dimension of a class of nonconformal repellers leaving invariant a strong unstable foliation. Related ideas were applied by Simon and Solomyak in [91] to compute the Hausdorff dimension of a class of hyperbolic sets in $\mathbb{R}^3$. Falconer also studied a class of limit sets of geometric constructions obtained from the composition of affine transformations that are not necessarily conformal [30]. In another direction, Bothe [17] and Simon [90] studied the dimension of solenoids. A solenoid is a hyperbolic set of the form

$$\Lambda = \bigcap_{n=1}^{\infty} f^n(T),$$

where $T \subset \mathbb{R}^3$ is diffeomorphic to a solid torus $S^1 \times D$ for some closed disk $D \subset \mathbb{R}^2$ and $f: T \to T$ is a diffeomorphism such that for each $x \in S^1$, the intersection $f(T) \cap (\{x\} \times D)$ is a disjoint union of p sets homeomorphic to a closed disk.

1.7   Multifractal Analysis and Irregular Sets

By Theorem 1.3, if $f \colon M \to M$ is a $C^{1+\alpha}$ diffeomorphism and μ is a finite f-invariant measure supported on a hyperbolic set, then the limit

$$\lim_{r \to 0} \frac{\log \mu(B(x,r))}{\log r}$$

exists for μ-almost every $x \in M$. Multifractal analysis studies the properties of the level sets

$$\left\{ x \in M : \lim_{r \to 0} \frac{\log \mu(B(x,r))}{\log r} = \alpha \right\}$$

for $\alpha \in \mathbb{R}$. We present in this section the main components of multifractal analysis.

Birkhoff's ergodic theorem—another basic but also fundamental result of ergodic theory—states that if $S \colon X \to X$ is a measurable transformation preserving a finite measure μ in X, then for each integrable function $\varphi \in L^1(X, \mu)$, the limit

$$\varphi_S(x) = \lim_{n \to \infty} \frac{1}{n} \sum_{k=0}^{n-1} \varphi(S^k(x))$$

exists for μ-almost every $x \in X$. Furthermore, if μ is ergodic, then

$$\varphi_S(x) = \frac{1}{\mu(X)} \int_X \varphi \, d\mu \qquad (1.13)$$

for μ-almost every $x \in X$. Of course, this does not mean that (1.13) holds for *every* $x \in X$ for which $\varphi_S(x)$ is well defined. Given $\alpha \in \mathbb{R}$, we consider the level set

$$K_\alpha(\varphi) = \left\{ x \in X : \lim_{n \to \infty} \frac{1}{n} \sum_{k=0}^{n-1} \varphi(S^k x) = \alpha \right\}.$$

We also consider the set

$$K(\varphi) = \left\{ x \in X : \liminf_{n \to \infty} \frac{1}{n} \sum_{k=0}^{n-1} \varphi(S^k x) < \limsup_{n \to \infty} \frac{1}{n} \sum_{k=0}^{n-1} \varphi(S^k x) \right\}.$$

Clearly,

$$X = K(\varphi) \cup \bigcup_{\alpha \in \mathbb{R}} K_\alpha(\varphi). \qquad (1.14)$$

We call the decomposition of X in (1.14) a *multifractal decomposition*.

One way to measure the complexity of the sets $K_\alpha(\varphi)$ is to compute their Hausdorff dimension. We define a function

$$\mathcal{D}: \{\alpha \in \mathbb{R} : K_\alpha(\varphi) \neq \varnothing\} \to \mathbb{R}$$

by

$$\mathcal{D}(\alpha) = \dim_H K_\alpha(\varphi).$$

We also consider the numbers

$$\underline{\alpha} = \inf_\mu \int_X \varphi \, d\mu \quad \text{and} \quad \overline{\alpha} = \sup_\mu \int_X \varphi \, d\mu,$$

where the infimum and supremum are taken over all S-invariant probability measures μ in X. It is easy to verify that $K_\alpha(\varphi) = \varnothing$ whenever $\alpha \notin [\underline{\alpha}, \overline{\alpha}]$. We also define a function $T: \mathbb{R} \to \mathbb{R}$ by

$$T(q) = P(q\varphi) - qP(\varphi),$$

where P denotes the topological pressure. For topological Markov chains (see Sect. 1.6), the function T is analytic (see [84]). Under the assumptions in Theorem 1.8 below, there exists a unique equilibrium measure ν_q of $q\varphi$ (see Sect. 1.5).

The following result shows that in the case of topological Markov chains, the set $K_\alpha(\varphi)$ is nonempty for every $\alpha \in (\underline{\alpha}, \overline{\alpha})$ and that the function $\mathcal{D}$ is analytic and strictly convex.

Theorem 1.8 (Multifractal analysis of Birkhoff averages). *If the topological Markov chain $\sigma|X$ is topologically mixing and $\varphi: X \to \mathbb{R}$ is Hölder continuous, then:*

1. *$K_\alpha(\varphi)$ is dense in X for each $\alpha \in (\underline{\alpha}, \overline{\alpha})$.*
2. *The function $\mathcal{D}: (\underline{\alpha}, \overline{\alpha}) \to \mathbb{R}$ is analytic and strictly convex.*
3. *The function $\mathcal{D}$ is the Legendre transform of T, that is,*

$$\mathcal{D}(-T'(q)) = T(q) - qT'(q)$$

   for each $q \in \mathbb{R}$.
4. *If $q \in \mathbb{R}$, then $\nu_q(K_{-T'(q)}(\varphi)) = 1$ and*

$$\lim_{r \to 0} \frac{\log \nu_q(B(x, r))}{\log r} = T(q) - qT'(q)$$

   for ν_q-almost every point $x \in K_{-T'(q)}(\varphi)$.

Theorem 1.8 reveals an enormous complexity of multifractal decompositions that is not foreseen by Birkhoff's ergodic theorem. In particular, it shows that the multifractal decomposition in (1.14) is composed of an *uncountable* number of (pairwise disjoint) dense invariant sets, each of them of positive Hausdorff dimension. Statement 1 is an exercise. The remaining statements in Theorem 1.8

follow from results of Pesin and Weiss in [76]. In [86], Schmeling showed that the domain of $\mathcal{D}$ coincides with $[\underline{\alpha}, \overline{\alpha}]$, that is, that $K_\alpha(\varphi) \neq \varnothing$ if and only if $\alpha \in [\underline{\alpha}, \overline{\alpha}]$.

The concept of multifractal analysis was suggested in [38]. The first rigorous approach is due to Collet, Lebowitz and Porzio in [25] for a class of measures invariant under one-dimensional Markov maps. Lopes [56] considered the measure of maximal entropy for hyperbolic Julia sets, and Rand [80] studied Gibbs measures for a class of repellers. We refer the reader to the books [8, 73] for detailed discussions and further references.

Now let M be a surface and let $\Lambda \subset M$ be a locally maximal hyperbolic set for a $C^{1+\alpha}$ diffeomorphism $f: M \to M$. We assume that f is topologically mixing on Λ. Consider an equilibrium measure μ of a Hölder continuous function $\varphi: \Lambda \to \mathbb{R}$. We define functions $T_s: \Lambda \to \mathbb{R}$ and $T_u: \Lambda \to \mathbb{R}$ by

$$T_s(q) = P(-q \log\|df|E^s\| + q\varphi) - qP(\varphi)$$

and

$$T_u(q) = P(q \log\|df|E^u\| + q\varphi) - qP(\varphi).$$

In [92], Simpelaere showed that

$$\dim_H \left\{ x \in M : \lim_{r\to 0} \frac{\log \mu(B(x,r))}{\log r} = \alpha \right\} = T_s(q) - qT_s'(q) + T_u(q) - qT_u'(q),$$

where $q \in \mathbb{R}$ is the unique real number such that

$$\alpha = -T_s'(q) - T_u'(q).$$

Again we observe an enormous complexity that is not precluded by the μ-almost everywhere existence of the pointwise dimension in Theorem 1.3.

Now we consider the irregular set $K(\varphi)$ in (1.14). When $\varphi: X \to \mathbb{R}$ is a continuous function, it follows from Birkhoff's ergodic theorem that the set $K(\varphi)$ has zero measure with respect to *any* S-invariant finite measure in X. Therefore, at least from the point of view of measure theory, the set $K(\varphi)$ is very small. Remarkably, from the point of view of dimension theory, this set is as large as the whole space. Let $S: X \to X$ be a continuous transformation of a topological space X. Two continuous functions $\varphi_1: X \to \mathbb{R}$ and $\varphi_2: X \to \mathbb{R}$ are said to be *cohomologous* if there exist a continuous function $\psi: X \to \mathbb{R}$ and a constant $c \in \mathbb{R}$ such that

$$\varphi_1 - \varphi_2 = \psi - \psi \circ S + c$$

on X. If the function φ is cohomologous to a constant, then $K(\varphi) = \varnothing$. The following result of Barreira and Schmeling in [14] shows that if φ is not cohomologous to a constant, then $K(\varphi)$ is as large as the whole space from the points of view of topological entropy and Hausdorff dimension. We recall that $h(f|X)$ denotes the topological entropy of $f|X$ (see Sect. 1.5).

Theorem 1.9 (Irregular sets). *If X is a repeller of a $C^{1+\alpha}$ transformation, for some $\alpha > 0$, such that f is conformal and topologically mixing on X, and $\varphi: X \to \mathbb{R}$ is a Hölder continuous function, then the following properties are equivalent:*

1. *φ is not cohomologous to a constant.*
2. *$K(\varphi)$ is a nonempty dense set with*

$$h(f|K(\varphi)) = h(\sigma|X) \quad and \quad \dim_H K(\varphi) = \dim_H X. \quad (1.15)$$

A priori, Property 1 in Theorem 1.9 could be rare. However, precisely the opposite happens. Let $C^\theta(X)$ be the space of Hölder continuous functions in X with Hölder exponent $\theta \in (0, 1]$ equipped with the norm

$$\|\varphi\|_\theta = \sup\{|\varphi(x)| : x \in X\} + \sup\left\{C > 0 : \frac{|\varphi(x) - \varphi(y)|}{d(x, y)^\theta} \le C \text{ for all } x, y \in X\right\}.$$

It is shown in [14] that for each $\theta \in (0, 1]$, the family of functions in $C^\theta(X)$ that are not cohomologous to a constant is an open dense set. Therefore, given $\theta \in (0, 1]$ and a generic function φ in $C^\theta(X)$, the set $K(\varphi)$ is dense and satisfies the identities in (1.15).

Now let $K = \bigcup_\varphi K(\varphi)$, with the union taken over all Hölder continuous functions $\varphi: X \to \mathbb{R}$. Under the hypotheses of Theorem 1.9, we have

$$h(\sigma|K) = h(\sigma|X) \quad and \quad \dim_H K = \dim_H X.$$

These identities were established by Pesin and Pitskel in [74] when σ is a Bernoulli shift with two symbols, that is, for the transition matrix $A = \left(\begin{smallmatrix} 1 & 1 \\ 1 & 1 \end{smallmatrix}\right)$. A related result of Shereshevsky in [88] shows that for a generic C^2 surface diffeomorphism with a locally maximal hyperbolic set Λ, and an equilibrium measure μ of a Hölder continuous function that is generic in the C^0 topology, the set

$$I = \left\{x \in \Lambda : \liminf_{r \to 0} \frac{\log \mu(B(x, r))}{\log r} < \limsup_{r \to 0} \frac{\log \mu(B(x, r))}{\log r}\right\}$$

has positive Hausdorff dimension. In fact, it follows from results in [14] that $\dim_H I = \dim_H \Lambda$ (under those generic assumptions).

1.8 Quantitative Recurrence and Dimension

Poincaré's recurrence theorem (Theorem 1.2) is one of the fundamental results of the theory of dynamical systems. Unfortunately, it only provides qualitative information. In particular, it does not consider:

1. With which frequency the orbit of a point visits a given set of positive measure.
2. With which rate the orbit of a point returns to an arbitrarily small neighborhood of the initial point.

Birkhoff's ergodic theorem gives a fairly complete answer to the first problem.

Now we briefly describe some results concerning the second problem. Given a transformation $f: M \rightarrow M$, the *(first) return time* of a point $x \in M$ to the ball $B(x, r)$ is given by

$$\tau_r(x) = \inf\{n \in \mathbb{N} : d(f^n(x), x) < r\}.$$

The *lower* and *upper recurrence rates* of x are defined by

$$\underline{R}(x) = \liminf_{r \to 0} \frac{\log \tau_r(x)}{-\log r} \quad \text{and} \quad \overline{R}(x) = \limsup_{r \to 0} \frac{\log \tau_r(x)}{-\log r}. \tag{1.16}$$

When $\underline{R}(x) = \overline{R}(x)$, we denote the common value by $R(x)$ and call it the *recurrence rate* of x. In the present context, the study of quantitative recurrence started with the work of Ornstein and Weiss [67], closely followed by the work of Boshernitzan [16]. In [67], the authors considered the case of symbolic dynamics (and thus the corresponding symbolic metric in (1.6)) and for an ergodic σ-invariant probability measure μ showed that $R(x) = h_\mu(\sigma)$ for μ-almost every x. On the other hand, Boshernitzan considered an arbitrary metric space M and showed in [16] that

$$\underline{R}(x) \leq \dim_H \mu \tag{1.17}$$

for μ-almost every $x \in M$. In the particular case of hyperbolic sets, the following result of Barreira and Saussol in [13] shows that (1.17) is often an identity.

Theorem 1.10 (Quantitative recurrence). *For a $C^{1+\alpha}$ diffeomorphism with a hyperbolic set Λ, for some $\alpha > 0$, if μ is an ergodic equilibrium measure of a Hölder continuous function, then*

$$R(x) = \lim_{r \to 0} \frac{\log \mu(B(x, r))}{\log r} \tag{1.18}$$

for μ-almost every $x \in \Lambda$.

We note that identity (1.18) relates two quantities of very different nature. In particular, $R(x)$ does not depend on the measure, and the pointwise dimension does not depend on the map. Putting together (1.16) and (1.18), we obtain

$$\lim_{r \to 0} \frac{\log \tau_r(x)}{-\log r} = \lim_{r \to 0} \frac{\log \mu(B(x, r))}{\log r}$$

for μ-almost every point $x \in \Lambda$. Therefore, the return time $\tau_r(x)$ is approximately equal to $1/\mu(B(x, r))$ when r is sufficiently small.

Part I
Ergodic Theory

Chapter 2
Basic Notions and Examples

We introduce in this chapter the basic notions and results of ergodic theory, starting with the notion of invariant measure with respect to a measurable transformation. In particular, we establish two basic but also fundamental results of ergodic theory: Poincaré's recurrence theorem and Birkhoff's ergodic theorem. We also discuss the notion of ergodicity as well as its consequences. All the notions and results are illustrated with a number of examples. These include rotations and translations of $\mathbb{R}^n$, rotations and expanding maps of the circle, toral endomorphisms, etc. We conclude this chapter with some applications of ergodic theory to number theory, namely, to fractional parts of polynomials and continued fractions. All the necessary material from measure theory is recalled in Appendix A.

2.1 Introduction

Ergodic theory can be described as the study of measurable maps and flows and more generally group actions, preserving a certain measure. Some emphasis is given to the study of the stochastic properties of the dynamics, such as ergodicity and mixing with respect to a given invariant measure.

The origins of ergodic theory go back to statistical mechanics with an attempt to apply probability theory to conservative mechanical systems with many degrees of freedom. For conservative systems defined by a Hamiltonian, it follows from Liouville's theorem that the volume in phase space is invariant under the dynamics. More precisely, let

$$q' = \frac{\partial H}{\partial p}, \quad p' = -\frac{\partial H}{\partial q} \tag{2.1}$$

be the Hamiltonian equations defined by an autonomous Hamiltonian $H = H(p, q)$ in $\mathbb{R}^n \times \mathbb{R}^n$, where q gives the positions of some particles and q gives their corresponding momenta. Since the divergence of the vector field $(\partial H / \partial p, -\partial H / \partial q)$ is zero (provided that H is of class C^2), the Hamiltonian flow φ_t defined by (2.1)

L. Barreira, *Ergodic Theory, Hyperbolic Dynamics and Dimension Theory*, Universitext,
DOI 10.1007/978-3-642-28090-0_2, © Springer-Verlag Berlin Heidelberg 2012

preserves volume, that is, $v(\varphi_t(A)) = v(A)$ for any $t \in \mathbb{R}$ and any measurable set A, where

$$v(A) = \int_A dp \, dq.$$

Due to the ubiquity of Hamiltonian equations in physics, this yields a large class of examples of measure-preserving flows and in fact also of measure-preserving maps, simply by taking the corresponding time-t maps. Since the Hamiltonian H is invariant under the Hamiltonian flow, that is,

$$H \circ \varphi_t = H \quad \text{for every} \quad t \in \mathbb{R},$$

it is sufficient to study the restriction of the dynamics to each level set of constant energy

$$L_c = \{(p, q) : H(p, q) = c\},$$

on which the volume v induces a corresponding invariant volume v_c.

Given a function F, Boltzmann's ergodic hypothesis corresponds to assume that points x in a given level set L_c (assuming that the volume v_c is finite) satisfy

$$\lim_{t \to +\infty} \frac{1}{t} \int_0^t F(\varphi_s(x)) \, ds = \frac{1}{v_c(L_c)} \int_{L_c} F \, dv_c \qquad (2.2)$$

or some appropriate version of this requirement. Rigorous versions of Boltzmann's ergodic hypothesis were established independently by Birkhoff and von Neumann. In particular, Birkhoff's ergodic theorem states that for an integrable function F, the limit in the left-hand side of (2.2) exists for v_c-almost every $x \in L_c$.

However, in general, the limit in (2.2) may depend on x (and may not exist). In order to obtain independence with respect to the point, we need to consider the notion of ergodicity, which means that from the point of view of ergodic theory, that is, from the point of view of an invariant measure, the space cannot be decomposed into invariant sets of positive measure. When the measure v_c is ergodic, it follows from Birkhoff's ergodic theorem that identity (2.2) holds for v_c-almost every $x \in L_c$.

In another direction, the existence of a finite invariant measure naturally gives rise to the concept of nontrivial recurrence. Let T be the time-t map of a Hamiltonian flow. It also preserves the Liouville volume v. For a level set L_c of finite v_c-volume, a simple yet striking property was established by Poincaré: the transformation T exhibits a nontrivial recurrence in any set $A \subset L_c$ of positive measure, in the sense that the trajectory of almost every point in A returns infinitely often to A. In other words,

$$v_c \left(\limsup_{n \to \infty} T^{-n} A \right) = v_c(A).$$

This is Poincaré's recurrence theorem. The result follows from the fact that due to the invariance and finiteness of the measure, it is impossible that successive iterations of a set of positive measure only intersect on a set of zero measure.

2.2 Invariant Measures

We introduce in this section the most basic notion of ergodic theory: the notion of invariant measure with respect to a dynamical system. Roughly speaking, a measure is invariant if the evolution of the dynamics does not change the value of the measure. We also illustrate the concept with several instructive examples.

We recall that a triple $(X, \mathcal{A}, \mu)$ is said to be a *measure space* if $\mathcal{A}$ is a σ-algebra of subsets of X and μ is a measure in $\mathcal{A}$. We say that the measure μ is *finite* if $\mu(X) < \infty$.

2.2.1 The Notion of Invariant Measure

Let $(X, \mathcal{A}, \mu)$ be a measure space. We say that a transformation $T: X \to X$ is $\mathcal{A}$-*measurable* if

$$T^{-1}B := \{x \in X : T(x) \in B\} \in \mathcal{A} \quad \text{for every} \quad B \in \mathcal{A},$$

that is, if the preimage of any set in $\mathcal{A}$ is also in $\mathcal{A}$. Whenever there is no danger of confusion, we simply say that T is *measurable* without any explicit reference to the σ-algebra.

Definition 2.1. Let $(X, \mathcal{A}, \mu)$ be a measure space and let $T: X \to X$ be an $\mathcal{A}$-measurable transformation. We say that μ is T-*invariant* and that T *preserves* μ if

$$\mu(T^{-1}B) = \mu(B) \quad \text{for every} \quad B \in \mathcal{A}. \tag{2.3}$$

When the transformation T is invertible and T^{-1} is $\mathcal{A}$-measurable, we note that $T^{-1}B \in \mathcal{A}$ if and only if $B \in \mathcal{A}$, and thus, in this case, condition (2.3) is equivalent to

$$\mu(T(B)) = \mu(B) \quad \text{for every} \quad B \in \mathcal{A}.$$

In order to present a characterization of the invariance of a finite measure, we start by recalling that the *characteristic function* $\chi_B: X \to \{0, 1\}$ of a set $B \subset X$ is defined by

$$\chi_{B(x)} = \begin{cases} 1 & \text{if } x \in B, \\ 0 & \text{if } x \notin B. \end{cases} \tag{2.4}$$

One can easily verify that

$$\chi_B \circ T = \chi_{T^{-1}B}, \tag{2.5}$$

and thus, property (2.3) is equivalent to

$$\int_X (\chi_B \circ T) \, d\mu = \int_X \chi_B \, d\mu \quad \text{for every} \quad B \in \mathcal{A}. \tag{2.6}$$

Now let $L^1(X, \mu)$ be the set of all μ-integrable functions, that is, the set of all measurable functions $\varphi \colon X \to \mathbb{R}$ such that

$$\int_X |\varphi| \, d\mu < \infty.$$

Proposition 2.1. *Let $T \colon X \to X$ be a measurable transformation and let μ be a finite measure in X. Then μ is T-invariant if and only if for every function $\varphi \in L^1(X, \mu)$, we have $\varphi \circ T \in L^1(X, \mu)$ and*

$$\int_X (\varphi \circ T) \, d\mu = \int_X \varphi \, d\mu. \tag{2.7}$$

Proof. We first assume that (2.7) holds. Given a measurable set $B \subset X$, we consider the function $\varphi = \chi_B$. Since $\int_X |\varphi| \, d\mu = \mu(B) < \infty$, we have $\chi_B \in L^1(X, \mu)$, and using (2.5), it follows from (2.7) that (2.3) holds.

Now we assume that μ is T-invariant. Then (2.6) holds. If s is a simple function, that is, a finite linear combination of characteristic functions, then it follows from (2.6) that

$$\int_X (s \circ T) \, d\mu = \int_X s \, d\mu. \tag{2.8}$$

Now we consider an arbitrary function $\varphi \in L^1(X, \mu)$. We can write it in the form $\varphi = \varphi^+ - \varphi^-$, where φ^+ and φ^- are the integrable functions

$$\varphi^+ = \max\{\varphi, 0\} \quad \text{and} \quad \varphi^- = \max\{-\varphi, 0\}.$$

Since $\varphi^+, \varphi^- \geq 0$, it is sufficient to prove the result for functions $\varphi \in L^1(X, \mu)$ with $\varphi \geq 0$. Thus, let s_n be a sequence of simple functions such that

$$0 \leq s_n \leq s_{n+1} \leq \varphi \quad \text{for each} \quad n \in \mathbb{N},$$

with $s_n \to \varphi$ pointwise and

$$\int_X s_n \, d\mu \to \int_X \varphi \, d\mu$$

when $n \to \infty$. Since $s_n \circ T \nearrow \varphi \circ T$ when $n \to \infty$, it follows from Fatou's lemma (Theorem A.1) and (2.8) that

$$\int_X (\varphi \circ T) \, d\mu \leq \liminf_{n \to \infty} \int_X (s_n \circ T) \, d\mu$$

$$= \liminf_{n \to \infty} \int_X s_n \, d\mu$$

$$= \int_X \varphi \, d\mu < \infty.$$

Fig. 2.1 Rotations and translations of $\mathbb{R}^n$

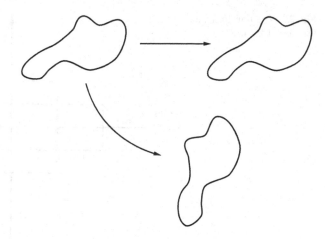

This shows that the function $\varphi \circ f$ is integrable. Moreover, it follows from the monotone convergence theorem (Theorem A.2) that

$$\int_X (\varphi \circ T)\, d\mu = \lim_{n \to \infty} \int_X (s_n \circ T)\, d\mu$$

$$= \lim_{n \to \infty} \int_X s_n\, d\mu = \int_X \varphi\, d\mu.$$

This completes the proof of the proposition. $\qquad\qquad\square$

2.2.2 Rotations and Translations of $\mathbb{R}^n$

Here and in the following sections, we present several examples of transformations with invariant measures.

We start by considering a class of linear transformations. Let $S(n, \mathbb{R})$ be the set of $n \times n$ matrices A with entries in $\mathbb{R}$ such that $|\det A| = 1$. Given $A \in S(n, \mathbb{R})$ and $b \in \mathbb{R}^n$, we define a transformation $T: \mathbb{R}^n \to \mathbb{R}^n$ by

$$T(x) = Ax + b \qquad\qquad (2.9)$$

for each $x \in \mathbb{R}^n$. For example, all rotations and translations are of this form. Moreover, any composition of rotations and translations (see Fig. 2.1) can also be written as in (2.9) for some orthogonal matrix A. This means that $A^*A = \mathrm{Id}$, where A^* is the transpose of A. Clearly, each transformation T in (2.9) is invertible and differentiable, with derivative $d_x T = A$ for each $x \in \mathbb{R}^n$.

Now let m be the Lebesgue measure in $\mathbb{R}^n$ (see Appendix A for the definition). If $B \subset \mathbb{R}^n$ is a measurable set, then

Fig. 2.2 The transformation
$(x, y) \mapsto (2x, y/2)$

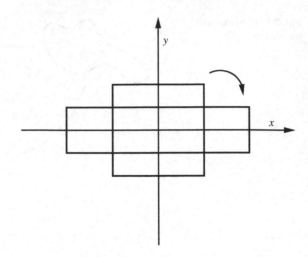

$$m(T(B)) = \int_B |\det d_x T| \, dm(x)$$

$$= \int_B |\det A| \, dm = m(B). \tag{2.10}$$

Since the transformation T is invertible, it follows from (2.10) that the Lebesgue measure m is T-invariant. In particular, any composition of rotations and translations preserves the Lebesgue measure. We note that the converse does not hold. For an example, consider the transformation $(x, y) \mapsto (2x, y/2)$ (see Fig. 2.2).

2.2.3 Circle Rotations and Interval Translations

Now we consider the rotations of the circle

$$S^1 = \{z \in \mathbb{C} : |z| = 1\}.$$

We can identify S^1 with the quotient $\mathbb{T} = [0, 1]/\{0, 1\}$ by the one-to-one transformation $h: \mathbb{T} \to S^1$ given by

$$h(\theta) = e^{2\pi i \theta}. \tag{2.11}$$

Definition 2.2. Given $w \in S^1$, we define the *circle rotation* $R_w: S^1 \to S^1$ by

$$R_w(z) = wz,$$

and given $\tau \in \mathbb{R}$, we define the *interval translation* $T_\tau: \mathbb{T} \to \mathbb{T}$ (see Fig. 2.3) by

$$T_\tau(x) = x + \tau \bmod 1 = x + \tau - \lfloor x + \tau \rfloor, \tag{2.12}$$

Fig. 2.3 An interval
translation

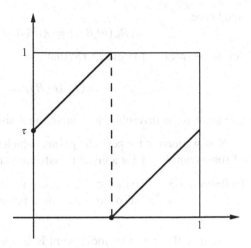

where $\lfloor x \rfloor$ denotes the integer part of x.

We note that if $w = e^{2\pi i \tau}$, then

$$(R_w \circ h)(\theta) = h(\theta + \tau) = (h \circ T_\tau)(\theta) \tag{2.13}$$

for every $\theta \in \mathbb{R}$.

Now let m be the probability measure in $\mathbb{T}$ induced by the Lebesgue measure in $[0, 1] \subset \mathbb{R}$. Since the latter is invariant under translations (see Sect. 2.2.2), we have

$$m(B + \tau) = m(B) \tag{2.14}$$

for every measurable set $B \subset [0, 1]$, where

$$B + \tau = \{\theta + \tau : \theta \in B\}.$$

Therefore, any interval translation preserves the Lebesgue measure. We also consider the *Lebesgue measure* λ in S^1 defined by

$$\lambda(h(B)) = m(B) \tag{2.15}$$

for each measurable set $B \subset [0, 1]$.

Proposition 2.2. *Any circle rotation preserves the Lebesgue measure.*

Proof. By (2.13), for any measurable set $B \subset [0, 1]$, we have

$$R_w(h(B)) = h(T_\tau(B)),$$

and hence,

$$\lambda(R_w(h(B))) = \lambda(h(B + \tau)) = m(B + \tau).$$

It follows from (2.14) and (2.15) that

$$\lambda(R_w(h(B))) = \lambda(h(B)),$$

and since R_w is invertible, we obtain the desired statement. $\square$

Now we look at the periodic points, which sometimes may indicate the existence of some complexity for a given transformation (see also Sect. 2.2.4).

Definition 2.3. Given $m \in \mathbb{N}$, we say that $x \in X$ is an *m-periodic point* of T if $T^m(x) = x$. We also say that x is a *periodic point* of T if it is m-periodic for some m.

We note that an m-periodic point is km-periodic for every $k \in \mathbb{N}$. Given an m-periodic point $x \in X$ of a transformation $T: X \to X$, we can define a T-invariant probability measure in X by

$$\mu = \frac{1}{m} \sum_{k=0}^{m-1} \delta_{T^k(x)}, \tag{2.16}$$

where δ_x is the probability measure such that $\delta_x(\{x\}) = 1$. The measure μ is concentrated on the *periodic orbit*

$$\mathcal{O}_T(x) = \{T^k(x) : k = 0, \ldots, m-1\},$$

in the sense that $\mu(X) = \mu(\mathcal{O}_T(x))$.

Now we study the periodic points in the particular case of the circle rotations.

Proposition 2.3. *Given $w = e^{2\pi i \tau} \in S^1$, the following properties hold:*

1. *If $\tau \notin \mathbb{Q}$, then R_w has no periodic points.*
2. *If $\tau \in \mathbb{Q}$, then R_w has m-periodic points if and only if $m\tau \in \mathbb{Z}$.*
3. *If there is a n-periodic point of R_w, then all points of S^1 are n-periodic.*

Proof. We note that a point $z \in S^1$ is m-periodic for R_w if and only if $e^{2\pi i m\tau} z = z$, that is, if and only if $e^{2\pi i m\tau} = 1$. Property 3 follows immediately from this observation. Furthermore, $e^{2\pi i m\tau} = 1$ if and only if $m\tau \in \mathbb{Z}$. This yields Properties 1 and 2. $\square$

Let $w = e^{2\pi i \tau} \in S^1$ and consider the interval translation T_τ in (2.12).

Definition 2.4. When $\tau \in \mathbb{Q}$, we say that R_w is a *rational circle rotation* and that T_τ is a *rational interval translation*. When $\tau \notin \mathbb{Q}$, we say that R_w is an *irrational circle rotation* and that T_τ is an *irrational interval translation*.

It follows from Proposition 2.3 and the construction in (2.16) that each rational circle rotation and each rational interval translation have uncountably many invariant probability measures.

2.2.4 Expanding Maps of the Circle

We consider in this section a class of transformations in the circle that expand distances.

Definition 2.5. Given $q \in \mathbb{Z}$, we define a transformation $E_q \colon S^1 \to S^1$ by $E_q(z) = z^q$ for each $z \in S^1$. When $|q| > 1$, we say that E_q is an *expanding map of the circle*.

Proposition 2.4. *Any expanding map of the circle preserves the Lebesgue measure.*

Proof. For each $\alpha \in (0, 1)$, we have

$$E_q^{-1}(h([0, \alpha])) = \bigcup_{i=1}^{q} h\left(\left[\frac{i}{q}, \frac{\alpha + i}{q}\right]\right) \tag{2.17}$$

whenever $q > 0$, and

$$E_q^{-1}(h([0, \alpha])) = \bigcup_{i=1}^{|q|} h\left(\left[\frac{\alpha + i}{q}, \frac{i}{q}\right]\right) \tag{2.18}$$

whenever $q < 0$. We note that both unions in (2.17) and (2.18) are disjoint. Hence,

$$\lambda\big(E_q^{-1}(h([0, \alpha]))\big) = \sum_{i=1}^{|q|} \frac{\alpha}{|q|} = \alpha$$

$$= \lambda(h([0, \alpha]))$$

for every $\alpha \in (0, 1)$. Since the intervals $[0, \alpha]$ generate the Borel σ-algebra of $[0, 1]$, we conclude that

$$\lambda(E_q^{-1}B) = \lambda(B)$$

for any measurable set $B \subset S^1$. Therefore, E_q preserves the Lebesgue measure. □

Now we study the periodic points of the expanding maps of the circle. Since $(E_q)^m = E_{q^m}$, a point $e^{2\pi i \theta} \in S^1$ is m-periodic for E_q if and only if

$$e^{2\pi i \theta(q^m - 1)} = 1,$$

that is, if and only if

$$\theta = j/|q^m - 1| \quad \text{for some} \quad j = 1, \dots, |q^m - 1|. \tag{2.19}$$

Fig. 2.4 The eight 2-periodic
points of E_3

Hence, the number of m-periodic points of E_q is equal to $|q^m - 1|$, and they
are uniformly distributed in S^1 (see Fig. 2.4 for an example). In particular, each
expanding map E_q has m-periodic points for every $m \in \mathbb{N}$.

Proposition 2.5. *The periodic points of an expanding map E_q are dense in S^1, that
is, the set*

$$\{z \in S^1 : z \text{ is } m\text{-periodic for } E_q \text{ for some } m \in \mathbb{N}\}$$

is dense in S^1.

Proof. The statement is an immediate consequence of (2.19).   □

In order to measure the complexity of a transformation from the point of view of
the periodic points, we now introduce the notion of periodic entropy.

Definition 2.6. Given a transformation $T\colon X \to X$, we define the *periodic entropy
of T* by

$$p(T) = \limsup_{m \to \infty} \frac{1}{m} \log^+ \operatorname{card} \{x \in X : T^m(x) = x\}, \qquad (2.20)$$

where $\log^+ a = \max\{0, \log a\}$ (with the convention that $\log 0 = -\infty$).

For example, for the expanding maps of the circle, we have

$$p(E_q) = \limsup_{m \to \infty} \frac{1}{m} \log^+ |q^m - 1| = \log|q| > 0. \qquad (2.21)$$

2.2.5 Toral Endomorphisms

We consider in this section a higher-dimensional version of the expanding maps of
the circle. Let $G(n, \mathbb{Z})$ be the set of $n \times n$ matrices A with entries in $\mathbb{Z}$ such that
$\det A \neq 0$. We note that

$$A(\mathbb{Z}^n) \subset \mathbb{Z}^n \qquad (2.22)$$

for each $A \in G(n, \mathbb{Z})$. Since A defines a linear transformation of $\mathbb{R}^n$, this is
equivalent to

$$Ay - Ax \in \mathbb{Z}^n \quad \text{whenever} \quad y - x \in \mathbb{Z}^n. \qquad (2.23)$$

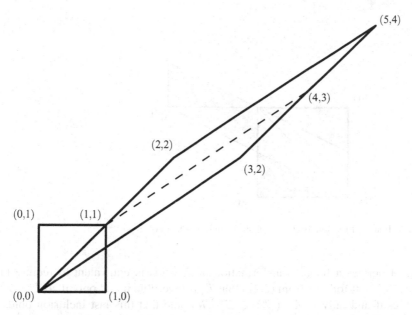

Fig. 2.5 Image of the square $[0, 1]^2$ under an endomorphism of $\mathbb{T}^2$ (the broken line divides the parallelogram into two identical parallelograms)

Now we consider the sets

$$[x] = \{y \in \mathbb{R}^n : y - x \in \mathbb{Z}^n\}$$

for each $x \in \mathbb{R}^n$ and the *n-torus*

$$\mathbb{T}^n = \mathbb{R}^n / \mathbb{Z}^n = \{[x] : x \in \mathbb{R}^n\}.$$

It follows from (2.23) that if $y \in [x]$, then $Ay \in [Ax]$. Therefore, one can define a map T_A of the *n*-torus by $T_A[x] = [Ax]$.

Definition 2.7. Given a matrix $A \in G(n, \mathbb{Z})$, we define the *toral endomorphism* $T_A : \mathbb{T}^n \to \mathbb{T}^n$ by

$$T_A[x] = [Ax] \quad \text{for each} \quad [x] \in \mathbb{T}^n.$$

We also say that T_A is the *toral endomorphism induced by* A.

We note that, in general, the map T_A need not be invertible, even though the matrix $A \in G(n, \mathbb{Z})$ is always invertible. As an example, we represent in Fig. 2.5 the noninvertible endomorphism of $\mathbb{T}^2$ induced by the matrix $\left(\begin{smallmatrix} 3 & 2 \\ 2 & 2 \end{smallmatrix}\right)$. A necessary and sufficient condition for the invertibility of the toral endomorphism T_A is that

$$Ay - Ax \in \mathbb{Z}^n \quad \text{if and only if} \quad y - x \in \mathbb{Z}^n.$$

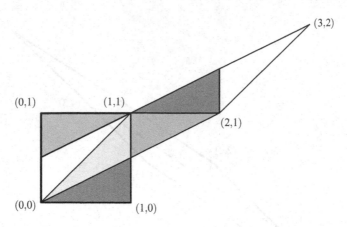

Fig. 2.6 Image of the square $[0, 1]^2$ under an automorphism of $\mathbb{T}^2$

Since A defines a linear transformation of $\mathbb{R}^n$, this is equivalent to require that $A(\mathbb{Z}^n) = \mathbb{Z}^n$. It follows from (2.22) that T_A is invertible if and only if $A(\mathbb{Z}^n) \supset \mathbb{Z}^n$ and thus if and only if $A^{-1}(\mathbb{Z}^n) \subset \mathbb{Z}^n$. We note that this last inclusion holds if and only if A^{-1} has only integer entries. In this case, both $\det A$ and $\det(A^{-1})$ are integers, which implies that $|\det A| = 1$. On the other, if $|\det A| = 1$, then clearly A^{-1} has only integer entries.

Let $S(n, \mathbb{Z}) \subset G(n, \mathbb{Z})$ be the set of $n \times n$ matrices A with $|\det A| = 1$. It follows from the former discussion that

$$T_A \text{ is invertible if and only if } A \in S(n, \mathbb{Z}). \tag{2.24}$$

Definition 2.8. For each matrix $A \in S(n, \mathbb{Z})$, the endomorphism $T_A \colon \mathbb{T}^n \to \mathbb{T}^n$ is also called a *toral automorphism*.

By (2.24), each toral automorphism is invertible. Moreover, the inverse of a toral automorphism T_A is also a toral automorphism, with $(T_A)^{-1} = T_{A^{-1}}$. As an example, we represent in Fig. 2.6 the automorphism of $\mathbb{T}^2$ induced by the matrix $\left(\begin{smallmatrix} 2 & 1 \\ 1 & 1 \end{smallmatrix}\right) \in S(2, \mathbb{Z})$. We note that the parallelogram $\left(\begin{smallmatrix} 2 & 1 \\ 1 & 1 \end{smallmatrix}\right) [0, 1]^2$ can be decomposed into four triangles, which can then be used to reassemble the square.

Now let m be the Lebesgue measure in $\mathbb{R}^n$. We define the Lebesgue measure λ in the torus $\mathbb{T}^n$ by

$$\lambda([B]) = m(B)$$

for each measurable set $B \subset [0, 1]^n$, where $[B] = \{[x] : x \in B\}$.

Proposition 2.6. *Any toral endomorphism preserves the Lebesgue measure.*

Proof. Let $A \in G(n, \mathbb{Z})$. Each point $x \in \mathbb{T}^n$ has a number $k = |\det A|$ of preimages under the toral endomorphism T_A. Moreover, for each sufficiently small $r > 0$ and each $x \in \mathbb{T}^n$, the preimage $T_A^{-1} B(x, r)$ of the open ball $B(x, r)$ of radius r centered at x consists of a number k of connected components $B_1, \ldots, B_k$. We consider the corresponding local inverses

$$S_i: B(x, r) \to B_i$$

of T_A for $i = 1, \ldots, k$. Then $T_A \circ S_i = \mathrm{Id}$ in $B(x, r)$, and thus, $d_y S_i = A^{-1}$ for each $y \in B(x, r)$ and $i = 1, \ldots, k$. Therefore,

$$\lambda(T_A^{-1} B(x, r)) = \sum_{i=1}^{k} \lambda(B_i) = \sum_{i=1}^{k} \lambda(S_i B(x, r))$$

$$= \sum_{i=1}^{k} \int_{B(x,r)} |\det d_y S_i| \, d\lambda(y)$$

$$= |\det A| \int_{B(x,r)} |\det A^{-1}| \, d\lambda$$

$$= \lambda(B(x, r)).$$

Since the family of open balls $B(x, r)$ with $x \in \mathbb{T}^n$ and $r > 0$ sufficiently small generates the Borel σ-algebra of $\mathbb{T}^n$, we conclude that λ is T_A-invariant. $\square$

2.2.6 Piecewise Linear Expanding Maps

Given $a \in (0, 1)$, we consider the piecewise linear transformation $T: [0, 1] \to [0, 1]$ defined by

$$T(x) = \begin{cases} x/a & \text{if } 0 \le x \le a, \\ (x - a)/(1 - a) & \text{if } a < x \le 1 \end{cases} \tag{2.25}$$

(see Fig. 2.7). For each $\alpha \in [0, 1]$ (see Fig. 2.8), we have

$$T^{-1}(0, \alpha) = (0, a\alpha) \cup (a, a + \alpha - a\alpha),$$

with a disjoint union. Therefore,

$$m(T^{-1}(0, \alpha)) = m((0, \alpha)) \quad \text{for every} \quad \alpha \in [0, 1].$$

Since the intervals of the form $(0, \alpha)$ generate the Borel σ-algebra of $[0, 1]$, we conclude that T preserves the Lebesgue measure.

We note that in the definition of the transformation T, the interval $[0, 1]$ is divided into two disjoint subintervals: $(0, a)$ and $(a, 1)$. More generally, we can divide the interval $[0, 1]$ into a (finite or infinite) countable number of disjoint subintervals $(a_k, b_k) \subset [0, 1]$ such that

Fig. 2.7 A piecewise linear
expanding map

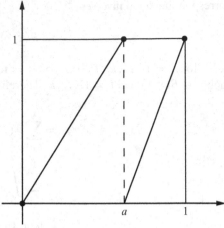

Fig. 2.8 Preimage of the
interval $[0, \alpha]$, where
$b = a + \alpha - a\alpha$

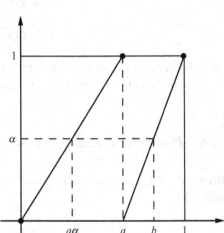

$$\sum_k (b_k - a_k) = 1. \qquad (2.26)$$

Definition 2.9. Given a (finite or infinite) countable number of disjoint subintervals
$(a_k, b_k) \subset [0, 1]$ satisfying (2.26), any transformation $T : [0, 1] \to [0, 1]$ such that

$$T(x) = \frac{x - a_k}{b_k - a_k}$$

whenever $x \in (a_k, b_k)$ for some k is called a *piecewise linear expanding map of the
interval*.

We note that for each $x \in \bigcup_k \{a_k, b_k\}$, the value $T(x)$ is arbitrary. Moreover, for
each piecewise linear expanding map of the interval, there exists $\tau > 1$ such that

$$T'(x) > \tau \quad \text{for every} \quad x \in \bigcup_k (a_k, b_k).$$

A similar argument to that for the particular linear expanding map of the interval in (2.25) establishes the following:

Proposition 2.7. *Any piecewise linear expanding map of the interval preserves the Lebesgue measure.*

Proof. For each $\alpha \in [0, 1]$, we have

$$T^{-1}(0, \alpha) = \bigcup_k (a_k, a_k + \alpha(b_k - a_k)),$$

with a disjoint union. Therefore, by (2.26),

$$m(T^{-1}(0, \alpha)) = \sum_k m\big((a_k, a_k + \alpha(b_k - a_k))\big)$$

$$= \sum_k \alpha(b_k - a_k) = \alpha$$

$$= m\big((0, \alpha)\big).$$

Since the intervals $(0, \alpha)$ generate the Borel σ-algebra of $[0, 1]$, we conclude that the map T preserves the Lebesgue measure. $\qquad \Box$

2.3 Poincaré's Recurrence Theorem

We establish in this section a basic but also fundamental result of ergodic theory—Poincaré's recurrence theorem. Essentially, it says that the existence of a finite invariant measure causes a nontrivial recurrence for the dynamics.

Let $T: X \to X$ be a measurable transformation and let μ be a finite T-invariant measure in X. Let also $A \subset X$ be a measurable set with $\mu(A) > 0$. If the sets $A, T^{-1}A, T^{-2}A, \ldots$ are pairwise disjoint (see Fig. 2.9), that is, if

$$T^{-m}A \cap T^{-n}A = \varnothing \quad \text{for every} \quad m \neq n,$$

then because μ is T-invariant, we have

$$\mu\left(\bigcup_{k=1}^{\infty} T^{-k}A\right) = \sum_{k=1}^{\infty} \mu(T^{-k}A) = \sum_{k=1}^{\infty} \mu(A) = \infty.$$

But this is impossible because μ is a finite measure. Therefore, there exist positive integers $m < n$ such that

Fig. 2.9 Pairwise disjoint
preimages

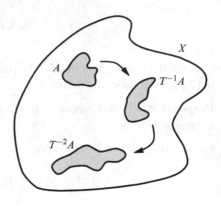

$$T^{-m}(A \cap T^{m-n} A) = T^{-m} A \cap T^{-n} A \neq \emptyset.$$

Hence, $A \cap T^{-(n-m)} A \neq \emptyset$, and some preimage of the set A must intersect A. In other words, whenever there exists a *finite* invariant measure, there is a nontrivial recurrence in each set of positive measure. Now we formulate a much stronger statement.

Theorem 2.1 (Poincaré's recurrence theorem). *Let $T: X \to X$ be a measurable transformation and let μ be a finite T-invariant measure in X. If $A \subset X$ is measurable, then the set*

$$B = \{x \in A : T^n(x) \in A \text{ for infinitely many integers } n \in \mathbb{N}\}$$

has measure $\mu(B) = \mu(A)$.

Proof. We have

$$B = A \cap \limsup_{n \to \infty} T^{-n} A$$

$$= A \cap \bigcap_{n=1}^{\infty} \bigcup_{k=n}^{\infty} T^{-k} A \qquad (2.27)$$

$$= A \setminus \bigcup_{n=1}^{\infty} \left(A \setminus \bigcup_{k=n}^{\infty} T^{-k} A \right).$$

Moreover,

$$A \setminus \bigcup_{k=n}^{\infty} T^{-k} A \subset \bigcup_{k=0}^{\infty} T^{-k} A \setminus \bigcup_{k=n}^{\infty} T^{-k}$$

$$= \bigcup_{k=0}^{\infty} T^{-k} A \setminus T^{-n} \bigcup_{k=0}^{\infty} T^{-k} A. \qquad (2.28)$$

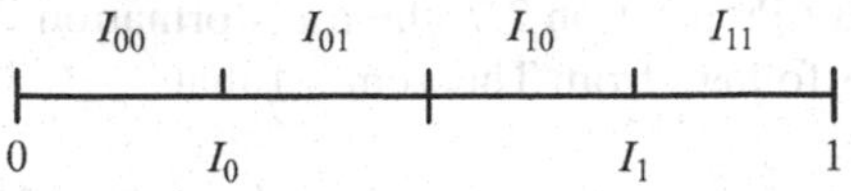

Fig. 2.10 Intervals I_{i_1} and
$I_{i_1 i_2}$ for the base-2
representation of numbers
in $[0, 1]$

On the other hand, since μ is T-invariant, we have

$$\mu\left(\bigcup_{k=0}^{\infty} T^{-k} A\right) = \mu\left(T^{-n} \bigcup_{k=0}^{\infty} T^{-k} A\right). \tag{2.29}$$

Moreover, since

$$\bigcup_{k=0}^{\infty} T^{-k} A \supset \bigcup_{k=n}^{\infty} T^{-k} A = T^{-n} \bigcup_{k=0}^{\infty} T^{-k} A,$$

and μ is finite, it follows from (2.28) and (2.29) that

$$\mu\left(A \setminus \bigcup_{k=n}^{\infty} T^{-k} A\right) \le \mu\left(\bigcup_{k=0}^{\infty} T^{-k} A\right) - \mu\left(T^{-n} \bigcup_{k=0}^{\infty} T^{-k} A\right) = 0$$

for every $n \in \mathbb{N}$. We conclude from (2.27) that $\mu(B) = \mu(A)$. $\qquad\square$

Exercises 2.7 and 2.8 show that the finiteness of an invariant measure is crucial for the existence of nontrivial recurrence.

The following is an application of Poincaré's recurrence theorem to base-2 representations:

Example 2.1. For each $n \in \mathbb{N}$ and $(i_1, \ldots, i_n) \in \{0, 1\}^n$, we consider the interval

$$I_{i_1 \cdots i_n} = \left(\sum_{j=1}^{n} \frac{i_j}{2^j}, \sum_{j=1}^{n} \frac{i_j}{2^j} + \frac{1}{2^n}\right) \subset [0, 1].$$

We note that $I_{i_1 \cdots i_n}$ has length $1/2^n$ and that for each fixed n, the intervals $I_{i_1 \cdots i_n}$ are pairwise disjoint (see Fig. 2.10). A point $x \in [0, 1]$ is in the closure of $I_{i_1 \cdots i_n}$ if and only if the first n digits of some base-2 representation of x are equal to $i_1 \cdots i_n$. Incidentally, we note that the base-2 representation is unique for Lebesgue-almost every point.

Now we consider the piecewise linear expanding map of the interval (see Sect. 2.2.6) defined by

$$T(x) = \begin{cases} 2x & \text{if } 0 \le x \le 1/2, \\ 2x - 1 & \text{if } 1/2 < x \le 1. \end{cases}$$

By Proposition 2.7, the transformation T preserves the Lebesgue measure. Hence, it follows from Theorem 2.1 that

$$\text{card}\left\{m \in \mathbb{N} : T^m(x) \in I_{i_1 \cdots i_n}\right\} = \infty \tag{2.30}$$

for Lebesgue-almost every $x \in I_{i_1 \cdots i_n}$. Now we observe that if $0.i_1 i_2 \cdots$ is a base-2 representation of x, then $0.i_{m+1} i_{m+2} \cdots$ is a base-2 representation of $T^m(x)$. Thus, it follows from (2.30) that for Lebesgue-almost every $x \in I_{i_1 \cdots i_n}$, the digits $i_1 \cdots i_n$ appear infinitely often, in this order, in some base-2 representation of x.

2.4 Invariant Sets and Invariant Functions

We already know that one of the basic notions of ergodic theory is the notion of invariant measure. We are also interested in other invariant objects. Here we introduce the notions of invariant set and invariant function.

Definition 2.10. Given a transformation $T: X \to X$, we say that:

1. A set $A \subset X$ is T-*invariant* if $T^{-1}A = A$.
2. A function $\varphi: X \to \mathbb{R}$ is T-*invariant* if $\varphi(T(x)) = \varphi(x)$ for every $x \in X$.

We note that if φ is a T-invariant function, then $\varphi(y) = \varphi(x)$ for every $y \in T^{-1}x$ since $T(y) = x$. We also note that when T is invertible, a set $A \subset X$ is T-invariant if and only if $T(A) = A$. For example, any periodic orbit of an invertible transformation is an invariant set.

Example 2.2. Consider the piecewise linear expanding map of the interval (see Sect. 2.2.6) defined by

$$T(x) = \begin{cases} 3x & \text{if } 0 \leq x \leq 1/3, \\ 3x - 1 & \text{if } 1/3 < x < 2/3, \\ 3x - 2 & \text{if } 2/3 \leq x \leq 1. \end{cases}$$

Clearly, the interval $[0, 1]$ is T-invariant. Now let

$$J = [0, 1/3] \cup [2/3, 1] \subset [0, 1],$$

and consider the set

$$A = \bigcap_{n=0}^{\infty} T^{-n} J.$$

One can easily verify that A is a (nonempty) Cantor set, that is, a closed set with empty interior and without isolated points. Moreover,

$$J \cap T^{-1}A = A. \tag{2.31}$$

Now we consider the transformation $S = T|J : J \to J$. It follows from (2.31) that $A \subset J$ is an S-invariant set.

We can easily verify that a set A is T-invariant if and only if its characteristic function χ_A is T-invariant (see (2.4) for the definition of χ_A). The following statement is a generalization of this observation for an arbitrary function:

Proposition 2.8. *A function φ is T-invariant if and only if the sets $\varphi^{-1}\alpha$ are T-invariant for every $\alpha \in \mathbb{R}$.*

Proof. The set $\varphi^{-1}\alpha$ is T-invariant if and only if

$$\varphi^{-1}\alpha = T^{-1}(\varphi^{-1}\alpha) = (\varphi \circ T)^{-1}\alpha.$$

This happens provided that $x \in \varphi^{-1}\alpha$ if and only if $x \in (\varphi \circ T)^{-1}\alpha$, which is the same as $\varphi(x) = \alpha$ if and only if $\varphi(T(x)) = \alpha$. We conclude that $\varphi^{-1}\alpha$ is T-invariant for every $\alpha \in \mathbb{R}$ if and only if $\varphi(T(x)) = \varphi(x)$ for every $x \in X$, that is, if and only if φ is T-invariant. $\square$

When the space X is equipped with a measure, it is often convenient to relax the notions of invariant set and invariant function up to sets of zero measure.

Definition 2.11. Given a measurable transformation $T : X \to X$ and a measure μ in X, we say that:

1. A measurable set $A \subset X$ is T-*invariant almost everywhere* if there is a T-invariant measurable subset $B \subset A$ with $\mu(A \setminus B) = 0$.
2. A measurable function $\varphi : X \to \mathbb{R}$ is T-*invariant almost everywhere* if there is a T-invariant measurable set $B \subset X$ with $\mu(X \setminus B) = 0$ such that the restriction $\varphi|B$ is $(T|B)$-invariant.

2.5   Birkhoff's Ergodic Theorem

Poincaré's recurrence theorem (Theorem 2.1) shows that the existence of a finite invariant measure causes a nontrivial recurrence. Unfortunately, this information is only of qualitative nature. In particular, the theorem says nothing about the frequency with which an orbit visits a given set. Birkhoff's ergodic theorem (Theorem 2.2) essentially gives a complete answer to this problem.

2.5.1   Formulation of the Theorem

Let $T : X \to X$ be a measurable transformation and let $A \subset X$ be a measurable set. Given $x \in X$ and $n \in \mathbb{N}$, we define

$$\tau_n(A, x) = \text{card} \{k \in \{0, \ldots, n - 1\} : T^k(x) \in A\}. \tag{2.32}$$

We note that $\tau_1(A, x) = \chi_A(x)$, and thus,

$$\tau_n(A, x) = \sum_{k=0}^{n-1} \tau_1(A, T^k(x)) = \sum_{k=0}^{n-1} \chi_A(T^k(x)).$$

When the limit

$$\lim_{n \to \infty} \frac{\tau_n(A, x)}{n} = \lim_{n \to \infty} \frac{1}{n} \sum_{k=0}^{n-1} \chi_A(T^k(x)) \tag{2.33}$$

exists, it gives the frequency with which the orbit of x visits the set A.

Now let μ be a finite T-invariant measure in X. Poincaré's recurrence theorem says that $\tau_n(A, x) \to \infty$ when $n \to \infty$, for μ-almost every $x \in A$, but it gives no information about the existence of the limit in (2.33). The following result essentially solves this problem.

Theorem 2.2 (Birkhoff's ergodic theorem). *Let $T: X \to X$ be a measurable transformation and let μ be a finite T-invariant measure in X. If $\varphi \in L^1(X, \mu)$, then the limit*

$$\varphi_T(x) = \lim_{n \to \infty} \frac{1}{n} \sum_{k=0}^{n-1} \varphi(T^k(x)) \tag{2.34}$$

exists for μ-almost every $x \in X$. Moreover, the following properties hold:

1. φ_T is T-invariant almost everywhere.
2. $\varphi_T \in L^1(X, \mu)$ and

$$\int_X \varphi_T \, d\mu = \int_X \varphi \, d\mu. \tag{2.35}$$

The proof of Theorem 2.2 is given in Sect. 2.5.3. Here we make some preliminary observations. Given a function $\varphi: X \to \mathbb{R}$, we have

$$\frac{1}{n} \sum_{k=0}^{n-1} \varphi(T^{k+1}(x)) = \frac{n+1}{n} \cdot \frac{1}{n+1} \sum_{k=0}^{n} \varphi(T^k(x)) - \frac{\varphi(x)}{n}$$

for every $n \in \mathbb{N}$. This shows that $\varphi_T(T(x))$ is well defined if and only if $\varphi_T(x)$ is well defined, in which case

$$\varphi_T(T(x)) = \varphi_T(x).$$

Therefore, the existence μ-almost everywhere of the limit in (2.34) yields Property 1 in the theorem.

2.5.2 Conditional Expectation

We introduce in this section the notion of conditional expectation of a function, which will allow us to give a somewhat expedite proof of Birkhoff's ergodic theorem.

Proposition 2.9. *Let $(X, \mathcal{A}, \mu)$ be a finite measure space and let $\mathcal{F} \subset \mathcal{A}$ be a σ-subalgebra. For each $\mathcal{A}$-measurable function $\varphi \in L^1(X, \mu)$, there exists an $\mathcal{F}$-measurable function $\varphi_{\mathcal{F}} \in L^1(X, \mu)$ such that*

$$\int_A \varphi_{\mathcal{F}} \, d\mu = \int_A \varphi \, d\mu \quad \text{for every} \quad A \in \mathcal{F}. \tag{2.36}$$

Proof. We define a finite measure ν in $\mathcal{F}$ by

$$\nu(A) = \int_A \varphi \, d\mu \quad \text{for every} \quad A \in \mathcal{F}.$$

If $\mu(A) = 0$ for some set $A \in \mathcal{F}$, then $\nu(A) = 0$. In other words, ν is absolutely continuous with respect to the restriction of μ to the σ-algebra $\mathcal{F}$. Therefore, by the Radon–Nikodym theorem (Theorem A.5) there exists an $\mathcal{F}$-measurable function $\varphi_{\mathcal{F}} \in L^1(X, \mu)$ (the Radon–Nikodym derivative of ν with respect to $\mu|\mathcal{F}$) such that

$$\nu(A) = \int_A \varphi_{\mathcal{F}} \, d\mu \quad \text{for every} \quad A \in \mathcal{F}.$$

This yields the desired result.   □

Definition 2.12. Any function $\varphi_{\mathcal{F}}$ as in Proposition 2.9 is called a *conditional expectation* of φ with respect to $\mathcal{F}$.

We note that, in general, $\varphi_{\mathcal{F}}$ is not unique since one can always change it on any set of zero measure in $\mathcal{F}$.

We give several examples of conditional expectations.

Example 2.3. Let $(X, \mathcal{A}, \mu)$ be a finite measure space and let $\varphi = \chi_B$ be the characteristic function of a set $B \in \mathcal{A}$. Given $E \in \mathcal{A}$ with $\mu(E) > 0$ and $\mu(X \setminus E) > 0$, we consider the σ-subalgebra

$$\mathcal{F} = \{\varnothing, E, X \setminus E, X\} \subset \mathcal{A},$$

and we find explicitly $\varphi_{\mathcal{F}}$. By Proposition 2.9, we have

$$\int_A \varphi_{\mathcal{F}} \, d\mu = \int_A \chi_B \, d\mu = \mu(A \cap B) \tag{2.37}$$

for every set $A \in \mathcal{F}$. We note that a function is $\mathcal{F}$-measurable if and only if it is constant on E and $X \setminus E$. Since $\varphi_{\mathcal{F}}$ is $\mathcal{F}$-measurable, letting $A = E$ and $A = X \setminus E$ in (2.37), we thus obtain

$$\varphi_{\mathcal{F}}(x) = \begin{cases} \mu(E \cap B)/\mu(E) & \text{if } x \in E, \\ \mu((X \setminus E) \cap B)/\mu(X \setminus E) & \text{if } x \in X \setminus E. \end{cases} \tag{2.38}$$

Now we consider the σ-subalgebra of invariant sets of a given transformation.

Example 2.4. Let $(X, \mathcal{A}, \mu)$ be a finite measure space and let $T \colon X \to X$ be an $\mathcal{A}$-measurable transformation. We consider the σ-subalgebra of T-invariant sets

$$\mathcal{F} = \{A \in \mathcal{A} : T^{-1}A = A\}. \tag{2.39}$$

Let also $\varphi \colon X \to \mathbb{R}$ be an $\mathcal{A}$-measurable function. We show that φ is $\mathcal{F}$-measurable if and only if the sets $\varphi^{-1}\alpha$ are T-invariant for any $\alpha \in \mathbb{R}$. Indeed, φ is $\mathcal{F}$-measurable if and only if $\varphi^{-1}B$ is T-invariant for every measurable set $B \subset \mathbb{R}$ and thus if and only if

$$(\varphi \circ T)^{-1}B = T^{-1}(\varphi^{-1}B) = \varphi^{-1}B \tag{2.40}$$

for every measurable set $B \subset \mathbb{R}$. Since $\{\alpha\} \subset \mathbb{R}$ is measurable for each $\alpha \in \mathbb{R}$, it follows from (2.40) that

$$(\varphi \circ T)^{-1}\alpha = \varphi^{-1}\alpha \tag{2.41}$$

for every $\alpha \in \mathbb{R}$. Now we assume that (2.41) holds for every $\alpha \in \mathbb{R}$. Then

$$\begin{aligned} (\varphi \circ T)^{-1}B &= (\varphi \circ T)^{-1} \bigcup_{\alpha \in B} \{\alpha\} \\ &= \bigcup_{\alpha \in B} (\varphi \circ T)^{-1}\alpha \\ &= \bigcup_{\alpha \in B} \varphi^{-1}\alpha \\ &= \varphi^{-1} \bigcup_{\alpha \in B} \{\alpha\} = \varphi^{-1}B \end{aligned}$$

for any set $B \subset \mathbb{R}$. This implies that φ is $\mathcal{F}$-measurable.

It follows from Proposition 2.8 that φ is $\mathcal{F}$-measurable if and only if is T-invariant. In particular, any conditional expectation $\varphi_{\mathcal{F}}$ of an $\mathcal{A}$-measurable function $\varphi \in L^1(X, \mu)$ is a T-invariant function.

2.5.3 Proof of Birkhoff's Ergodic Theorem

We are now ready to prove Birkhoff's ergodic theorem (Theorem 2.2).

Proof. Given $\psi \in L^1(X, \mu)$, for each $n \in \mathbb{N}$, we consider the function

$$\psi_n = \max\left\{\sum_{k=0}^{\ell-1} \psi \circ T^k : 1 \le \ell \le n\right\}.$$

Clearly, $\psi_{n+1} \ge \psi_n$ for each n. Since

$$|\psi_n| \le \sum_{k=0}^{n-1} |\psi \circ T^k|$$

and μ is T-invariant, it follows from Proposition 2.1 that

$$\int_X |\psi_n|\, d\mu \le \sum_{k=0}^{n-1} \int_X |\psi \circ T^k|\, d\mu$$

$$= n \int_X |\psi|\, d\mu < \infty.$$

Therefore, $(\psi_n)_{n\in\mathbb{N}}$ is a nondecreasing sequence of μ-integrable functions. Moreover,

$$\psi_{n+1} = \max\left\{\sum_{k=0}^{\ell} \psi \circ T^k : 1 \le \ell \le n\right\}$$

$$= \psi + \max\left\{0, \sum_{k=1}^{\ell} \psi \circ T^k : 1 \le \ell \le n\right\} \tag{2.42}$$

$$= \psi + \max\{0, \psi_n \circ T\}.$$

This implies that

$$\lim_{n\to\infty} \psi_{n+1}(x) = +\infty \quad \text{if and only if} \quad \lim_{n\to\infty} \psi_n(T(x)) = +\infty,$$

and hence,

$$A = \left\{x \in X : \lim_{n\to\infty} \psi_n(x) = +\infty\right\}$$

is a T-invariant set. It also follows from (2.42) that

$$\psi_{n+1} - \psi_n \circ T = \psi + \max\{0, \psi_n \circ T\} - \psi_n \circ T$$

$$= \psi - \min\{0, \psi_n \circ T\}, \tag{2.43}$$

and hence,

$$\psi_{n+1} - \psi_n \circ T \searrow \psi \text{ in } A \text{ when } n \to \infty.$$

Moreover, again by (2.43), we have

$$0 \le \psi_{n+1} - \psi_n \circ T - \psi \le \psi_2 - \psi_1 \circ T - \psi$$

for every $n \in \mathbb{N}$. Since $\psi_2 - \psi_1 \circ T - \psi$ is μ-integrable, it follows from the dominated convergence theorem (Theorem A.3) and Proposition 2.1 that

$$0 \le \int_A (\psi_{n+1} - \psi_n) \, d\mu$$
$$= \int_A (\psi_{n+1} - \psi_n \circ T) \, d\mu \to \int_A \psi \, d\mu$$

when $n \to \infty$, and thus,

$$\int_A \psi \, d\mu \ge 0. \tag{2.44}$$

Now let $\mathcal{F}$ be the σ-subalgebra $\mathcal{F}$ of T-invariant sets in (2.39). Given $\varepsilon > 0$, we consider the function

$$\psi = \varphi - \varphi_{\mathcal{F}} - \varepsilon.$$

By (2.36), since A is T-invariant, we have

$$\int_A \psi \, d\mu = -\varepsilon \mu(A),$$

and it follows from (2.44) that $\mu(A) = 0$. Hence,

$$\lim_{n \to \infty} \psi_n(x) < +\infty$$

for μ-almost every $x \in X$. Since

$$\frac{1}{n} \sum_{k=0}^{n-1} \psi \circ T^k \le \frac{\psi_n}{n},$$

we obtain

$$\limsup_{n \to \infty} \frac{1}{n} \sum_{k=0}^{n-1} \psi \circ T^k \le 0$$

μ-almost everywhere in X, and since $\varphi_{\mathcal{F}}$ is T-invariant (see Example 2.4), we conclude that

$$\limsup_{n \to \infty} \frac{1}{n} \sum_{k=0}^{n-1} \varphi \circ T^k \le \varphi_{\mathcal{F}} + \varepsilon \tag{2.45}$$

μ-almost everywhere in X. Now we observe that $(-\varphi)_{\mathscr{F}} = -\varphi_{\mathscr{F}}$. Therefore, replacing φ by $-\varphi$ in (2.45), we obtain

$$\liminf_{n\to\infty} \frac{1}{n} \sum_{k=0}^{n-1} \varphi \circ T^k \geq \varphi_{\mathscr{F}} - \varepsilon \qquad (2.46)$$

μ-almost everywhere in X. By (2.45) and (2.46), there is a set $X_\varepsilon \subset X$ with measure $\mu(X_\varepsilon) = \mu(X)$ on which

$$\varphi_{\mathscr{F}} - \varepsilon \leq \liminf_{n\to\infty} \frac{1}{n} \sum_{k=0}^{n-1} \varphi \circ T^k \leq \limsup_{n\to\infty} \frac{1}{n} \sum_{k=0}^{n-1} \varphi \circ T^k \leq \varphi_{\mathscr{F}} + \varepsilon.$$

Therefore,

$$\lim_{n\to\infty} \frac{1}{n} \sum_{k=0}^{n-1} \varphi \circ T^k = \varphi_{\mathscr{F}} \qquad (2.47)$$

in the set of full μ-measure

$$\bigcap_{m=1}^{\infty} X_{1/m} \subset X.$$

By (2.47), the number $\varphi_T(x)$ in (2.34) is well defined for μ-almost every $x \in X$, and $\varphi_T = \varphi_{\mathscr{F}}$ μ-almost everywhere. Since $\varphi_{\mathscr{F}} \in L^1(X, \mu)$, we conclude that $\varphi_T \in L^1(X, \mu)$, and it follows from (2.36) that

$$\int_X \varphi_T \, d\mu = \int_X \varphi_{\mathscr{F}} \, d\mu = \int_X \varphi \, d\mu.$$

This completes the proof of the theorem.   □

2.6 Ergodicity

In the theory of dynamical systems, one is often interested in decomposing the phase space into invariant sets that are indecomposable in some appropriate sense. In ergodic theory, we are also interested in a corresponding notion.

2.6.1 Basic Notions

We first introduce the notion of ergodic measure.

Definition 2.13. Let $T: X \to X$ be a measurable transformation. A measure μ in X is said to be *ergodic* with respect to T if the measure of any T-invariant measurable subset $A \subset X$ satisfies $\mu(A) = 0$ or $\mu(X \setminus A) = 0$. In this case, we also say that T is ergodic with respect to μ.

We note that an ergodic measure is not necessarily invariant. We start with a simple example.

Example 2.5. Given an invertible measurable transformation $T: X \to X$, let $x \in X$ be an m-periodic point which is not k-periodic for any other $k \mid m$. Then the T-invariant measure μ in (2.16) is ergodic since the only nonempty T-invariant set contained in the orbit of x is the orbit itself.

Now we present a criterion for ergodicity in terms of invariant functions. More precisely, we use the notion of T-invariance almost everywhere introduced in Definition 2.11.

Proposition 2.10. *If $T: X \to X$ be a measurable transformation and μ is a measure in X, then the following properties are equivalent:*

1. *The measure μ is ergodic with respect to T.*
2. *If $\varphi: X \to \mathbb{R}$ is a measurable function that is T-invariant almost everywhere, then it is constant almost everywhere, that is, there exists $\alpha \in \mathbb{R}$ such that $\varphi(x) = \alpha$ for μ-almost every $x \in X$.*

Proof. We first assume that μ is ergodic. Let $\varphi: X \to \mathbb{R}$ be a measurable function that is T-invariant almost everywhere. Proceeding in a similar manner to that in the proof of Proposition 2.8, we can show that $\varphi^{-1}A$ is T-invariant almost everywhere for every measurable subset $A \subset \mathbb{R}$. Indeed, let $B \subset X$ be a T-invariant measurable set with $\mu(X \setminus B) = 0$ such that $\varphi|B$ is $(T|B)$-invariant. If $A \subset \mathbb{R}$ is measurable, then by Proposition 2.8, we have

$$B \cap \varphi^{-1}A = B \cap \bigcup_{\alpha \in A} \varphi^{-1}\alpha$$

$$= \bigcup_{\alpha \in A} (\varphi|B)^{-1}\alpha$$

$$= \bigcup_{\alpha \in A} (T|B)^{-1}(\varphi|B)^{-1}\alpha$$

$$= \bigcup_{\alpha \in A} (\varphi|B \circ T|B)^{-1}\alpha$$

$$= \left((\varphi \circ T)|B\right)^{-1}A = B \cap (\varphi \circ T)^{-1}A.$$

Therefore, by ergodicity,

$$\mu(\varphi^{-1}A) = 0 \quad \text{or} \quad \mu(X \setminus \varphi^{-1}A) = 0. \tag{2.48}$$

For each $n \in \mathbb{Z}$ and $m \in \mathbb{N}$, we consider the interval $A_{mn} = [n/2^m, (n+1)/2^m)$. For a fixed m, the sets

$$\varphi^{-1} A_{mn} = \{x \in X : n/2^m \le \varphi(x) < (n+1)/2^m\}$$

are pairwise disjoint and satisfy $\bigcup_{n \in \mathbb{Z}} \varphi^{-1} A_{mn} = X$. It follows from (2.48) that there exists a unique $n = n_m \in \mathbb{Z}$ such that $\mu(X \setminus \varphi^{-1} A_{mn_m}) = 0$. Now we observe that the intersection $A = \bigcap_{m \in \mathbb{N}} A_{mn_m}$ contains at most one point since each interval A_{mn_m} has length $1/2^m$. Moreover,

$$\mu(X \setminus \varphi^{-1} A) = \mu \left(X \setminus \bigcap_{m \in \mathbb{N}} \varphi^{-1} A_{mn_m} \right)$$

$$= \mu \left(\bigcup_{m \in \mathbb{N}} (X \setminus A_{mn_m}) \right) = 0.$$

This shows that A is nonempty, and thus, it consists of a single point say α. In other words, there exists $\alpha \in \mathbb{R}$ such that $\varphi(x) = \alpha$ for μ-almost every $x \in X$.

Now we assume that all measurable functions that are T-invariant almost everywhere are constant almost everywhere. In particular, if a set $A \subset X$ is measurable and T-invariant, then its characteristic function χ_A is also measurable and T-invariant, and thus, $\chi_A = 0$ μ-almost everywhere or $\chi_A = 1$ μ-almost everywhere in X. We conclude that $\mu(A) = 0$ or $\mu(X \setminus A) = 0$. Therefore, μ is ergodic. $\square$

The following is a simple consequence of the proof of Proposition 2.10 for finite measures:

Proposition 2.11. *If* $T : X \to X$ *be a measurable transformation and* μ *is a finite measure in* X, *then the following properties are equivalent:*

1. *The measure* μ *is ergodic with respect to* T.
2. *If* $\varphi \in L^1(X, \mu)$ *is* T*-invariant almost everywhere, then it is constant almost everywhere.*

2.6.2 Fourier Coefficients and Ergodicity

We illustrate in this section how the Fourier coefficients can be used to establish the ergodicity of some transformations.

Given a function $\varphi \in L^1(\mathbb{T}, m)$, where m is the Lebesgue measure in $[0, 1]$, we define the *Fourier coefficients* of φ by

$$a_k(\varphi) = \int_0^1 e^{-2\pi i k x} \varphi(x) \, dm(x)$$

for each $k \in \mathbb{Z}$. We briefly recall some properties of these numbers. Two functions $\varphi, \psi \in L^1(\mathbb{T}, m)$ are equal Lebesgue-almost everywhere if and only if $a_k(\varphi) = a_k(\psi)$ for every $k \in \mathbb{Z}$. In other words, the Fourier coefficients of a function determine it completely on a set of full Lebesgue measure. In particular:

1. A function $\varphi \in L^1(\mathbb{T}, m)$ is zero Lebesgue-almost everywhere if and only if $a_k(\varphi) = 0$ for every $k \in \mathbb{Z}$.
2. A function $\varphi \in L^1(\mathbb{T}, m)$ is constant almost everywhere if and only if $a_k(\varphi) = 0$ for every $k \in \mathbb{Z} \setminus \{0\}$, in which case $\varphi(x) = a_0(\varphi)$ for Lebesgue-almost every $x \in \mathbb{T}$.

Furthermore, $a_k(\varphi) \to 0$ when $|k| \to \infty$, by the Riemann–Lebesgue lemma.

Example 2.6. Given $\alpha \in \mathbb{R} \setminus \mathbb{Q}$, let $T_\alpha : \mathbb{T} \to \mathbb{T}$ be the irrational interval translation defined by $T_\alpha(x) = x + \alpha \mod 1$ (see Sect. 2.2.3). For each $\varphi \in L^1(\mathbb{T}, m)$ and $k \in \mathbb{Z}$, we have

$$\begin{aligned}
a_k(\varphi \circ T_\alpha) &= \int_0^1 e^{-2\pi i k x} \varphi(T_\alpha(x)) \, dm(x) \\
&= \int_0^1 e^{-2\pi i k(x-\alpha)} \varphi(x) \, dm(x) \qquad (2.49) \\
&= e^{2\pi i k \alpha} a_k(\varphi).
\end{aligned}$$

Hence, the function φ is T_α-invariant almost everywhere if and only if

$$a_k(\varphi \circ T_\alpha) = a_k(\varphi) \quad \text{for every} \quad k \in \mathbb{Z}.$$

Since $\alpha \in \mathbb{R} \setminus \mathbb{Q}$, we have $e^{2\pi i k \alpha} \neq 1$ for every $k \neq 0$, and thus, by (2.49), the function φ is T_α-invariant almost everywhere if and only if $a_k(\varphi) = 0$ for every $k \neq 0$. This shows that all functions in $L^1(\mathbb{T}, m)$ that are T_α-invariant almost everywhere are constant almost everywhere. By Proposition 2.11, the Lebesgue measure is ergodic with respect to T_α.

One the other hand, the Lebesgue measure is not ergodic with respect to the rational translations of the interval. Indeed, consider the interval translation T_α with $\alpha = p/q \in \mathbb{Q}$ and $p, q \in \mathbb{Z}$. It is sufficient to observe that the function $\varphi \colon \mathbb{T} \to \mathbb{R}$ defined by $\varphi(x) = \cos(2\pi q x)$ is T_α-invariant, that is, $\varphi \circ T_\alpha = \varphi$, but it is clearly not constant almost everywhere.

In conclusion, the Lebesgue measure is ergodic with respect to an interval translation T_α if and only if $\alpha \in \mathbb{R} \setminus \mathbb{Q}$.

We can also consider a multidimensional version of the interval translations. Given a function $\varphi \in L^1(\mathbb{T}^n, \lambda)$, where λ is the Lebesgue measure in $\mathbb{T}^n$, we define the *Fourier coefficients* of φ by

$$a_k(\varphi) = \int_{\mathbb{T}^n} e^{-2\pi i \langle k, x \rangle} \varphi(x) \, d\lambda(x)$$

for each $k \in \mathbb{Z}^n$, where

$$\langle k, x \rangle = k_1 x_1 + \cdots + k_n x_n \tag{2.50}$$

is the standard inner product in $\mathbb{R}^n$. We recall that two functions $\varphi, \psi \in L^1(\mathbb{T}^n, \lambda)$ are equal Lebesgue-almost everywhere if and only if $a_k(\varphi) = a_k(\psi)$ for every $k \in \mathbb{Z}^n$. Moreover, $a_k(\varphi) \to 0$ when $\|k\| \to \infty$ by the Riemann–Lebesgue lemma.

Example 2.7. Let T_A be the automorphism of the torus $\mathbb{T}^n$ induced by a matrix A. We assume that

$$\det(A^m - \mathrm{Id}) \neq 0 \quad \text{for every} \quad m \in \mathbb{Z} \setminus \{0\}. \tag{2.51}$$

Now let $\varphi \in L^1(\mathbb{T}^n, \lambda)$. Since $|\det A| = 1$, for each $k \in \mathbb{Z}^n$, we have

$$\begin{aligned}
a_k(\varphi \circ T_A) &= \int_{\mathbb{T}^n} e^{-2\pi i \langle k, x \rangle} \varphi(Ax) \, d\lambda(x) \\
&= \int_{\mathbb{T}^n} e^{-2\pi i \langle k, A^{-1}x \rangle} \varphi(x) \, d\lambda(x) \\
&= \int_{\mathbb{T}^n} e^{-2\pi i \langle (A^{-1})^* k, x \rangle} \varphi(x) \, d\lambda(x) = a_{(A^{-1})^* k}(\varphi).
\end{aligned}$$

The function φ is T_A-invariant almost everywhere (with respect to λ) if and only if

$$a_k(\varphi) = a_{(A^{-1})^* k}(\varphi) \tag{2.52}$$

for every $k \in \mathbb{Z}^n$. Now we observe that if $k = (A^*)^m k$ for some $m, k \in \mathbb{Z}$, with $k \neq 0$, then $(A^*)^m$ and thus also A^m would have 1 as eigenvalue, which contradicts (2.51). This implies that the function $m \mapsto (A^*)^m k$ is injective for each $k \neq 0$, and hence,

$$\|(A^*)^m k\| \to \infty \quad \text{when} \quad m \to \infty.$$

It follows from the Riemann–Lebesgue lemma that $a_{(A^*)^m k}(\varphi) \to 0$ when $m \to \infty$, and using (2.52), we conclude that $a_k(\varphi) = 0$ for every $k \neq 0$. Therefore, $\varphi(x) = a_0(\varphi)$ for λ-almost every $x \in \mathbb{T}^n$. By Proposition 2.11, the measure λ is ergodic with respect to T_A.

Example 2.8. If a matrix $A \in S(n, \mathbb{Z})$ has no eigenvalues in S^1, such as $A = \begin{pmatrix} 2 & 1 \\ 1 & 1 \end{pmatrix}$, then no power of A has 1 as eigenvalue, and (2.51) holds. It follows from Example 2.7 that if $A \in S(n, \mathbb{Z})$ has no eigenvalues in S^1, then λ is ergodic with respect to the toral automorphism T_A.

2.6.3   The Case of Invariant Measures

Now we consider the particular case of *invariant* ergodic measures, and we obtain additional information to that given by Birkhoff's ergodic theorem (Theorem 2.2).

Proposition 2.12. *Let $T: X \to X$ be a measurable transformation and let μ be a T-invariant ergodic measure in X with $\mu(X) < \infty$. If $\varphi \in L^1(X, \mu)$, then*

$$\lim_{n \to \infty} \frac{1}{n} \sum_{k=0}^{n-1} \varphi(T^k(x)) = \frac{1}{\mu(X)} \int_X \varphi \, d\mu$$

for μ-almost every $x \in X$.

Proof. By Theorem 2.2, the function φ_T defined μ-almost everywhere by (2.34) is T-invariant almost everywhere. Therefore, by Proposition 2.11, we conclude that φ_T is constant almost everywhere. The desired result follows now readily from (2.35). □

Under the assumptions of Proposition 2.12, using (2.32) and (2.33), we find that if $A \subset X$ is a measurable set, then

$$\lim_{n \to \infty} \frac{\tau_n(A, x)}{n} = \frac{\mu(A)}{\mu(X)}$$

for μ-almost every $x \in X$. In other words, with respect to an invariant ergodic measure, the frequency with which the orbit of a "typical" point visits a given set is proportional to the measure of the set.

2.7   Applications to Number Theory

We present briefly in this section some applications of ergodic theory to problems of number theory. In particular, we consider fractional parts of polynomials and continued fractions.

2.7.1   Fractional Parts of Polynomials

Consider the polynomial

$$P(t) = a_0 t^r + \cdots + a_r, \tag{2.53}$$

where $r \in \mathbb{N}$ and $a_0, \ldots, a_r \in \mathbb{R}$.

Definition 2.14. The numbers $P_n = P(n) \bmod 1 \in [0, 1)$, for $n \in \mathbb{N}$, are called the *fractional parts* of the polynomial P.

Now we introduce the notion of uniformly distributed sequence.

Definition 2.15. We say that a sequence $(y_n)_{n \in \mathbb{N}} \subset [0, 1]$ is *uniformly distributed* if

$$\lim_{n \to \infty} \frac{1}{n} \sum_{k=1}^{n} \varphi(y_k) = \int_0^1 \varphi(x) \, dx$$

for every continuous function $\varphi \colon [0, 1] \to \mathbb{R}$.

We want to illustrate how ergodic theory can be used to study the distribution of the fractional parts of a polynomial.

Example 2.9. Given $r = 1$, $a_0 = \alpha \in \mathbb{R} \setminus \mathbb{Q}$, and $a_1 = \beta \in \mathbb{R}$, we have $P_n = T^n(\beta)$, where $T(x) = x + \alpha \bmod 1$. By Example 2.6, the irrational translation T is ergodic. Therefore, by Proposition 2.12, for any continuous function $\varphi \colon [0, 1] \to \mathbb{R}$, we have

$$\lim_{n \to \infty} \frac{1}{n} \sum_{k=1}^{n} \varphi(P_k) = \lim_{n \to \infty} \frac{1}{n} \sum_{k=1}^{n} \varphi(T^k(\beta)) = \int_0^1 \varphi(x) \, dx$$

for Lebesgue-almost every $\beta \in [0, 1]$. This shows that if $\alpha \notin \mathbb{Q}$, then the sequence $(P_n)_{n \in \mathbb{N}}$ of fractional parts of the polynomial $P(t) = \alpha t + \beta$ is uniformly distributed for Lebesgue-almost every $\beta \subset \mathbb{R}$.

Now we consider polynomials of degree 2.

Proposition 2.13. *Given $\alpha \in \mathbb{R} \setminus \mathbb{Q}$, the sequence $(P_n)_{n \in \mathbb{N}}$ of fractional parts of the polynomial $P(t) = \alpha t^2 + \beta t + \gamma$ is uniformly distributed for Lebesgue-almost every $(\beta, \gamma) \in \mathbb{R}^2$.*

Proof. Let $T \colon \mathbb{T}^2 \to \mathbb{T}^2$ be defined by

$$T(x, y) = (x + \alpha, y + 2x + \alpha) \bmod 1.$$

Since $\det d_{(x,y)} T = 1$ for $(x, y) \in \mathbb{T}^2$, the transformation T preserves the Lebesgue measure. Moreover, one can use induction to show that

$$T^n(x, y) = (x + n\alpha, \alpha n^2 + 2nx + y) \bmod 1$$

$$= (x + n\alpha, P_n) \bmod 1$$

for every $n \in \mathbb{N}$, with $\beta = 2x$ and $\gamma = y$. Now we show that the Lebesgue measure λ in $\mathbb{T}^2$ is ergodic with respect to T. Take $\varphi \in L^1(\mathbb{T}^2, \lambda)$. The Fourier coefficients of the function $\varphi \circ T$ are given by

$$a_{k\ell}(\varphi \circ T) = \int_{\mathbb{T}^2} e^{-2\pi i(kx+\ell y)} \varphi(x+\alpha, y+2x+\alpha) \, d\lambda(x,y)$$

$$= \int_{\mathbb{T}^2} e^{-2\pi i[(k-2\ell)x+\ell y]} e^{2\pi i(k-\ell)\alpha} \varphi(x,y) \, d\lambda(x,y)$$

$$= e^{2\pi i(k-\ell)\alpha} a_{k-2\ell,\ell}(\varphi).$$

If φ is T-invariant almost everywhere (with respect to λ), then

$$e^{2\pi i(k-\ell)\alpha} a_{k-2\ell,\ell}(\varphi) = a_{k\ell}(\varphi) \tag{2.54}$$

for every $k, \ell \in \mathbb{Z}$. In particular, if $\ell = 0$, then

$$e^{2\pi ik\alpha} a_{k0}(\varphi) = a_{k0}(\varphi) \quad \text{for every} \quad k \in \mathbb{Z}.$$

Since $\alpha \in \mathbb{R} \setminus \mathbb{Q}$, we have $e^{2\pi ik\alpha} \neq 1$, and hence, $a_{k0}(\varphi) = 0$ for every $k \in \mathbb{Z} \setminus \{0\}$.
On the other hand, if $a_{k\ell}(\varphi) \neq 0$ for some $\ell \neq 0$, then it follows from (2.54) that

$$|a_{k-2\ell,\ell}(\varphi)| = |a_{k\ell}(\varphi)| \neq 0 \quad \text{for every} \quad k \in \mathbb{Z},$$

and hence,

$$|a_{k-2n\ell,\ell}(\varphi)| = |a_{k\ell}(\varphi)| \neq 0$$

for every $k \in \mathbb{Z}$ and $n \in \mathbb{N}$. But this contradicts the Riemann–Lebesgue lemma, and
we conclude that

$$a_{k\ell}(\varphi) = 0 \quad \text{for every} \quad (k, \ell) \in \mathbb{Z}^2 \setminus \{(0,0)\}.$$

Therefore, $\varphi = a_{00}(\varphi)$ Lebesgue-almost everywhere, and so φ is constant almost
everywhere. This shows that the Lebesgue measure is ergodic with respect to T.

For each continuous function $\psi: [0, 1] \to \mathbb{R}$, we define a continuous function
in $\mathbb{T}^2$ by $\varphi(x, y) = \psi(y)$. By Proposition 2.12, we have

$$\lim_{n \to \infty} \frac{1}{n} \sum_{k=1}^{n} \psi(P_k) = \lim_{n \to \infty} \frac{1}{n} \sum_{k=1}^{n} \varphi(T^k(x,y))$$

$$= \int_{\mathbb{T}^2} \varphi \, d\lambda$$

$$= \int_0^1 \psi(\tau) \, d\tau$$

for λ-almost every $(x, y) \in \mathbb{T}^2$ and thus for Lebesgue-almost every $(\beta, \gamma) \in \mathbb{R}^2$.
This yields the desired result. $\qquad\square$

Fig. 2.11 Gauss transformation

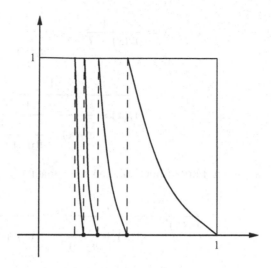

2.7.2 Continued Fractions

Now we consider the relation between ergodic theory and the theory of continued fractions. We first introduce a transformation that is closely related to the theory of continued fractions.

Definition 2.16. We define the *Gauss transformation* $T: [0, 1) \to [0, 1)$ by

$$T(x) = \begin{cases} 1/x \bmod 1 & \text{if } x \neq 0, \\ 0 & \text{if } x = 0 \end{cases}$$

(see Fig. 2.11).

Given an irrational number $x \in (0, 1)$, we can define positive integers

$$n_j(x) = \left\lfloor \frac{1}{T^{j-1}(x)} \right\rfloor \tag{2.55}$$

for each $j \in \mathbb{N}$. Since

$$T^j(x) = T\big(T^{j-1}(x)\big) = \frac{1}{T^{j-1}(x)} - n_j(x),$$

we obtain

$$T^{j-1}(x) = \frac{1}{n_j(x) + T^j(x)}$$

for each $j \in \mathbb{N}$. Therefore,

$$x = \frac{1}{n_1(x) + T(x)} = \frac{1}{n_1(x) + \dfrac{1}{n_2(x) + T^2(x)}}$$

$$= \frac{1}{n_1(x) + \dfrac{1}{n_2(x) + \dfrac{1}{n_3(x) + T^3(x)}}}$$

and so on. Furthermore, one can show that the sequence

$$\frac{1}{n_1(x)}, \quad \frac{1}{n_1(x) + \dfrac{1}{n_2(x)}}, \quad \frac{1}{n_1(x) + \dfrac{1}{n_2(x) + \dfrac{1}{n_3(x)}}}, \quad \ldots$$

converges to x. So we simply write

$$x = \frac{1}{n_1(x) + \dfrac{1}{n_2(x) + \cdots}}. \tag{2.56}$$

The right-hand side of (2.56) is called the *continued fraction* of x.

Now we consider the problem of computing the frequency with which an integer $k \in \mathbb{N}$ occurs in the sequence $n_1(x), n_2(x), \ldots$. We first show how to translate this problem into the language of ergodic theory. It follows from (2.55) that $n_j(x) = k$ if and only if

$$T^{j-1}(x) \in \left(\frac{1}{k+1}, \frac{1}{k} \right].$$

Therefore, given an irrational number $x \in (0, 1)$, the frequency with which $k \in \mathbb{N}$ occurs in the sequence $n_1(x), n_2(x), \ldots$ is given by

$$\eta_k(x) = \lim_{n \to \infty} \frac{1}{n} \, \mathrm{card} \left\{ j \in \{1, \ldots, n\} : n_j(x) = k \right\}$$

$$= \lim_{n \to \infty} \frac{1}{n} \sum_{j=1}^{n} \chi_{\left(\frac{1}{k+1}, \frac{1}{k} \right]}(T^{j-1}(x)),$$

whenever the limit exists. We will show that $\eta_k(x)$ is well defined for Lebesgue-almost every $x \in (0, 1)$.

We start by considering an appropriate measure.

Definition 2.17. The *Gauss measure* is the probability measure μ in $[0, 1)$ defined by

$$\mu(A) = \frac{1}{\log 2} \int_A \frac{dx}{1+x}$$

for each measurable set $A \subset [0, 1)$.

We note that

$$\frac{1}{2} < \frac{1}{1+x} < 1 \quad \text{for every} \quad x \in [0, 1).$$

This implies that

$$\frac{m(A)}{2 \log 2} \le \mu(A) \le \frac{m(A)}{\log 2} \tag{2.57}$$

for any measurable set $A \subset [0, 1)$, where m is the Lebesgue measure in $\mathbb{R}$. In particular, the Gauss measure μ is equivalent to the Lebesgue measure, that is, μ is absolutely continuous with respect to m, and vice versa. We also have the following property:

Proposition 2.14. *The Gauss transformation preserves the Gauss measure.*

Proof. It is sufficient to consider the intervals of the form $[0, b)$ for $b \in (0, 1)$ since these generate the Borel σ-algebra of $[0, 1)$. We have

$$T^{-1}[0, b) = \bigcup_{n=1}^{\infty} \left[\frac{1}{n+b}, \frac{1}{n} \right),$$

with a disjoint union. Therefore,

$$
\begin{aligned}
\mu(T^{-1}[0, b)) &= \sum_{n=1}^{\infty} \mu\left(\left[\frac{1}{n+b}, \frac{1}{n} \right) \right) \\
&= \sum_{n=1}^{\infty} \frac{1}{\log 2} \int_{1/(n+b)}^{1/n} \frac{dx}{1+x} \\
&= \sum_{n=1}^{\infty} \frac{1}{\log 2} \log \frac{1+1/n}{1+1/(n+b)} \\
&= \frac{1}{\log 2} \sum_{n=1}^{\infty} \left(\log \frac{n+1}{n+1+b} - \log \frac{n}{n+b} \right) \\
&= -\frac{1}{\log 2} \log \frac{1}{1+b} \\
&= \frac{1}{\log 2} \int_0^b \frac{dx}{1+x} = \mu([0, b)).
\end{aligned}
$$

This yields the desired statement. $\qquad\square$

Now we show that the Gauss measure is ergodic.

Theorem 2.3. *The Gauss measure is ergodic with respect to the Gauss transformation.*

Proof. Consider the open intervals

$$I_k = \left(\frac{1}{k+1}, \frac{1}{k} \right)$$

for $k \in \mathbb{N}$. Given $n \in \mathbb{N}$ and $i_0, \ldots, i_{n-1} \in \mathbb{N}$, we also consider the open interval

$$I_{i_0 \cdots i_{n-1}} = \bigcap_{k=0}^{n-1} T^{-k} I_{i_k}.$$

Clearly, the transformation T^n is of class C^∞ and is strictly decreasing on each interval $I_{i_0 \cdots i_{n-1}}$. Furthermore, $T^n(I_{i_0 \cdots i_{n-1}}) = (0, 1)$. We denote by

$$\psi_{i_0 \cdots i_{n-1}} \colon (0, 1) \to I_{i_0 \cdots i_{n-1}}$$

the corresponding local inverse of T^n, that is, the unique C^∞ function in $(0, 1)$ such that

$$\psi_{i_0 \cdots i_{n-1}}((0, 1)) = I_{i_0 \cdots i_{n-1}}, \tag{2.58}$$

and

$$T^n(\psi_{i_0 \cdots i_{n-1}}(x)) = x \quad \text{for every} \quad x \in (0, 1).$$

On the other hand, for each $x \in I_k$, we have $|T'(x)| = 1/x^2 \geq 1$, and

$$(T^2)'(x) = \left(1 - x \left\lfloor \frac{1}{x} \right\rfloor \right)^{-2} > (k+1)^2 \geq 4.$$

When n is even, we obtain

$$|(T^n)'(x)| = \left| [(T^2)^{n/2}]'(x) \right| \geq 4^{n/2} = 2^n,$$

and when n is odd,

$$\begin{aligned}
|(T^n)'(x)| &= \left| (T^{n-1} \circ T)'(x) \right| \\
&= \left| (T^{n-1})'(T(x)) \right| \cdot |T'(x)| \\
&\geq 4^{(n-1)/2} = 2^{n-1}.
\end{aligned}$$

Therefore, $|(T^n)'(x)| \geq 2^{n-1}$ for every $x \in \bigcup_{k=1}^{\infty} I_k$.

Now take $x, y \in [0, 1]$, $n \in \mathbb{N}$, and $i_0, \ldots, i_{n-1} \in \mathbb{N}$. By the mean value theorem, we obtain

$$
\begin{aligned}
|x - y| &= \left| T^n(\psi_{i_0 \cdots i_{n-1}}(x)) - T^n(\psi_{i_0 \cdots i_{n-1}}(y)) \right| \\
&= \left| (T^n)'(z) \cdot |\psi_{i_0 \cdots i_{n-1}}(x) - \psi_{i_0 \cdots i_{n-1}}(y)| \right. \\
&\geq 2^{n-1} |\psi_{i_0 \cdots i_{n-1}}(x) - \psi_{i_0 \cdots i_{n-1}}(y)|
\end{aligned}
$$

for some point $z \in I_{i_0 \cdots i_{n-1}}$, and thus,

$$
|\psi_{i_0 \cdots i_{n-1}}(x) - \psi_{i_0 \cdots i_{n-1}}(y)| \leq 2^{1-n} |x - y|. \tag{2.59}
$$

By (2.58), it follows from (2.59) that the interval $I_{i_0 \cdots i_{n-1}}$ has length at most 2^{1-n}. Therefore, given a measurable set $B \subset [0, 1)$ and $\varepsilon > 0$, there is a disjoint union J of intervals in the family

$$
\mathcal{F} = \left\{ I_{i_0 \cdots i_{n-1}} : n \in \mathbb{N} \text{ and } i_0, \ldots, i_{n-1} \in \mathbb{N} \right\} \tag{2.60}
$$

such that

$$
\mu(B \setminus J) < \varepsilon \quad \text{and} \quad \mu(J \setminus B) < \varepsilon.
$$

Now we observe that

$$
\begin{aligned}
\frac{|\psi'_{i_0 \cdots i_{n-1}}(x)|}{|\psi'_{i_0 \cdots i_{n-1}}(y)|} &= \frac{|(T^n)'(\psi_{i_0 \cdots i_{n-1}}(y))|}{|(T^n)'(\psi_{i_0 \cdots i_{n-1}}(x))|} \\
&= \prod_{j=0}^{n-1} \frac{|T'(\psi_{i_j \cdots i_{n-1}}(y))|}{|T'(\psi_{i_j \cdots i_{n-1}}(x))|} \\
&\leq \prod_{j=0}^{n-1} \left(1 + \frac{|T'(\psi_{i_j \cdots i_{n-1}}(y)) - T'(\psi_{i_j \cdots i_{n-1}}(x))|}{|T'(\psi_{i_j \cdots i_{n-1}}(x))|} \right) \\
&\leq \prod_{j=0}^{n-1} \left(1 + \frac{|T''(z_j)| \cdot |\psi_{i_j \cdots i_{n-1}}(y) - \psi_{i_j \cdots i_{n-1}}(x)|}{|T'(\psi_{i_j \cdots i_{n-1}}(x))|} \right)
\end{aligned}
$$

for some $z_j \in I_{i_j \cdots i_{n-1}}$. Moreover,

$$
|\psi_{i_{j+1} \cdots i_{n-1}}(y) - \psi_{i_{j+1} \cdots i_{n-1}}(x)| = |T'(w_j)| \cdot |\psi_{i_j \cdots i_{n-1}}(y) - \psi_{i_j \cdots i_{n-1}}(x)|
$$

for some $w_j \in I_{i_j \cdots i_{n-1}}$, and since $I_{i_j \cdots i_{n-1}} \subset I_{i_j}$, we obtain

$$\frac{|T''(z_j)|}{|T'(\psi_{i_j \cdots i_{n-1}}(x))| \cdot |T'(w_j)|} = \frac{2\psi_{i_j \cdots i_{n-1}}(x)^2 w_j^2}{z_j^3}$$

$$\leq \frac{2(i_j + 1)^3}{i_j^4}$$

$$= \frac{2}{i_j}\left(1 + \frac{1}{i_j}\right)^3 \leq 16$$

Therefore, it follows from (2.59) that

$$\frac{|\psi'_{i_0 \cdots i_{n-1}}(x)|}{|\psi'_{i_0 \cdots i_{n-1}}(y)|} \leq \prod_{j=0}^{n-1}(1 + 16 \cdot 2^{2-n+j}|x - y|)$$

$$\leq \prod_{j=-4}^{\infty}(1 + 2^{-j})$$

$$= \exp \sum_{j=-4}^{\infty} \log(1 + 2^{-j})$$

$$\leq \exp \sum_{j=-4}^{\infty} 2^{-j} = e^{2^5} =: c$$

since $\log(1 + x) \leq x$ for every $x \geq 0$.

Let again m be the Lebesgue measure in $[0, 1]$. For any measurable set $A \subset [0, 1]$, it follows from (2.57) that

$$\mu(I_{i_0 \cdots i_{n-1}} \cap T^{-n}A) = \frac{1}{\log 2}\int_{I_{i_0 \cdots i_{n-1}} \cap T^{-n}A} \frac{dx}{1 + x}$$

$$\geq \frac{1}{2\log 2}m(I_{i_0 \cdots i_{n-1}} \cap T^{-n}A)$$

$$= \frac{1}{2\log 2}m(\psi_{i_0 \cdots i_{n-1}}(A))$$

$$= \frac{1}{2\log 2}\int_A |\psi'_{i_0 \cdots i_{n-1}}|\, dm$$

$$\geq \frac{1}{2\log 2}m(A)\inf\{|\psi'_{i_0 \cdots i_{n-1}}(x)| : x \in (0, 1)\}$$

$$\geq \frac{1}{2c\log 2}m(A)\int_0^1 |\psi'_{i_0 \cdots i_{n-1}}|\, dm$$

$$= \frac{1}{2c \log 2} m(A) \int_{I_{i_0 \cdots i_{n-1}}} 1 \, dm$$

$$\geq \frac{\log 2}{2c} \mu(A) \mu(I_{i_0 \cdots i_{n-1}}).$$

Now we assume that the set A is T-invariant. By former inequality, we obtain

$$\mu(I_{i_0 \cdots i_{n-1}} \cap A) \geq \frac{\log 2}{2c} \mu(A) \mu(I_{i_0 \cdots i_{n-1}}). \tag{2.61}$$

As observed above, given $\varepsilon > 0$, there is a disjoint union J of intervals in the family $\mathcal{F}$ (see (2.60)) such that setting $X = [0, 1)$,

$$\mu(J \setminus (X \setminus A)) < \varepsilon \quad \text{and} \quad \mu((X \setminus A) \setminus J) < \varepsilon.$$

By (2.61), we conclude that

$$\mu(J \cap A) \geq \frac{\log 2}{2c} \mu(A) \mu(J),$$

and since $J \setminus (X \setminus A) = J \cap A$, we obtain

$$\varepsilon > \frac{\log 2}{2c} \mu(A)(\mu(X \setminus A) - \varepsilon).$$

Letting $\varepsilon \to 0$ yields $\mu(A)\mu(X \setminus A) = 0$, and thus, the measure μ is ergodic. $\qquad \square$

With slight changes, the method of proof of Theorem 2.3 can be applied to a large class of transformations with an invariant measure that is absolutely continuous with respect to the Lebesgue measure. See in particular Exercise 3.28.

It follows from Theorem 2.3 and Proposition 2.12 that

$$\eta_k(x) = \lim_{n \to \infty} \frac{1}{n} \sum_{j=1}^{n} \chi_{\left(\frac{1}{k+1}, \frac{1}{k}\right]}(T^{j-1}(x))$$

$$= \int_{[0,1)} \chi_{\left(\frac{1}{k+1}, \frac{1}{k}\right]} \, d\mu$$

$$= \frac{1}{\log 2} \int_{1/(k+1)}^{1/k} \frac{dx}{1 + x}$$

$$= \frac{1}{\log 2} \log \frac{(k+1)^2}{k(k+2)}$$

for μ-almost every $x \in [0, 1)$. Moreover, since

$$\left(\log \frac{(x+1)^2}{x(x+2)}\right)' = -\frac{2}{x(x+1)(x+2)} < 0$$

for $x > 0$, all frequencies $\eta_k(x)$ are distinct, and

$$\eta_1(x) > \eta_2(x) > \eta_3(x) > \cdots$$

for μ-almost every $x \in [0, 1)$.

2.8 Exercises

Exercise 2.1. Show that if R_w is an irrational circle rotation, then for each $z \in S^1$, the set $\{R_w^n(z) : n \in \mathbb{N}\}$ is dense in S^1.

Exercise 2.2. Given an integer $q > 1$, show that if T is a piecewise linear expanding map of the interval obtained from the subintervals

$$\left(0, \frac{1}{q}\right), \left(\frac{1}{q}, \frac{2}{q}\right), \ldots, \left(\frac{q-1}{q}, 1\right),$$

then $h \circ T \circ h^{-1} = E_q$ in the set

$$S^1 \setminus \{e^{2\pi i k/q} : k = 0, \ldots, q-1\},$$

where $h: \mathbb{T} \to S^1$ is the one-to-one map in (2.11).

Exercise 2.3. Show that if T is a piecewise linear expanding map of the interval, obtained from a (finite or infinite) number q of subintervals, then $p(T) - \log q$.

Exercise 2.4. For the toral endomorphism of $\mathbb{T}^2$ induced by the matrix $\left(\begin{smallmatrix} 3 & 1 \\ 1 & 1 \end{smallmatrix}\right)$, show that:

1. The orbit of each point with rational coordinates contains at least one periodic point.
2. Not all points with rational coordinates are periodic.
3. $(1/3, 1/3)$ is a periodic point of period 8.

Exercise 2.5. Compute the periodic entropy of the automorphism of $\mathbb{T}^2$ induced by the matrix $\left(\begin{smallmatrix} 2 & 1 \\ 1 & 1 \end{smallmatrix}\right)$.

Exercise 2.6. Find all periodic points of the automorphism of $\mathbb{T}^2$ induced by the matrix $\left(\begin{smallmatrix} 2 & 1 \\ 1 & 1 \end{smallmatrix}\right)$. Hint: consider the set of points with rational coordinates $\mathbb{Q}^2/\mathbb{Z}^2 \subset \mathbb{T}^2$.

Exercise 2.7. Show that the statement in Theorem 2.1 in general does not hold for infinite measures. Hint: consider the map $T: \mathbb{R} \to \mathbb{R}$ given by $T(x) = x + 1$.

Exercise 2.8. Find a continuous transformation $T: X \to X$ of a compact metric space and an infinite measure μ in X for which the statement in Theorem 2.1 does not hold.

Exercise 2.9. Let $T: X \to X$ be a measurable transformation and let μ be a T-invariant probability measure in X. Show that:

1. μ is ergodic if and only if

$$\mu\left(\bigcup_{n=1}^{\infty} T^{-n} A\right) = 1$$

   for every measurable set $A \subset X$ with $\mu(A) > 0$.

2. If μ is ergodic and $\varphi: X \to \mathbb{R}_0^+$ is a measurable function such that

$$\limsup_{n\to\infty} \frac{1}{n} \sum_{k=0}^{n-1} \varphi(T^k(x)) < \infty$$

   for μ-almost every $x \in X$, then $\varphi \in L^1(X, \mu)$. Hint: use Theorem A.1.

Exercise 2.10. Let $x \in X$ be an m-periodic point of a transformation $T: X \to X$. Show that the limit

$$\lim_{n\to\infty} \frac{1}{n} \sum_{k=0}^{n-1} \varphi(T^k(x))$$

exists for any function $\varphi: X \to \mathbb{R}$. Also, compute the limit explicitly and express it in terms of the measure μ in (2.16). Hint: verify that $\int_X \varphi \, d\delta_x = \varphi(x)$.

Exercise 2.11. Let $T: X \to X$ be a measurable transformation preserving an ergodic finite measure μ in X. Show that if $\varphi \in L^1(X, \mu)$ has integral $\int_X \varphi \, d\mu \neq 0$, then

$$\lim_{n\to\infty} \frac{1}{\log n} \log \left| \sum_{k=0}^{n-1} \varphi(T^k(x)) \right| = 1$$

for μ-almost every $x \in X$.

Exercise 2.12. Show that for Lebesgue-almost every $x \in [0, 1]$, the frequency of the number 1 in the base-2 representation of x is equal to $1/2$. Hint: show that the transformation $x \mapsto 2x \bmod 1$ is ergodic with respect to the Lebesgue measure.

Exercise 2.13. Show that if $T: X \to X$ is a measurable transformation preserving a finite measure μ in X and $\varphi \in L^1(X, \mu)$, then

$$\lim_{n\to\infty} \frac{\varphi(T^n(x))}{n} = 0$$

for μ-almost every $x \in X$.

Exercise 2.14. Given $\alpha_1, \ldots, \alpha_n \in \mathbb{R}$, let $T_{\alpha_1, \ldots, \alpha_n} : \mathbb{T}^n \to \mathbb{T}^n$ be defined by

$$T_{\alpha_1, \ldots, \alpha_n}(x_1, \ldots, x_n) = (x_1 + \alpha_1, \ldots, x_n + \alpha_n) \bmod 1. \tag{2.62}$$

The transformation $T_{\alpha_1, \ldots, \alpha_n}$ is called a *toral translation*. Describe a necessary and sufficient condition in terms of $\alpha_1, \ldots, \alpha_n$ so that the Lebesgue measure in $\mathbb{T}^n$ is ergodic with respect to $T_{\alpha_1, \ldots, \alpha_n}$.

Exercise 2.15. Let $(X, \mathcal{A}, \mu)$ be a finite measure space and let $\mathcal{F} \subset \mathcal{A}$ be a finite σ-subalgebra. Given an $\mathcal{A}$-measurable function $\varphi \in L^1(X, \mu)$ and a set $A \in \mathcal{F}$ with $\mu(A) > 0$ such that no proper subset of A in $\mathcal{F}$ has positive measure, show that

$$\varphi_{\mathcal{F}}(x) = \frac{\int_A \varphi \, d\mu}{\mu(A)}$$

for μ-almost every $x \in A$. This yields a generalization of formula (2.38) for arbitrary functions.

Exercise 2.16. Let $T : X \to X$ be a measurable transformation preserving a σ-finite measure μ in X with $\mu(X) = \infty$ (see Appendix A for the definition of σ-finite measure). Show that if $\varphi \in L^1(X, \mu)$, then:

1. The limit $\varphi_T(x)$ in (2.34) exists for μ-almost every $x \in X$.
2. The function φ_T is measurable and T-invariant almost everywhere.

Hint: first obtain a version of Proposition 2.9 for σ-finite measures (see Theorems A.5 and A.6). This yields a generalization of Birkhoff's ergodic theorem to some infinite measure spaces.

Exercise 2.17. Let $T : X \to X$ be a measurable transformation preserving an ergodic measure μ in X which is infinite but σ-finite. Show that if $\varphi \in L^1(X, \mu)$, then

$$\lim_{n \to \infty} \frac{1}{n} \sum_{k=0}^{n-1} \varphi(T^k(x)) = 0$$

for μ-almost every $x \in X$. Hint: use Exercise 2.16.

Exercise 2.18. Let $\varphi : S^1 \to \mathbb{R}$ be the function defined by $\varphi = \chi_{h([0, 1/2])}$ (see Fig. 2.12), with h as in (2.11). For the Lebesgue measure in S^1, find the conditional expectation $\varphi_{\mathcal{F}}$ with respect to the σ-subalgebra $\mathcal{F}$ of E_2-invariant sets.

Exercise 2.19. Show that for each integer $n > 1$, the Lebesgue measure m in $\mathbb{T}$ is ergodic with respect to the transformation $T(x) = nx \bmod 1$. Hint: show that the Fourier coefficients of a function $\varphi \in L^1(\mathbb{T}, m)$ satisfy

$$a_k(\varphi \circ T) = \begin{cases} a_{k/n}(\varphi) & \text{if } n \mid k, \\ 0 & \text{if } n \nmid k. \end{cases}$$

Fig. 2.12 Conditional
expectation in the circle

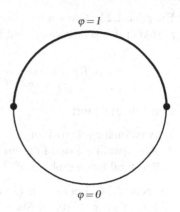

Exercise 2.20. Show that for each integer $n > 1$, the Lebesgue measure in S^1 is ergodic with respect to the transformation $z \mapsto z^n$.

Exercise 2.21. Let $T: \mathbb{R}^n \to \mathbb{R}^n$ be a C^1 diffeomorphism. Show that:

1. T preserves the Lebesgue measure if and only if $|\det d_x T| = 1$ for every $x \in \mathbb{R}^n$.
2. For a measure μ in $\mathbb{R}^n$ that is absolutely continuous with respect to the Lebesgue measure, with Radon–Nikodym derivative ρ (see Appendix A for the definition), the diffeomorphism T preserves μ if and only if

$$\rho = (\rho \circ T)|\det dT| \tag{2.63}$$

Lebesgue-almost everywhere in $\mathbb{R}^n$.
3. If ρ is defined everywhere and is continuous, then T preserves μ if and only if (2.63) holds everywhere.

Exercise 2.22. Let $T: \mathbb{R}^n \to \mathbb{R}^n$ be a C^1 diffeomorphism preserving a measure that is absolutely continuous with respect to the Lebesgue measure, with a continuous Radon–Nikodym derivative ρ defined everywhere. We also assume that

$$0 < \inf\{\rho(x) : x \in \mathbb{R}^n\} \le \sup\{\rho(x) : x \in \mathbb{R}^n\} < \infty.$$

1. Show that $|\det d_x T^m| = 1$ for each m-periodic point $x \in \mathbb{R}^n$.
2. Show that the set

$$\{\det d_x T^n : x \in \mathbb{R}^n \text{ and } n \in \mathbb{N}\}$$

is bounded.

Hint: use Exercise 2.21.

Exercise 2.23. Given a transformation $T: X \to X$, we say that a point $x \in X$ has *period m* if x is m-periodic, but is not k-periodic for any other $k \mid m$. Let

$$q(T) = \limsup_{m \to \infty} \frac{1}{m} \log^+ \operatorname{card} \{x \in X : x \text{ has period } m\}.$$

Compute $q(T)$ for:

1. A rational circle rotation.
2. An expanding map of the circle.
3. The toral automorphism of $\mathbb{T}^2$ induced by the matrix $\left(\begin{smallmatrix} 2 & 1 \\ 1 & 1 \end{smallmatrix}\right)$.

Exercise 2.24. Let $n_i = n_i(x)$ for $i \in \mathbb{N}$ be the integers in the continued fraction of a number $x \in (0, 1)$. Show that the geometric average $\sqrt[k]{n_1 \cdots n_k}$ converges Lebesgue-almost everywhere to a constant when $k \to \infty$, and compute this constant as explicitly as possible.

Exercise 2.25. Let $T: X \to X$ be a measurable transformation preserving a probability measure μ in X. We also assume that T is invertible and that its inverse is measurable. By Poincaré's recurrence theorem (Theorem 2.1), given a measurable set $A \subset X$ with $\mu(A) > 0$, the integer

$$n_A(x) = \min \{n \in \mathbb{N} : T^n(x) \in A\}$$

is well defined for μ-almost every $x \in A$. It is called the *(first) return time* of the orbit of x to A. Define a transformation $S_A: A \to A$ for μ-almost every $x \in A$ by

$$S_A(x) = T^{n_A(x)}(x).$$

This is called the *induced transformation* by T on A. Show that:

1. S_A preserves the probability measure $\mu_A = \mu/\mu(A)$ in A, called the *induced invariant measure*.
2. The function $x \mapsto n_A(x)$ defined μ-almost everywhere in the set A is μ_A-integrable.
3. If μ is ergodic, then $\int_A n_A \, d\mu = 1$.

Hint: consider the decomposition of A, up to a set of zero measure, into the pairwise disjoint subsets $\{x \in A : n_A(x) = n\}$ for $n \in \mathbb{N}$.

Exercise 2.26. Draw the graph of the transformation $S_{[0,1/2)}$ induced by the interval translation $T(x) = x + \alpha \mod 1$ on the interval $[0, 1/2)$.

Exercise 2.27. Show that any transformation induced by an ergodic transformation is ergodic with respect to the induced invariant measure in Exercise 2.25.

Exercise 2.28. Consider the transformation $T: [0, 1) \to [0, 1)$ given by

Fig. 2.13 The transformation
in Exercise 2.28

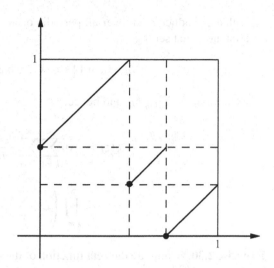

$$T(x) = \begin{cases} x + 1/2 & \text{if } 0 \le x < 1/2, \\ x + 1/2 - 1/\sqrt{2} & \text{if } 1/2 \le x < 1/\sqrt{2}, \\ x - 1/\sqrt{2} & \text{if } 1/\sqrt{2} \le x < 1 \end{cases}$$

(see Fig. 2.13). Show that:

1. T preserves the Lebesgue measure.
2. T is ergodic but T^2 is not ergodic. Hint: the interval $[0, 1/2)$ is T^2-invariant.

Exercise 2.29. Given a transformation $T: X \to X$ with

$$a_n := \text{card}\{x \in X : T^n(x) = x\} < \infty$$

for every $n \in \mathbb{N}$, we define its *zeta function* by

$$\zeta(z) = \exp \sum_{n=1}^{\infty} \frac{a_n z^n}{n} \tag{2.64}$$

for each $z \in \mathbb{C}$ with a convergent power series. Show that:

1. If $a_n > 0$ for every $n \in \mathbb{N}$, then the radius of convergence of the power series is
   equal to $e^{-p(T)}$ with $p(T)$ as in (2.20).
2. If $|z| < e^{-p(T)}$, then

$$\zeta(z) = \prod_{\gamma} \frac{1}{1 - z^{\ell(\gamma)}},$$

with the product taken over all periodic orbits γ, where $\ell(\gamma)$ is the period of γ.
Hint: note that setting

$$b_k = \text{card}\left\{x \in X : x \text{ has period } k\right\},$$

we have $a_n = \sum_{k|n} b_k$, and hence,

$$\zeta(z) = \exp \sum_{k=1}^{\infty} \sum_{j=1}^{\infty} \frac{b_k z^{jk}}{jk}$$

$$= \prod_{k=1}^{\infty} \left(\exp \sum_{j=1}^{\infty} \frac{z^{jk}}{j}\right)^{b_k/k}.$$

Exercise 2.30. Compute the zeta function of the toral automorphism of $\mathbb{T}^2$ induced by the matrix $\left(\begin{smallmatrix} 2 & 1 \\ 1 & 1 \end{smallmatrix}\right)$.

Notes

The books [26, 44, 57, 103] are excellent sources for ergodic theory. A slightly modified version of Theorem 2.1 was established by Poincaré in his seminal work on the three-body problem [78]. Incidentally, Lebesgue's theory of integration [53] only appeared afterwards. Theorem 2.2 is due to Birkhoff [15], and our proof is based on [44]. The statements in Example 2.9 and Proposition 2.13 are special cases of a result of Weyl [104] saying that if $a_k \in \mathbb{R} \setminus \mathbb{Q}$ for some $k \in \{0, \ldots, r\}$, for the polynomial P in (2.53), then the sequence $(P_n)_{n \in \mathbb{N}}$ of fractional parts of P is uniformly distributed. See, for example, [26, Sect. 7.2] for a proof using ergodic theory. The zeta function in (2.64) was introduced by Artin and Mazur in [6]. We refer to [70] for a detailed description.

Chapter 3
Further Topics

We discuss in this chapter several additional topics of ergodic theory. In particular, we establish the existence of finite invariant measures for any continuous transformation of a compact metric space. We also discuss the notions of unique ergodicity and mixing. In particular, we show how unique ergodicity can be characterized in terms of uniform convergence in Birkhoff's ergodic theorem. We conclude this chapter with brief introductions to symbolic dynamics and topological dynamics, with emphasis on their relations to ergodic theory. This includes the construction of Markov measures and Bernoulli measures, as well as the relation between Markov measures and topological Markov chains. We also show that each ergodic property implies a corresponding topological property.

3.1 Introduction

We already described several classes of measure-preserving transformations, but we discussed no procedure to find invariant measures of a given transformation. There is however a simple yet powerful procedure due to Krylov and Bogolubov to find finite invariant measures. For a brief description, let $T: X \to X$ be a continuous transformation of a compact metric space. Given a finite measure μ in X, we consider the sequence of measures μ_n defined by

$$\mu_n(A) = \frac{1}{n} \sum_{k=0}^{n-1} \mu(T^{-k} A)$$

obtained from averaging over the pullbacks of the measure μ under the dynamics. One can show that any sublimit of the sequence μ_n (which always has sublimits) is a finite T-invariant measure.

In another direction, the notion of ergodicity was already discussed in the former chapter. It turns out that it can be strengthen in several ways. In particular,

L. Barreira, *Ergodic Theory, Hyperbolic Dynamics and Dimension Theory*, Universitext, DOI 10.1007/978-3-642-28090-0_3, © Springer-Verlag Berlin Heidelberg 2012

if a given transformation has a single invariant probability measure (such as an irrational rotation of the circle), then this measure is ergodic. This justifies why such transformation is called uniquely ergodic. The existence of a single invariant probability measure turns out to imply that the convergence in Birkhoff's ergodic theorem is uniform. In particular, this property has applications to number theory, among many other applications of the theory of dynamical systems to number theory.

Still another way to strengthen the notion of ergodicity is motivated by an equivalent description: a finite T-invariant measure μ is ergodic if and only if

$$\lim_{n\to\infty} \frac{1}{n} \sum_{k=0}^{n-1} \mu(T^{-k} A \cap B) = \mu(A)\mu(B)$$

for any measurable sets A and B. This can be described as an asymptotic independence in the average of the events $T^{-n} A$ and B when $n \to \infty$. The stronger notion of mixing is obtained by discarding the average: we say that a measure μ is mixing if

$$\lim_{n\to\infty} \mu(T^{-n} A \cap B) = \mu(A)\mu(B)$$

for any measurable sets A and B. This can be described as an asymptotic independence of the events $T^{-n} A$ and B when $n \to \infty$. There are several other possibilities of strengthening the notion of ergodicity, although a comprehensive description falls out of the scope of this book.

We emphasize that ergodic theory is a confluence of many areas and not only from the theory of dynamical systems. Among them is symbolic dynamics. Our related primary interest is the possibility of coding repellers and hyperbolic sets, which in many situations allows to give simpler proofs. The symbolic coding is effected by dividing the phase space into appropriate pieces, forming what is called a Markov partition, and then coding each trajectory by enumerating successively the pieces of the partition containing the elements of the trajectory. It turns out that the symbolic coding of a measure-preserving dynamics induces shift-invariant measures on the symbolic dynamics, such as Markov measures and Bernoulli measures.

3.2   Existence of Invariant Measures

In Chap. 2, we described several examples of measure-preserving transformations. However, we did not discuss the problem of the existence of (finite) invariant measures for a given transformation. For example, this is crucial for Poincaré's recurrence theorem (Theorem 2.1) and Birkhoff's ergodic theorem (Theorem 2.2). We address in this section the problem of the existence of invariant measures for continuous transformations of compact metric spaces.

3.2.1 Preliminary Results

Given a compact metric space X, we denote by $C(X)$ the vector space of continuous functions $\varphi\colon X \to \mathbb{R}$. We define a norm in $C(X)$ by

$$\|\varphi\|_\infty = \max\{|\varphi(x)| : x \in X\}. \tag{3.1}$$

Proposition 3.1. *If X is a compact metric space, then with the norm in (3.1), the space $C(X)$ is separable, that is, it contains a countable dense set.*

Proof. Since X is compact, there exists a sequence $(x_n)_{n\in\mathbb{N}}$ dense in X. Now for each $m, n \in \mathbb{N}$, we consider the continuous function

$$\chi_{nm}(x) = \begin{cases} 1/m - \rho(x, x_n) & \text{if } \rho(x, x_n) \le 1/m, \\ 0 & \text{if } \rho(x, x_n) > 1/m, \end{cases}$$

where ρ is the distance in X. Let also $\mathcal{A} \subset C(X)$ be the smallest algebra containing the constant function 1 and all the functions χ_{nm}. This means that $\mathcal{A}$ is the smallest subset of $C(X)$ which contains these functions such that $\varphi + \psi, \varphi\psi, \lambda\varphi \in \mathcal{A}$ whenever $\varphi, \psi \in \mathcal{A}$ and $\lambda \in \mathbb{R}$. Now we observe that given two distinct points $x, y \in X$, there exist $n, m \in \mathbb{N}$ such that $\chi_{nm}(x) \ne \chi_{nm}(y)$. Indeed, taking

$$\delta = \rho(x, y), \quad x_n \in B(x, \delta/N), \quad \text{and} \quad m \in (N/((N-1)\delta), N/\delta) \cap \mathbb{N}$$

(the interval is nonempty provided that N is sufficiently large), we have

$$\chi_{nm}(x) = 1/m - \rho(x, x_n) > 0 \quad \text{and} \quad \chi_{nm}(y) = 0.$$

Hence, the functions χ_{nm} and thus also those in $\mathcal{A}$ distinguish between the points of X. Now we can apply Stone–Weierstrass theorem which (for our purposes) says that if X is a compact metric space and $\mathcal{A} \subset C(X)$ is an algebra containing the constant function 1 and distinguishing between the points of X, then $\mathcal{A} = C(X)$. Finally, we consider the countable subset $\mathcal{B} \subset \mathcal{A}$ formed by all functions of the form

$$c_0 + \sum_{i=1}^{N} c_i \chi_{n_i m_i}$$

with $c_0, c_1, \ldots, c_N \in \mathbb{Q}$ and $n_i, m_i \in \mathbb{N}$ for $i = 1, \ldots, N$, and all their finite products. One can verify that $\mathcal{B}$ is dense in $\mathcal{A}$ and thus also in $C(X)$. This shows that $C(X)$ is separable. $\square$

Now we denote by $\mathcal{M}(X)$ the set of all Borel probability measures in X. We want to show that $\mathcal{M}(X)$ is metrizable. Proposition 3.1 guarantees the existence of a sequence of continuous functions $(\varphi_n)_{n\in\mathbb{N}}$ with closure equal to the ball

$$\{\varphi \in C(X) : \|\varphi\|_\infty \leq 1\}.$$

We define a function $d \colon \mathcal{M}(X) \times \mathcal{M}(X) \to \mathbb{R}_0^+$ by

$$d(\mu, \nu) = \sum_{n=1}^{\infty} \frac{1}{2^n} \left| \int_X \varphi_n \, d\mu - \int_X \varphi_n \, d\nu \right|. \tag{3.2}$$

Clearly,

$$d(\mu, \nu) \leq \sum_{n=1}^{\infty} \frac{2\|\varphi_n\|_\infty}{2^n} \leq 2$$

for every $\mu, \nu \in \mathcal{M}(X)$, and thus, d is well defined.

Proposition 3.2. *If X is a compact metric space, then d is a distance in $\mathcal{M}(X)$. Moreover, given a sequence of measures $(\mu_n)_{n \in \mathbb{N}}$ in $\mathcal{M}(X)$ and $\mu \in \mathcal{M}(X)$, we have $d(\mu_n, \mu) \to 0$ when $n \to \infty$ if and only if*

$$\lim_{n \to \infty} \int_X \varphi \, d\mu_n = \int_X \varphi \, d\mu$$

for every function $\varphi \in C(X)$.

Proof. For the first property, we note that $d(\mu, \nu) \geq 0$ and $d(\mu, \nu) = d(\nu, \mu)$ for any measures μ and ν. It is also immediate that d satisfies the triangle inequality. Thus, to show that d is a distance, it remains to verify that $\mu = \nu$ whenever $d(\mu, \nu) = 0$. If $d(\mu, \nu) = 0$, then

$$\int_X \varphi_m \, d\mu = \int_X \varphi_m \, d\nu$$

for every $m \in \mathbb{N}$. By the choice of the sequence $(\varphi_m)_{m \in \mathbb{N}}$, given a nonzero function $\varphi \in C(X)$ and $\varepsilon > 0$, there exists $m \in \mathbb{N}$ such that

$$\left\| \frac{\varphi}{\|\varphi\|_\infty} - \varphi_m \right\|_\infty < \varepsilon. \tag{3.3}$$

Therefore,

$$\frac{1}{\|\varphi\|_\infty} \left| \int_X \varphi \, d\mu - \int_X \varphi \, d\nu \right| \leq \left| \int_X \varphi_m \, d\mu - \int_X \varphi_m \, d\nu \right| + 2\varepsilon = 2\varepsilon.$$

Since ε is arbitrary, we conclude that

$$\int_X \varphi \, d\mu = \int_X \varphi \, d\nu$$

for every $\varphi \in C(X)$. Hence, $\mu = \nu$, and d is a distance.

For the second property, we first assume that $d(\mu_n, \mu) \to 0$ when $n \to \infty$. Then

$$\lim_{n \to \infty} \int_X \varphi_m \, d\mu_n = \int_X \varphi_m \, d\mu$$

for every $m \in \mathbb{N}$. Again, given a nonzero function $\varphi \in C(X)$ and $\varepsilon > 0$, there exists $m \in \mathbb{N}$ such that inequality (3.3) holds. Thus, for each $n, m \in \mathbb{N}$, we have

$$\frac{1}{\|\varphi\|_\infty} \left| \int_X \varphi \, d\mu_n - \int_X \varphi \, d\mu \right| \leq \left| \int_X \varphi_m \, d\mu_n - \int_X \varphi_m \, d\mu \right| + 2\varepsilon.$$

Letting $n \to \infty$, we obtain

$$\frac{1}{\|\varphi\|_\infty} \limsup_{n \to \infty} \left| \int_X \varphi \, d\mu_n - \int_X \varphi \, d\mu \right| \leq 2\varepsilon,$$

and since ε is arbitrary, we conclude that

$$\lim_{n \to \infty} \int_X \varphi \, d\mu_n = \int_X \varphi \, d\mu. \tag{3.4}$$

Now we assume that (3.4) holds for any function $\varphi \in C(X)$. We want to show that $d(\mu_n, \mu) \to 0$ when $n \to \infty$. Given $\varepsilon > 0$, for each $m \in \mathbb{N}$, there exists $k_m \in \mathbb{N}$ such that if $n > k_m$, then

$$\left| \int_X \varphi_i \, d\mu_n - \int_X \varphi_i \, d\mu \right| < \varepsilon$$

for $i = 1, \ldots, m$. Therefore,

$$d(\mu_n, \mu) < \varepsilon + \sum_{i=m+1}^{\infty} \frac{1}{2^i} \left| \int_X \varphi_i \, d\mu_n - \int_X \varphi_i \, d\mu \right|$$

$$\leq \varepsilon + \sum_{i=m+1}^{\infty} \frac{2}{2^i} = \varepsilon + \frac{1}{2^{m-1}}$$

for every $n > k_m$. Taking $m = m(\varepsilon) \in \mathbb{N}$ such that $1/2^{m-1} < \varepsilon$, we obtain

$$d(\mu_n, \mu) < 2\varepsilon \quad \text{for every} \quad n > k_{m(\varepsilon)}.$$

Letting $\varepsilon \to 0$, we conclude that $d(\mu_n, \mu) \to 0$ when $n \to \infty$. This completes the proof of the proposition. $\square$

We also show that with the distance d in (3.2), the space $\mathcal{M}(X)$ is compact.

Theorem 3.1. *If X is a compact metric space, then $\mathcal{M}(X)$ is compact.*

Proof. Since $\mathcal{M}(X)$ is a metric space, it is sufficient to show that each sequence $(\mu_n)_{n \in \mathbb{N}}$ in $\mathcal{M}(X)$ has a convergent subsequence. Let $(\varphi_n)_{n \in \mathbb{N}}$ be again the sequence used to define the distance d in (3.2). For each $n \in \mathbb{N}$, we consider the bounded sequence $\beta_n \colon \mathbb{N} \to [-1, 1]$ defined by

$$\beta_n(k) = \int_X \varphi_k \, d\mu_n.$$

By Tychonoff's theorem, the space of sequences $[-1, 1]^{\mathbb{N}}$ with the product topology is compact, and there exists a subsequence $(\beta_{m_n})_{n \in \mathbb{N}}$ converging in this topology. This implies that the sequence $(\int_X \varphi_k \, d\mu_{m_n})_{n \in \mathbb{N}}$ converges for each $k \in \mathbb{N}$. On the other hand, given a nonzero function $\varphi \in C(X)$ and $\varepsilon > 0$, there exists $k \in \mathbb{N}$ such that

$$\left\| \frac{\varphi}{\|\varphi\|_\infty} - \varphi_k \right\|_\infty < \varepsilon.$$

Then

$$\frac{1}{\|\varphi\|_\infty} \left| \int_X \varphi \, d\mu_{m_n} - \int_X \varphi \, d\mu_{m_\ell} \right| \le \left| \int_X \varphi_k \, d\mu_{m_n} - \int_X \varphi_k \, d\mu_{m_\ell} \right| + 2\varepsilon$$

for every $n, \ell \in \mathbb{N}$. We conclude that if n and ℓ are sufficiently large, then

$$\left| \int_X \varphi \, d\mu_{m_n} - \int_X \varphi \, d\mu_{m_\ell} \right| \le 3\varepsilon \|\varphi\|_\infty,$$

and $(\int_X \varphi \, d\mu_{m_n})_{n \in \mathbb{N}}$ is a Cauchy sequence. Therefore, one can define a linear functional in $C(X)$ by

$$J(\varphi) = \lim_{n \to \infty} \int_X \varphi \, d\mu_{m_n}. \tag{3.5}$$

Clearly, $|J(\varphi)| \le \|\varphi\|_\infty$, and in particular, J is continuous. Moreover, $J(1) = 1$, and J is positive, that is, $J(\varphi) \ge 0$ whenever $\varphi \ge 0$. By Riesz's representation theorem (Theorem A.4), there exists a unique measure $\mu \in \mathcal{M}(X)$ such that

$$J(\varphi) = \int_X \varphi \, d\mu$$

for every function $\varphi \in C(X)$. By Proposition 3.2 and (3.5), we conclude that $d(\mu_{m_n}, \mu) \to 0$ when $n \to \infty$. This shows that the space $\mathcal{M}(X)$ is compact. $\qquad \square$

3.2.2   Existence of Invariant Measures

The following result shows that $\mathcal{M}(X)$ contains at least one T-invariant probability measure when T is continuous:

Theorem 3.2 (Krylov–Bogolubov). *Let X be a compact metric space. If the transformation $T \colon X \to X$ is continuous, then there exists a T-invariant probability measure in X.*

Proof. Consider the transformation $T_* \colon \mathcal{M}(X) \to \mathcal{M}(X)$ defined by

$$(T_* \mu)(A) = \mu(T^{-1} A). \tag{3.6}$$

We note that a measure μ is T-invariant if and only if it is a fixed point of T_*. Moreover, one can easily verify that

$$\int_X \varphi \, d(T_* \mu) = \int_X (\varphi \circ T) \, d\mu$$

for every function $\varphi \in C(X)$. By Proposition 3.2, if the sequence of measures $(\mu_n)_{n \in \mathbb{N}} \subset \mathcal{M}(X)$ converges to μ, then for each $\varphi \in C(X)$, we have

$$\int_X \varphi \, d\mu_n \to \int_X \varphi \, d\mu$$

when $n \to \infty$. Since T is continuous, we conclude that if $\varphi \in C(X)$, then

$$\int_X \varphi \, d(T_* \mu_n) = \int_X (\varphi \circ T) \, d\mu_n$$

$$\to \int_X (\varphi \circ T) \, d\mu = \int_X \varphi \, d(T_* \mu)$$

when $n \to \infty$, that is, the sequence $(T_* \mu_n)_{n \in \mathbb{N}}$ converges to $T_* \mu$. This shows that the transformation T_* is continuous.

Now given $\mu \in \mathcal{M}(X)$, we consider the sequence of measures

$$\mu_n = \frac{1}{n} \sum_{k=0}^{n-1} T_*^k \mu \in \mathcal{M}(X).$$

By Theorem 3.1, there exists a convergent subsequence $(\mu_{m_n})_{n \in \mathbb{N}}$, say with limit $\nu \in \mathcal{M}(X)$. Since

$$T_* \mu_{m_n} = \frac{1}{m_n} \sum_{k=1}^{m_n} T_*^k \mu = \mu_{m_n} + \frac{T_*^{m_n} \mu}{m_n} - \frac{\mu}{m_n}, \tag{3.7}$$

and T_* is continuous, letting $n \to \infty$ in (3.7), we obtain $T_* \nu = \nu$. $\square$

3.3 Unique Ergodicity

We consider in this section the transformations having a unique invariant probability measure—the so-called uniquely ergodic transformations—and we study their ergodic properties. In particular, we show how unique ergodicity leads to uniform convergence in Birkhoff's ergodic theorem.

3.3.1 Basic Notions and Uniform Convergence

Clearly, if μ is an invariant measure, then $c\mu$ is also an invariant measure for any $c > 0$. In the following definition, we normalize the measure in order to avoid this duplication.

Definition 3.1. We say that a transformation $T\colon X \to X$ is *uniquely ergodic* if T has exactly one T-invariant probability measure in X.

For example, rational circle rotations, rational interval translations, expanding maps of the circle, and linear expanding maps of the interval are not uniquely ergodic (see Exercise 3.3).

The following statement justifies the expression "uniquely ergodic":

Proposition 3.3. *The unique invariant probability measure of a uniquely ergodic transformation is ergodic.*

Proof. Let μ be the unique invariant probability measure of a uniquely ergodic transformation in X. If μ is not ergodic, then there is an invariant measurable set $A \subset X$ with $0 < \mu(A) < 1$. We define new invariant probability measures μ_1 and μ_2 by

$$\mu_1(B) = \frac{\mu(B \cap A)}{\mu(A)} \quad \text{and} \quad \mu_2(B) = \frac{\mu(B \cap (X \setminus A))}{\mu(X \setminus A)}.$$

Clearly, $\mu_1 \neq \mu_2$ (e.g., $\mu_1(A) = 1$ and $\mu_2(A) = 0$). This contradicts the unique ergodicity of the transformation, and we obtain the desired statement. $\square$

The following statement shows that Birkhoff's ergodic theorem can be considerably strengthened for the uniquely ergodic continuous transformations of a compact metric space. We recall that $C(X)$ denotes the space of all continuous functions $\varphi\colon X \to \mathbb{R}$.

Theorem 3.3. *Let $T\colon X \to X$ be a uniquely ergodic continuous transformation of a compact metric space. If $\varphi \in C(X)$, then the sequence of functions*

$$\frac{1}{n} \sum_{k=0}^{n-1} \varphi \circ T^k \tag{3.8}$$

converges uniformly when $n \to \infty$ to the constant $\int_X \varphi \, d\mu$, where μ is the unique T-invariant probability measure in X.

Proof. Given $x \in X$, we consider the sequence of probability measures

$$\mu_n = \frac{1}{n} \sum_{k=0}^{n-1} \delta_{T^k(x)}.$$

By Exercise 3.1, each accumulation point of this sequence is a T-invariant probability measure, and hence, it is equal to μ. This shows that $(\mu_n)_{n \in \mathbb{N}}$ converges to the measure μ. Given a function $\varphi \in C(X)$, since

$$\int_X \varphi \, d\mu_n = \frac{1}{n} \sum_{k=0}^{n-1} \varphi(T^k(x)),$$

we thus obtain

$$\lim_{n \to \infty} \frac{1}{n} \sum_{k=0}^{n-1} \varphi(T^k(x)) = \int_X \varphi \, d\mu. \tag{3.9}$$

If the convergence in (3.9) was not uniform, then for some sufficiently small $\varepsilon > 0$, it would exist a sequence $(m_n)_{n \in \mathbb{N}} \subset \mathbb{N}$ with $m_n \nearrow \infty$ such that

$$\left\| \frac{1}{m_n} \sum_{k=0}^{m_n - 1} \varphi \circ T^k - \int_X \varphi \, d\mu \right\|_\infty \geq \varepsilon$$

for every $n \in \mathbb{N}$. Therefore, since X is compact, there would exist a sequence $(x_n)_{n \in \mathbb{N}} \subset X$ such that

$$\left| \frac{1}{m_n} \sum_{k=0}^{m_n - 1} \varphi(T^k(x_n)) - \int_X \varphi \, d\mu \right| \geq \varepsilon \tag{3.10}$$

for every $n \in \mathbb{N}$. On the other hand, it follows from Theorem 3.1 that the sequence of measures

$$\nu_n = \frac{1}{m_n} \sum_{k=0}^{m_n - 1} \delta_{T^k(x_n)} \tag{3.11}$$

has a subsequence $(\nu_{\ell_n})_{n \in \mathbb{N}}$ converging to a measure $\nu \in \mathcal{M}(X)$. We show that ν is T-invariant. Given $\psi \in C(X)$, we have

$$\int_X (\psi \circ T)\, dv = \lim_{n \to \infty} \int_X (\psi \circ T)\, dv_{\ell_n}$$

$$= \lim_{n \to \infty} \frac{1}{m_{\ell_n}} \sum_{k=0}^{m_{\ell_n}-1} \psi(T^{k+1}(x_{\ell_n}))$$

$$= \lim_{n \to \infty} \frac{1}{m_{\ell_n}} \left(\sum_{k=0}^{m_{\ell_n}-1} \psi(T^k(x_{\ell_n})) + \psi(T^{m_{\ell_n}}(x_{\ell_n})) - \psi(x_{\ell_n}) \right)$$

$$= \lim_{n \to \infty} \frac{1}{m_{\ell_n}} \sum_{k=0}^{m_{\ell_n}-1} \psi(T^k(x_{\ell_n})),$$

and thus, by (3.11), we obtain

$$\int_X (\psi \circ T)\, dv = \lim_{n \to \infty} \int_X \psi\, dv_{\ell_n} = \int_X \psi\, dv.$$

It follows from Proposition 2.1 that v is T-invariant. But since T is uniquely ergodic, we have $v = \mu$, and in view of (3.10), we obtain the contradiction

$$\varepsilon \le \left| \frac{1}{m_{\ell_n}} \sum_{k=0}^{m_{\ell_n}-1} \varphi(T^k(x_{\ell_n})) - \int_X \varphi\, d\mu \right| = \left| \int_X \varphi\, dv_{\ell_n} - \int_X \varphi\, d\mu \right| \to 0$$

when $n \to \infty$. This completes the proof of the theorem. $\square$

3.3.2   Criteria for Unique Ergodicity and Examples

In order to determine whether a given transformation is uniquely ergodic, it is convenient to have the following criterion:

Theorem 3.4. *Let $T: X \to X$ be a continuous transformation of a compact metric space. If for each $\varphi \in C(X)$ the sequence of functions in (3.8) converges uniformly to a constant when $n \to \infty$, then T is uniquely ergodic.*

Proof. We define a linear functional $J: C(X) \to \mathbb{R}$ by

$$J(\varphi) = \lim_{n \to \infty} \frac{1}{n} \sum_{k=0}^{n-1} \varphi \circ T^k. \tag{3.12}$$

One can easily verify that J is continuous and positive (i.e., $J(\varphi) \ge 0$ whenever $\varphi \ge 0$), and that $J(1) = 1$. By Riesz's representation theorem (Theorem A.4), there is a unique probability measure μ in X such that

$$J(\varphi) = \int_X \varphi \, d\mu \quad \text{for every} \quad \varphi \in C(X).$$

On the other hand, it follows readily from (3.12) that

$$J(\varphi \circ T) = J(\varphi) \quad \text{for every} \quad \varphi \in C(X).$$

Therefore, by Proposition 2.1, we conclude that μ is T-invariant.

Now we consider the sequence of continuous functions $\varphi_n \colon X \to \mathbb{R}$ in (3.8). Since by hypothesis

$$\varphi_n \to \int_X \varphi \, d\mu \quad \text{uniformly when} \quad n \to \infty,$$

if ν is a T-invariant probability measure in X, then

$$\int_X \varphi_n \, d\nu \to \int_X \int_X \varphi \, d\mu \, d\nu = \int_X \varphi \, d\mu$$

when $n \to \infty$. On the other hand, since ν is T-invariant, by Proposition 2.1, we have

$$\int_X \varphi_n \, d\nu = \frac{1}{n} \sum_{k=0}^{n-1} \int_X (\varphi \circ T^k) \, d\nu = \int_X \varphi \, d\nu.$$

Therefore,

$$\int_X \varphi \, d\nu = \int_X \varphi \, d\mu$$

for every function $\varphi \in C(X)$, and $\nu = \mu$. This shows that T is uniquely ergodic. $\square$

It follows from Theorems 3.3 and 3.4 that if $T \colon X \to X$ a continuous transformation of a compact metric space, then the following properties are equivalent:

1. T is uniquely ergodic.
2. For each $\varphi \in C(X)$, the sequence of functions in (3.8) converges uniformly to a constant when $n \to \infty$.

In fact, in order to show that a given transformation is uniquely ergodic, it is sufficient to consider certain subfamilies of continuous functions for which the sequence in (3.8) converges uniformly to a constant. Given a family of functions $\Phi \subset C(X)$, we denote by $L(\Phi)$ the set of all finite linear combinations of functions in Φ.

Theorem 3.5. *Let $T \colon X \to X$ be a continuous transformation of a compact metric space, and let $\Phi \subset C(X)$ be a family of functions such that $L(\Phi)$ is dense in $C(X)$.*

If for each $\varphi \in \Phi$ the sequence in (3.8) converges uniformly to a constant when $n \to \infty$, then T is uniquely ergodic.

Proof. Given $\varphi \in C(X)$, there exist functions $\psi_m \in L(\Phi)$ for $m \in \mathbb{N}$ such that

$$\|\varphi - \psi_m\|_\infty \to 0 \quad \text{when} \quad m \to \infty.$$

Furthermore, for each $m \in \mathbb{N}$, there exists $c_m \in \mathbb{R}$ such that

$$\|\psi_{mn} - c_m\|_\infty \to 0 \quad \text{when} \quad n \to \infty, \tag{3.13}$$

where

$$\psi_{mn} = \frac{1}{n} \sum_{k=0}^{n-1} \psi_m \circ T^k.$$

Indeed, if

$$\psi_m = a_{m1} \eta_1 + \cdots + a_{mp} \eta_p$$

for some $\eta_1, \ldots, \eta_p \in \Phi$ and $a_{m1}, \ldots, a_{mp} \in \mathbb{R}$, then we can take

$$c_m = a_{m1} d_1 + \cdots + a_{mp} d_p,$$

where the constants $d_1, \ldots, d_p \in \mathbb{R}$ are such that

$$\frac{1}{n} \sum_{j=0}^{n-1} \eta_k \circ T^j \to d_k$$

uniformly when $n \to \infty$. On the other hand, given $m, \ell \in \mathbb{N}$, we have

$$\|\psi_{mn} - \psi_{\ell n}\|_\infty \leq \frac{1}{n} \sum_{k=0}^{n-1} \|\psi_m \circ T^k - \psi_\ell \circ T^k\|_\infty$$

$$\leq \frac{1}{n} \sum_{k=0}^{n-1} \|\psi_m - \psi_\ell\|_\infty = \|\psi_m - \psi_\ell\|_\infty,$$

and hence,

$$|c_m - c_\ell| \leq \|c_m - \psi_{mn}\|_\infty + \|\psi_{mn} - \psi_{\ell n}\|_\infty + \|\psi_{\ell n} - c_\ell\|_\infty$$

$$\leq \|c_m - \psi_{mn}\|_\infty + \|\psi_m - \psi_\ell\|_\infty + \|\psi_{\ell n} - c_\ell\|_\infty.$$

Letting $n \to \infty$, it follows from (3.13) that

$$|c_m - c_\ell| \leq \|\psi_m - \psi_\ell\|_\infty,$$

and hence, $(c_m)_{m \in \mathbb{N}}$ is a Cauchy sequence, say with limit c. Now let $(\varphi_n)_{n \in \mathbb{N}}$ be the sequence of functions in (3.8). We have

$$\|\varphi_n - c\|_\infty \leq \|\varphi_n - \psi_{mn}\|_\infty + \|\psi_{mn} - c_m\|_\infty + |c_m - c|$$
$$\leq \|\varphi - \psi_m\|_\infty + \|\psi_{mn} - c_m\|_\infty + |c_m - c|. \tag{3.14}$$

On the other hand, given $\varepsilon > 0$, for any sufficiently large m, we have

$$\|\varphi - \psi_m\|_\infty < \varepsilon \quad \text{and} \quad |c_m - c| < \varepsilon.$$

Hence, for any such m, letting $n \to \infty$ in (3.13), it follows from (3.14) that

$$\limsup_{n \to \infty} \|\varphi_n - c\|_\infty \leq 2\varepsilon.$$

Since ε is arbitrary, this implies that $\|\varphi_n - c\|_\infty \to 0$ when $n \to \infty$. Now we can apply Theorem 3.4 to obtain the desired statement. $\square$

We describe two families of functions Φ that can be used in Theorem 3.5.

Example 3.1. For the space $X = [a, b]$, let

$$\Phi = \{1\} \cup \{\varphi_k : k \in \mathbb{N}\} \subset C([a, b]),$$

where $\varphi_k(x) = x^k$ for each $k \in \mathbb{N}$. Then the set of polynomials $L(\Phi)$ in $[a, b]$ is dense in $C([a, b])$.

Example 3.2. For the space $X = [0, 1]$, let

$$\Phi = \{1\} \cup \{\varphi_k : k \in \mathbb{N}\} \cup \{\psi_k : k \in \mathbb{N}\} \subset C([0, 1]),$$

where

$$\varphi_k(x) = \cos(2\pi k x) \quad \text{and} \quad \psi_k(x) = \sin(2\pi k x) \tag{3.15}$$

for each $k \in \mathbb{N}$. Then $L(\Phi)$ is dense in $C([0, 1])$.

Now we apply the criterion for unique ergodicity in Theorem 3.5 to circle rotations.

Example 3.3. Consider the circle rotation R_w (see Sect. 2.2.3) and the functions $\chi_k(z) = z^k$ for each $k \in \mathbb{Z}$. We have $\chi_0 = 1$,

$$\chi_k(e^{2\pi i x}) + \chi_{-k}(e^{2\pi i x}) = 2\varphi_k(x),$$

and

$$\chi_k(e^{2\pi i x}) - \chi_{-k}(e^{2\pi i x}) = 2i\psi_k(x),$$

with φ_k and ψ_k as in (3.15). By Example 3.2, the set of all finite linear combinations of the functions χ_k is dense in $C(S^1)$. Moreover,

$$\chi_k(R_w(z)) = w^k \chi_k(z).$$

If R_w is an *irrational rotation*, that is, if $w = e^{2\pi i \tau}$ for some $\tau \notin \mathbb{Q}$, then $w^k \neq 1$ for every $k \in \mathbb{Z} \setminus \{0\}$, and hence,

$$\left| \frac{1}{n} \sum_{j=0}^{n-1} \chi_k(R_w{}^j(z)) \right| = \left| \frac{1}{n} \sum_{j=0}^{n-1} w^{kj} \chi_k(z) \right|$$

$$= \frac{|1 - w^{kn}|}{n|1 - w^k|}$$

$$\leq \frac{2}{n|1 - w^k|} \to 0$$

when $n \to \infty$. On the other hand,

$$\frac{1}{n} \sum_{j=0}^{n-1} \chi_0(R_w{}^j(z)) = 1$$

for every $n \in \mathbb{N}$. Thus, by Theorem 3.5, the rotation R_w is uniquely ergodic. In other words,

any irrational circle rotation is uniquely ergodic.   (3.16)

Moreover, since the circle rotations preserve the Lebesgue measure, the unique R_w-invariant probability measure is the Lebesgue measure.

   In addition, one can easily verify that the one-to-one map $h: \mathbb{T} \to S^1$ defined by $h(\theta) = e^{2\pi i \theta}$ (see Sect. 2.2.3) is a homeomorphism when $\mathbb{T}$ has the topology induced from the standard topology in $\mathbb{R}$. Hence, it follows from (2.13) and (3.16) that

any irrational interval translation is uniquely ergodic,   (3.17)

and again the unique invariant probability measure of an irrational interval translation is the Lebesgue measure.

   The following is an application of property (3.17):

Example 3.4. Consider the sequence $\ell_1 = 2, \ell_2 = 4, \ell_3 = 8, \ell_4 = 1, \ldots$ of the first digits in base 10 of the powers of 2. We want to find the frequency with which each $k \in \{1, \ldots, 9\}$ occurs in the sequence ℓ_n. More precisely, we want to show that the limit

$$\lim_{n \to \infty} \frac{\operatorname{card} \{ j \in \{1, \ldots, n\} : \ell_j = k \}}{n} \tag{3.18}$$

exists for each k, and to compute its value explicitly.

Consider the interval translation $T\colon \mathbb{T} \to \mathbb{T}$ given by

$$T(x) = x + \log_{10} 2 \bmod 1.$$

Since $\log_{10} 2$ is irrational, it follows from (3.17) that T is uniquely ergodic. On the other hand,

$$2^n = 10^{n \log_{10} 2 - \lfloor n \log_{10} 2 \rfloor} 10^{\lfloor n \log_{10} 2 \rfloor}$$
$$= 10^{T^n 0} 10^{\lfloor n \log_{10} 2 \rfloor}.$$

Therefore, $\ell_n = k$ if and only if

$$T^n(0) \in [\log_{10} k, \log_{10}(k + 1)).$$

Now, let

$$\varphi = \chi_{[\log_{10} k, \log_{10}(k+1))}.$$

We consider continuous functions $a_p, b_p \colon \mathbb{T} \to [0, 1]$ for $p \in \mathbb{N}$ with

$$a_p \le \varphi \le b_p, \quad p \in \mathbb{N},$$

such that

$$\int_0^1 a_p \, dm \to \int_0^1 \varphi \, dm \quad \text{and} \quad \int_0^1 b_p \, dm \to \int_0^1 \varphi \, dm \qquad (3.19)$$

when $p \to \infty$, where m is the Lebesgue measure in $[0, 1]$. We observe that

$$\frac{1}{n} \sum_{j=0}^{n-1} a_p(T^j(0)) \le \frac{1}{n} \sum_{j=0}^{n-1} \varphi(T^j(0)) \le \frac{1}{n} \sum_{j=0}^{n-1} b_p(T^j(0)). \qquad (3.20)$$

Since T is uniquely ergodic, it follows from Theorem 3.3 that

$$\frac{1}{n} \sum_{j=0}^{n-1} a_p(T^j(0)) \to \int_0^1 a_p \, dm \qquad (3.21)$$

and

$$\frac{1}{n} \sum_{j=0}^{n-1} b_p(T^j(0)) \to \int_0^1 b_p \, dm$$

when $n \to \infty$. By (3.20) and (3.21), we obtain

$$\int_0^1 a_p \, dm \le \liminf_{n\to\infty} \frac{1}{n} \sum_{j=0}^{n-1} \varphi(T^j(0))$$

$$\le \limsup_{n\to\infty} \frac{1}{n} \sum_{j=0}^{n-1} \varphi(T^j(0)) \le \int_0^1 b_p \, dm.$$

Letting $p \to \infty$, it follows from (3.19) that

$$\lim_{n\to\infty} \frac{1}{n} \sum_{j=0}^{n-1} \varphi(T^j(0)) = \int_0^1 \varphi \, dm.$$

Therefore, the frequency with which the integer k occurs in the sequence ℓ_n is given by

$$\int_0^1 \varphi \, dm = \int_0^1 \chi_{[\log_{10} k, \log_{10}(k+1))} \, dm$$

$$= \log_{10} \left(1 + \frac{1}{k} \right).$$

3.4   Mixing

We consider in this section an ergodic property which is stronger than the ergodicity of an invariant measure.

We first give an alternative characterization of ergodicity.

Proposition 3.4. *Let* $T : X \to X$ *be a measurable transformation preserving a probability measure* μ *in* X. *The measure* μ *is ergodic if and only if*

$$\lim_{n\to\infty} \frac{1}{n} \sum_{k=0}^{n-1} \mu(T^{-k} A \cap B) = \mu(A)\mu(B) \tag{3.22}$$

for any measurable sets $A, B \subset X$.

Proof. We first assume that μ is ergodic. Let $A, B \subset X$ be measurable sets. By Proposition 2.12, we obtain

$$\lim_{n\to\infty} \frac{1}{n} \sum_{k=0}^{n-1} \chi_A(T^k(x)) = \int_A \chi_A \, d\mu = \mu(A)$$

for μ-almost every $x \in X$. Therefore, by the dominated convergence theorem (Theorem A.3), we conclude that

$$\mu(A)\mu(B) = \mu(A) \int_X \chi_B \, d\mu$$

$$= \int_X \mu(A)\chi_B \, d\mu$$

$$= \int_X \lim_{n\to\infty} \frac{1}{n} \sum_{k=0}^{n-1} \chi_A(T^k(x))\chi_B(x) \, d\mu(x)$$

$$= \lim_{n\to\infty} \frac{1}{n} \sum_{k=0}^{n-1} \int_X \chi_A(T^k(x))\chi_B(x) \, d\mu(x).$$

Since

$$\chi_A(T^k(x))\chi_B(x) = 1$$

if and only if $T^k(x) \in A$ and $x \in B$, that is, if and only if $x \in T^{-k}A \cap B$, we have

$$(\chi_A \circ T^k)\chi_B = \chi_{T^{-k}A \cap B}, \tag{3.23}$$

and hence,

$$\mu(A)\mu(B) = \lim_{n\to\infty} \frac{1}{n} \sum_{k=0}^{n-1} \mu(T^{-k}A \cap B).$$

Now we assume that property (3.22) holds. Let $A \subset X$ be a T-invariant measurable set. Then

$$\mu(A)\mu(X \setminus A) = \lim_{n\to\infty} \frac{1}{n} \sum_{k=0}^{n-1} \mu(T^{-k}A \cap (X \setminus A))$$

$$= \lim_{n\to\infty} \frac{1}{n} \sum_{k=0}^{n-1} \mu(A \cap (X \setminus A)) = 0,$$

and thus, $\mu(A) = 0$ or $\mu(X \setminus A) = 0$. This shows that μ is ergodic. $\qquad \square$

Whenever we consider a convergence stronger than that in (3.22), we obtain a property stronger than ergodicity. In particular, we consider the following notion:

Definition 3.2. Let $T: X \to X$ be a measurable transformation and let μ be a T-invariant probability measure in X. We say that μ is *mixing* with respect to T if

$$\lim_{n\to\infty} \mu(T^{-n}A \cap B) = \mu(A)\mu(B)$$

for any measurable sets $A, B \subset X$.

One can easily verify that if μ is mixing, then μ is ergodic.

In order to describe a criterion for mixing, given a probability measure μ in X we consider the space $L^2(X, \mu)$ of all measurable functions $\varphi: X \to \mathbb{R}$ such that

$$\|\varphi\|_2 := \left(\int_X \varphi^2 \, d\mu \right)^{1/2} < \infty,$$

equipped with the norm $\varphi \mapsto \|\varphi\|_2$. We note that $L^2(X, \mu) \subset L^1(X, \mu)$. Indeed, by the Cauchy–Schwarz inequality, we have

$$\int_X |\varphi| \, d\mu = \int_X 1 \cdot |\varphi| \, d\mu$$

$$\leq \mu(X)^{1/2} \|\varphi\|_2 = \|\varphi\|_2 < \infty. \tag{3.24}$$

Given a family of functions $\Phi \subset L^2(X, \mu)$ we denote by $L(\Phi)$ the set of all finite linear combinations of functions in Φ.

Theorem 3.6. *Let $T: X \to X$ be a measurable transformation preserving a probability measure μ in X, and let $\Phi \subset L^2(X, \mu)$ be a family of functions such that $L(\Phi)$ is dense in $L^2(X, \mu)$. If*

$$\lim_{n\to\infty} \int_X (\varphi \circ T^n)\psi \, d\mu = \int_X \varphi \, d\mu \int_X \psi \, d\mu \tag{3.25}$$

for any $\varphi, \psi \in \Phi$, then the measure μ is mixing.

Proof. We note that by (3.24), the integrals in the right-hand side of (3.25) are well defined. Moreover, by the Cauchy–Schwarz inequality and Proposition 2.1, we have

$$\int_X |(\varphi \circ T^n)\psi| \, d\mu \leq \left(\int_X (\varphi \circ T^n)^2 \, d\mu \right)^{1/2} \left(\int_X \psi^2 \, d\mu \right)^{1/2}$$

$$= \left(\int_X \varphi^2 \, d\mu \right)^{1/2} \left(\int_X \psi^2 \, d\mu \right)^{1/2} < \infty,$$

and hence, the integral in the right-hand of (3.25) side is also well defined.

Now we proceed with the proof of the theorem. We first observe that since (3.25) holds for any $\varphi, \psi \in \Phi$, this property also holds for any $\varphi, \psi \in L(\Phi)$. Given $\bar{\varphi}, \bar{\psi} \in L^2(X, \mu)$ and $\varepsilon > 0$, we consider functions $\varphi, \psi \in L(\Phi)$ such that

$$\|\varphi - \bar{\varphi}\|_2 < \varepsilon \quad \text{and} \quad \|\psi - \bar{\psi}\|_2 < \varepsilon \tag{3.26}$$

(which is always possible because $L(\Phi)$ is dense in $L^2(X, \mu)$). Using the Cauchy–Schwarz inequality and (3.24), we obtain

$$D_n = \left| \int_X (\bar{\varphi} \circ T^n) \bar{\psi} \, d\mu - \int_X \bar{\varphi} \, d\mu \int_X \bar{\psi} \, d\mu \right|$$

$$\leq \left| \int_X (\bar{\varphi} \circ T^n)(\bar{\psi} - \psi) \, d\mu \right| + \left| \int_X (\bar{\varphi} \circ T^n - \varphi \circ T^n) \psi \, d\mu \right|$$

$$+ \left| \int_X (\varphi \circ T^n) \psi \, d\mu - \int_X \varphi \, d\mu \int_X \psi \, d\mu \right|$$

$$+ \left| \int_X (\varphi - \bar{\varphi}) \, d\mu \int_X \psi \, d\mu \right| + \left| \int_X \bar{\varphi} \, d\mu \int_X (\psi - \bar{\psi}) \, d\mu \right|$$

$$\leq \| \bar{\varphi} \circ T^n \|_2 \| \bar{\psi} - \psi \|_2 + \| (\bar{\varphi} - \varphi) \circ T^n \|_2 \| \psi \|_2$$

$$+ \left| \int_X (\varphi \circ T^n) \psi \, d\mu - \int_X \varphi \, d\mu \int_X \psi \, d\mu \right|$$

$$+ \| \bar{\varphi} - \varphi \|_2 \| \psi \|_2 + \| \bar{\varphi} \|_2 \| \psi - \bar{\psi} \|_2.$$

Hence, it follows from Proposition 2.1 and (3.26) that

$$D_n \leq 2\varepsilon \| \bar{\varphi} \|_2 + 2\varepsilon \| \psi \|_2 + \left| \int_X (\varphi \circ T^n) \psi \, d\mu - \int_X \varphi \, d\mu \int_X \psi \, d\mu \right|.$$

Letting $n \to \infty$, we conclude that

$$\limsup_{n \to \infty} D_n \leq 2\varepsilon \big(\| \bar{\varphi} \|_2 + \| \bar{\psi} \|_2 + \varepsilon \big),$$

and it follows from the arbitrariness of ε that (3.25) holds for $\varphi, \psi \in L^2(X, \mu)$.

Now we consider measurable sets $A, B \subset X$. Clearly, $\chi_A, \chi_B \in L^2(X, \mu)$. By identity (3.23), setting $\varphi = \chi_A$ and $\psi = \chi_B$, it follows from (3.25) that μ is mixing. $\square$

We use this criterion to show that some toral automorphisms have the mixing property.

Proposition 3.5. *Let T_A be the automorphism of the torus $\mathbb{T}^n$ induced by a matrix A such that (2.51) holds. Then the Lebesgue measure is mixing with respect to T_A.*

Proof. By Theorem 3.6, it is sufficient to consider the family of functions

$$\Phi = \{ \chi_k : k \in \mathbb{Z}^n \},$$

where $\chi_k(x) = e^{2\pi i \langle k, x \rangle}$ with $\langle k, x \rangle$ as in (2.50). Indeed, one can show that for this family, the set $L(\Phi)$ is dense in $L^2(X, \lambda)$, where λ is the Lebesgue measure in $\mathbb{T}^n$.

Now we study the integrals

$$\int_{\mathbb{T}^n} (\chi_k \circ T^p) \chi_\ell \, d\lambda.$$

We first note that

$$\int_{\mathbb{T}^n} \chi_k \overline{\chi_\ell} \, d\lambda = \delta_{k_1 \ell_1} \cdots \delta_{k_n \ell_n}, \tag{3.27}$$

where

$$\delta_{ab} = \begin{cases} 1 & \text{if} \quad a = b, \\ 0 & \text{if} \quad a \ne b. \end{cases}$$

If $k = \ell = 0$, then

$$\int_{\mathbb{T}^n} (\chi_k \circ T^p) \chi_\ell \, d\lambda = 1 = \int_{\mathbb{T}^n} \chi_k \, d\lambda \int_{\mathbb{T}^n} \chi_\ell \, d\lambda.$$

Now we assume that $k \ne 0$. We have

$$(\chi_k \circ T^p)(x) = e^{2\pi i \langle k, A^p x \rangle} = \chi_{(A^*)^p k}(x).$$

On the other hand, it follows readily from (2.51) that $(A^*)^{p_1} k \ne (A^*)^{p_2} k$ for any positive integers $p_1 > p_2$ since otherwise

$$(A^*)^{p_1 - p_2} (A^*)^{p_2} k = (A^*)^{p_2} k,$$

and $(A^*)^{p_2} k$ would be a nonzero eigenvector of $(A^*)^{p_1 - p_2}$ with eigenvalue 1. In particular, $(A^*)^p k \ne -\ell$ for any sufficiently large p, and it follows from (3.27) that

$$\int_{\mathbb{T}^n} (\chi_k \circ T^p) \chi_\ell \, d\lambda = \int_{\mathbb{T}^n} \chi_{(A^*)^p k} \chi_\ell \, d\lambda = 0.$$

Since $k \ne 0$, we also have

$$\int_{\mathbb{T}^n} \chi_k \, d\lambda \int_{\mathbb{T}^n} \chi_\ell \, d\lambda = 0.$$

This completes the proof of the proposition. $\square$

3.5   Symbolic Dynamics

We give in this section a brief introduction to symbolic dynamics and to its relations to ergodic theory. In particular, we construct Markov measures and Bernoulli measures, which are shift-invariant probability measures. We also describe the relation between Markov measures and topological Markov chains.

Fig. 3.1 Graph associated to a topological Markov chain

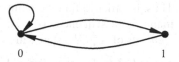

$$0 \qquad\qquad\qquad 1$$

3.5.1 Basic Notions

We first develop the theory of one-sided sequences. See Sect. 3.5.4 for the case of two-sided sequences.

For each $k \in \mathbb{N}$, we consider the set $X_k^+ = \{1, \ldots, k\}^{\mathbb{N}}$ of one-sided sequences of numbers in $\{1, \ldots, k\}$ that we write in the form

$$\omega = (i_1(\omega)i_2(\omega)\cdots).$$

Definition 3.3. We define the *shift map* $\sigma\colon X_k^+ \to X_k^+$ by

$$\sigma(i_1 i_2 \cdots) = (i_2 i_3 \cdots).$$

Clearly, the map $\sigma\colon X_k^+ \to X_k^+$ is not invertible. We also consider a particular class of shift-invariant set in X_k^+. Namely, given a $k \times k$ matrix $A = (a_{ij})$ with $a_{ij} \in \{0, 1\}$ for each i and j, we consider the set

$$X_A^+ = \{\omega \in X_k^+ : a_{i_n(\omega)i_{n+1}(\omega)} = 1 \text{ for every } n \in \mathbb{N}\}.$$

It is easy to verify that $\sigma(X_A^+) \subset X_A^+$.

Definition 3.4. Given a $k \times k$ matrix A with entries in $\{0, 1\}$, the restriction of the shift map $\sigma|X_A^+\colon X_A^+ \to X_A^+$ is called the *(one-sided) topological Markov chain* or *subshift of finite type* with *transition matrix A*.

Example 3.5. If $k = 2$ and $A = \left(\begin{smallmatrix} 1 & 1 \\ 1 & 0 \end{smallmatrix}\right)$, then

$$X_A^+ = \{\omega \in X_2^+ : (i_n(\omega)i_{n+1}(\omega)) \neq (11) \text{ for every } n \in \mathbb{N}\}$$

is the set of sequences in which the symbol 1 is always isolated (when it occurs). This set can be represented by the graph in Fig. 3.1, where an arrow indicates the possibility of a transition from i to j.

In particular, we want to find the number of m-periodic points of a topological Markov chain. We first observe that $\omega \in X_A^+$ is m-periodic if and only if $\sigma^m(\omega) = \omega$, that is, if and only if

$$i_{n+m}(\omega) = i_n(\omega) \quad \text{for every} \quad n \in \mathbb{N}.$$

Thus, in order to find the number of m-periodic points, it is sufficient to find how many finite sequences $(i_1 \cdots i_m)$ of length m with $a_{i_m i_1} = 1$ can occur in the elements of X_A^+. We start with an example.

Example 3.6. For the matrix $A = \left(\begin{smallmatrix} 1 & 1 \\ 1 & 0 \end{smallmatrix} \right)$ considered in Example 3.5, we note that the possible transitions from i to the same symbol i trough j are

$$1 \to 1 \to 1,\ 1 \to 2 \to 1,\ \text{and}\ 2 \to 1 \to 2.$$

Therefore, the shift map σ has exactly the 2-periodic points

$$(\cdots 11111 \cdots),\ (\cdots 21212 \cdots),\ \text{and}\ (\cdots 12121 \cdots).$$

Now we translate this observation into a more algebraic approach. We note that there is a transition from i to the same symbol i trough j if and only if $a_{ij} a_{ji} = 1$. Therefore, the number of 2-periodic points in X_A^+ is equal to

$$\sum_{i=1}^{2} \sum_{j=1}^{2} a_{ij} a_{ji} = \sum_{i=1}^{2} (A^2)_{ii} = \operatorname{tr}(A^2) = 3.$$

Now we find the number of m-periodic points of an arbitrary topological Markov chain. We recall that the *spectral radius* of a square matrix A is given by

$$\rho(A) = \max \{ |\tau| : \tau \in \operatorname{Sp}(A) \},$$

where $\operatorname{Sp}(A)$ denotes the set of eigenvalues of A. We also recall the notion of periodic entropy introduced in Definition 2.6.

Proposition 3.6. *For the one-sided topological Markov chain $\sigma | X_A^+$, we have:*

1. For each $m \in \mathbb{N}$,

$$\operatorname{card} \{ \omega \in X_A^+ : \sigma^m(\omega) = \omega \} = \operatorname{tr}(A^m).$$

2. $p(\sigma | X_A^+) = \log \rho(A)$.

Proof. In a similar manner to that in Example 3.6, the number of m-periodic points of the topological Markov chain $\sigma | X_A^+$ is given by

$$\sum_{i_1,\ldots,i_m=1}^{k} a_{i_1 i_2} a_{i_2 i_3} \cdots a_{i_m i_1} = \sum_{i=1}^{k} (A^m)_{ii} = \operatorname{tr}(A^m).$$

This establishes the first statement. The second statement follows immediately from the first. $\square$

Example 3.7. Consider the one-sided topological Markov chain with $k \times k$ transition matrix A. In view of Proposition 3.6, the zeta function of $\sigma | X_A^+$ (see Exercise 2.29) is given by

$$\zeta(z) = \exp \sum_{n=1}^{\infty} \frac{\operatorname{tr}(A^n)}{n} z^n.$$

Now let $\lambda_1, \ldots, \lambda_k$ be the eigenvalues of A, counted with their multiplicities. We have

$$\operatorname{tr}(A^n) = \lambda_1^n + \cdots + \lambda_k^n,$$

and hence,

$$\zeta(z) = \exp \sum_{j=1}^{k} \sum_{n=1}^{\infty} \frac{\lambda_j^n}{n} z^n$$

$$= \exp \sum_{j=1}^{k} \log \frac{1}{1 - \lambda_j z}$$

$$= \prod_{j=1}^{k} \frac{1}{1 - \lambda_j z}.$$

Therefore,

$$\zeta(z) = \frac{1}{\det(\operatorname{Id} - zA)}.$$

In particular, this shows that the zeta function of a topological Markov chain is always a rational function.

3.5.2 Markov Measures and Bernoulli Measures

We discuss in this section some relations between symbolic dynamics and ergodic theory. We start by introducing the notion of stochastic matrix. Let $P = (p_{ij})$ be a nonnegative $k \times k$ matrix. This means that $p_{ij} \geq 0$ for every $i, j = 1, \ldots, k$.

Definition 3.5. We say that P is a *stochastic matrix* if:

1. $\sum_{j=1}^{k} p_{ij} = 1$ for $i = 1, \ldots, k$.
2. There exist numbers $p_1, \ldots, p_k \in (0, 1)$ such that

$$\sum_{i=1}^{k} p_i = 1 \quad \text{and} \quad \sum_{i=1}^{k} p_i p_{ij} = p_j \tag{3.28}$$

for $j = 1, \ldots, k$.

We then say that $p = (p_1, \ldots, p_k)$ is a *probability vector* associated to P, and that (P, p) is a *stochastic pair*.

If P is a stochastic matrix, then 1 is an eigenvalue of P: simply note that $(1, \ldots, 1)$ is an eigenvector with eigenvalue 1. We also note that the second condition in (3.28) can be written in matrix form as $pP = p$.

Now to each stochastic pair, we associate a measure in X_k^+. We first need to introduce a σ-algebra in X_k^+. Given $m \in \mathbb{N}$ and $i_1, \ldots, i_m \in \{1, \ldots, k\}$, we define the *cylinder set*

$$C_{i_1 \cdots i_m} = \{(j_1 j_2 \cdots) \in X_k^+ : j_\ell = i_\ell \text{ for } \ell = 1, \ldots, m\},$$

and we consider the σ-algebra in X_k^+ generated by these sets.

Definition 3.6. Given a stochastic pair (P, p), we define a measure μ in X_k^+ by requiring that

$$\mu(C_{i_1 \cdots i_m}) = p_{i_1} p_{i_1 i_2} \cdots p_{i_{m-1} i_m}$$

for each $m \in \mathbb{N}$ and $i_1, \ldots, i_m \in \{1, \ldots, k\}$. We then say that μ is the *(one-sided) Markov measure* associated to the pair (P, p).

One can easily verify that any Markov measure in X_k^+ is well defined. Indeed, since the cylinder sets generate the σ-algebra of X_k^+, and since

$$C_{i_1 \cdots i_m} = \bigcup_{j=1}^{k} C_{i_1 \cdots i_m j} \tag{3.29}$$

is a disjoint union, it is sufficient to note that

$$\sum_{j=1}^{k} p_{i_1} p_{i_1 i_2} \cdots p_{i_{m-1} i_m} p_{i_m j} = p_{i_1} p_{i_1 i_2} \cdots p_{i_{m-1} i_m} \sum_{j=1}^{k} p_{i_m j} \tag{3.30}$$

$$= p_{i_1} p_{i_1 i_2} \cdots p_{i_{m-1} i_m}.$$

Proposition 3.7. *Any one-sided Markov measure is σ-invariant.*

Proof. Let μ be the Markov measure associated to a stochastic pair (P, p). Since

$$\sigma^{-1} C_{i_1 \cdots i_m} = \bigcup_{j=1}^{k} C_{j i_1 \cdots i_m},$$

we obtain

$$\mu(\sigma^{-1}C_{i_1\cdots i_m}) = \sum_{j=1}^{k}\mu(C_{ji_1\cdots i_m}) = \sum_{j=1}^{k} p_j\, p_{ji_1} p_{i_1 i_2}\cdots p_{i_{m-1} i_m}$$

$$= p_{i_1} p_{i_1 i_2}\cdots p_{i_{m-1} i_m} = \mu(C_{i_1\cdots i_m}).$$

Since the cylinder sets generate the σ-algebra this yields the desired result.   □

Example 3.8. Consider the matrix

$$P = \begin{pmatrix} 1/2 & 1/2 \\ 1 & 0 \end{pmatrix}.$$

Then $p = (2/3, 1/3)$ is an eigenvector of P^* with eigenvalue 1. Therefore, $p = pP$, and (P, p) is a stochastic pair. If μ is the corresponding Markov measure in X_2^+, then, for example,

$$\mu(C_{121}) = p_1 p_{12} p_{21} = \frac{2}{3}\cdot\frac{1}{2}\cdot 1 = \frac{1}{3},$$

and

$$\mu(C_{211}) = p_2 p_{21} p_{11} = \frac{1}{3}\cdot 1 \cdot\frac{1}{2} = \frac{1}{6}.$$

Now we consider a particular class of Markov measures.

Definition 3.7. We say that a measure μ in X_k^+ is a *(one-sided) Bernoulli measure* if there exist numbers $p_1,\ldots, p_k \in (0, 1)$ such that:

1. $\sum_{i=1}^{k} p_i = 1$.
2. $\mu(C_{i_1\cdots i_m}) = p_{i_1}\cdots p_{i_m}$ for each $m \in \mathbb{N}$ and $i_1,\ldots, i_m \in \{1,\ldots, k\}$.

To see that any Bernoulli measure is indeed a Markov measure, it is sufficient to take $p_{ij} = p_j$ for each i and j. In particular, by Proposition 3.7, all Bernoulli measures are invariant under the shift map. Moreover, the following property holds:

Proposition 3.8. *Any one-sided Bernoulli measure is mixing.*

Proof. Again it is sufficient to consider cylinder sets. We have

$$\mu(\sigma^{-n}C_{i_1\cdots i_k}\cap C_{j_1\cdots j_\ell}) = \mu\left(\bigcup_{m_1\cdots m_{n-\ell}} C_{j_1\cdots j_\ell m_1\cdots m_{n-\ell} i_1\cdots i_k}\right)$$

$$= \sum_{m_1\cdots m_{n-\ell}} p_{j_1}\cdots p_{j_\ell} p_{m_1}\cdots p_{m_{n-\ell}} p_{i_1}\cdots p_{i_k}$$

$$= p_{j_1}\cdots p_{j_\ell} p_{i_1}\cdots p_{i_k}$$

$$= \mu(C_{j_1\cdots j_\ell})\mu(C_{i_1\cdots i_k})$$

for any sufficiently large $n \in \mathbb{N}$. This yields the desired result.   □

Example 3.9. Let μ be an ergodic Markov measure (e.g., by Proposition 3.8, all Bernoulli measures are ergodic). By Proposition 2.12, for each $i = 1, \ldots, k$, we have

$$\lim_{n \to \infty} \frac{1}{n} \sum_{j=0}^{n-1} \chi_{C_i}(\sigma^j(\omega)) = \int_{X_k^+} \chi_{C_i} \, d\mu = \mu(C_i) = p_i$$

for μ-almost every $\omega \in X_k^+$.

We also give a criterion for a nonnegative matrix to be a stochastic matrix.

Theorem 3.7 (Perron–Frobenius). *Let $P = (p_{ij})$ be a nonnegative $k \times k$ matrix such that:*

1. *$\sum_{j=1}^{k} p_{ij} = 1$ for $i = 1, \ldots, k$.*
2. *For each $i, j = 1, \ldots, k$, there exists $m = m(i, j) \in \mathbb{N}$ such that the entry (i, j) of P^m is positive.*

Then there exists one and only one $p \in \mathbb{R}^k$ such that (P, p) is a stochastic pair.

Proof. Let

$$S = \{v \in (\mathbb{R}_0^+)^k : \|v\| = 1\},$$

with the norm $\|v\| = \sum_{i=1}^{k} |v_i|$. By the first condition, we can define a transformation $F \colon S \to S$ by

$$F(v) = P^* v / \|P^* v\|.$$

Since S is homeomorphic to the closed unit ball in $\mathbb{R}^{k-1}$ and F is continuous, it follows from Brouwer's fixed point theorem that F has a fixed point $p \in S$. We thus obtain $P^* p = \lambda p$, where

$$\lambda = \|P^* p\| = \sum_{j=1}^{k} (P^* p)_j$$

$$= \sum_{j=1}^{k} \sum_{i=1}^{k} p_i p_{ij} = \sum_{i=1}^{k} p_i \sum_{j=1}^{k} p_{ij}$$

$$= \sum_{i=1}^{k} p_i = 1.$$

Therefore, to show that (P, p) is a stochastic pair, it remains to verify that p has only positive entries. Since $p \in S$, there exists $j \in \{1, \ldots, k\}$ such that $p_j > 0$. On the other hand, by the second condition, for each $i = 1, \ldots, k$, there exists $m = m(i, j)$ such that the entry (i, j) of $(P^*)^m$ is positive. Since $(P^*)^m p = p$, we thus obtain

$$p_i = \sum_{\ell=1}^{k} (P^*)_{i\ell}^m p_\ell \geq (P^*)_{ij}^m p_j > 0 \qquad (3.31)$$

for $i = 1, \ldots, k$. Now we consider another vector $q \in \mathbb{R}^k$ with

$$\|q\| = 1 \quad \text{and} \quad \min\{q_i : i = 1, \ldots, k\} > 0$$

such that $P^*q = q$. We want to show that $q = p$. Otherwise, there would exist $t \in \mathbb{R}$ for which $p + tq$ would be a nonzero eigenvector with nonnegative entries, at least with one entry equal to zero. On the other hand, one can repeat the argument in (3.31) to show that all entries of $p + tq$ must be positive. This contradiction shows that $q = p$. □

3.5.3 Topological Markov Chains and Markov Measures

Now we introduce a distance and thus also a topology in X_k^+. Namely, given $\beta > 1$, for each $\omega, \omega' \in X_k^+$, we set

$$d_\beta(\omega, \omega') = \begin{cases} \beta^{-n} & \text{if } \omega \neq \omega', \\ 0 & \text{if } \omega = \omega', \end{cases} \tag{3.32}$$

where $n = n(\omega, \omega') \in \mathbb{N}$ is the smallest integer such that $i_n(\omega) \neq i_n(\omega')$. One can easily verify that d_β is indeed a distance in X_k^+. For additional properties, we refer to Exercises 3.14–3.16.

Now we describe the relation between Markov measures and topological Markov chains. Given a stochastic matrix $P = (p_{ij})$, we consider the topological Markov chain $\sigma|X_A^+ : X_A^+ \to X_A^+$ where the $k \times k$ transition matrix $A = (a_{ij})$ is defined by

$$a_{ij} = \begin{cases} 1 & \text{if } p_{ij} > 0, \\ 0 & \text{if } p_{ij} = 0. \end{cases} \tag{3.33}$$

We recall that the *support* of a measure μ in X_k^+ is the set $\operatorname{supp} \mu \subset X_k^+$ formed by all points $x \in X_k^+$ such that $\mu(U) > 0$ for any open set U containing x (see also Proposition 3.12 below).

Proposition 3.9. *If μ is the one-sided Markov measure associated to a stochastic pair (P, p), then $\operatorname{supp} \mu = X_A^+$, where A is the transition matrix defined by (3.33).*

Proof. For each $\omega = (i_1 i_2 \cdots) \in X_A^+$ and $m \in \mathbb{N}$, we have

$$\mu(C_{i_1 \cdots i_m}) = p_{i_1} p_{i_1 i_2} \cdots p_{i_{m-1} i_m} > 0.$$

Since the cylinder sets generate the topology, this shows that $\omega \in \operatorname{supp} \mu$.

Now we assume that $\omega = (i_1 i_2 \cdots) \notin X_A^+$. Then there exists $m \in \mathbb{N}$ such that $a_{i_m i_{m+1}} = 0$. Therefore,

$$p_{i_m i_{m+1}} = 0 \quad \text{and} \quad \mu(C_{i_1 \cdots i_{m+1}}) = 0.$$

Since $\omega \in C_{i_1 \cdots i_{m+1}}$, this shows that $\omega \notin \operatorname{supp} \mu$. $\qquad \qquad \square$

By Proposition 3.9, the support of any Bernoulli measure is X_k^+.

3.5.4 The Case of Two-Sided Sequences

We develop in this section a corresponding theory of two-sided sequences, highlighting the differences with respect to the theory of one-sided sequences.

For each $k \in \mathbb{N}$, we consider the set $X_k = \{1, \ldots, k\}^{\mathbb{Z}}$ of two-sided sequences of numbers in $\{1, \ldots, k\}$. Given a sequence $\omega \in X_k$, we write it in the form

$$\omega = (\cdots i_{-1}(\omega) i_0(\omega) i_1(\omega) \cdots).$$

Definition 3.8. We define the *shift map* $\sigma \colon X_k \to X_k$ by $\sigma(\omega) = \omega'$, where

$$i_n(\omega') = i_{n+1}(\omega) \quad \text{for each} \quad n \in \mathbb{Z}.$$

Clearly, the map $\sigma \colon X_k \to X_k$ is invertible. Given a $k \times k$ matrix $A = (a_{ij})$ with $a_{ij} \in \{0, 1\}$ for each i and j, we consider the set

$$X_A = \{\omega \in X_k : a_{i_n(\omega) i_{n+1}(\omega)} = 1 \text{ for every } n \in \mathbb{Z}\}.$$

It is easy to verify that $\sigma(X_A) = X_A$.

Definition 3.9. The restriction $\sigma|X_A \colon X_A \to X_A$ is called the *(two-sided) topological Markov chain* or *subshift of finite type* with *transition matrix* A.

We note that the statement in Proposition 3.6 also holds for two-sided topological Markov chains, that is,

$$\operatorname{card}\{\omega \in X_A : \sigma^m(\omega) = \omega\} = \operatorname{tr}(A^m) \tag{3.34}$$

for each $m \in \mathbb{N}$, and hence,

$$p(\sigma|X_A) = \log \rho(A). \tag{3.35}$$

Moreover, to each stochastic pair we associate a measure in X_k. Given $m \in \mathbb{N}$ and $i_{-m}, \ldots, i_m \in \{1, \ldots, k\}$, we define the *cylinder set*

$$C_{i_{-m} \cdots i_m} = \{(\cdots j_0 \cdots) \in X_k : j_\ell = i_\ell \text{ for } \ell = -m, \ldots, m\}, \tag{3.36}$$

and we consider the σ-algebra in X_k generated by these sets.

Definition 3.10. Given a stochastic pair (P, p), we define a measure μ in X_k by requiring that
$$\mu(C_{i_{-m}\cdots i_m}) = p_{i_{-m}} p_{i_{-m}i_{-m+1}} \cdots p_{i_{m-1}i_m}$$
for each $m \in \mathbb{N}$ and $i_{-m}, \ldots, i_m \in \{1, \ldots, k\}$. We then say that μ is the *(two-sided) Markov measure* associated to the pair (P, p).

Proceeding as in (3.29) and (3.30), one can easily verify that any Markov measure in X_k is well defined.

Proposition 3.10. *Any two-sided Markov measure is σ-invariant.*

Proof. Let μ be the Markov measure associated to a stochastic pair (P, p). In view of the identity
$$\sigma^{-1}C_{i_{-m}\cdots i_m} = \bigcup_{j=1}^{k}\bigcup_{\ell=1}^{k} C_{j\ell i_{-m}\cdots i_m},$$
we obtain
$$\mu(\sigma^{-1}C_{i_{-m}\cdots i_m}) = \sum_{j=1}^{k}\sum_{\ell=1}^{k} \mu(C_{j\ell i_{-m}\cdots i_m})$$
$$= \sum_{j=1}^{k}\sum_{\ell=1}^{k} p_j\, p_{j\ell} p_{\ell i_{-m}} \cdots p_{i_{m-1}i_m}$$
$$= p_{i_{-m}} p_{i_{-m}i_{-m+1}} \cdots p_{i_{m-1}i_m}$$
$$= \mu(C_{i_{-m}\cdots i_m}).$$

Since the cylinder sets generate the σ-algebra, this yields the desired result. □

We also consider the class of Bernoulli measures.

Definition 3.11. We say that a measure μ in X_k is a *(two-sided) Bernoulli measure* if there exist numbers $p_1, \ldots, p_k \in (0, 1)$ such that:

1. $\sum_{i=1}^{k} p_i = 1$.
2. $\mu(C_{i_{-m}\cdots i_m}) = p_{i_{-m}} \cdots p_{i_m}$ for each $m \in \mathbb{N}$ and $i_{-m}, \ldots, i_m \in \{1, \ldots, k\}$.

Now we introduce a distance and thus also a topology in X_k. Namely, given $\beta > 1$, for each $\omega, \omega' \in X_k$, we set
$$d_\beta(\omega, \omega') = \begin{cases} \beta^{-n} & \text{if } \omega \neq \omega' \\ 0 & \text{if } \omega = \omega', \end{cases} \tag{3.37}$$

where $n = n(\omega, \omega') \in \mathbb{N} \cup \{0\}$ is the smallest integer such that $i_n(\omega) \neq i_n(\omega')$ or $i_{-n}(\omega) \neq i_{-n}(\omega')$. Then d_β is a distance in X_k.

Finally, we describe the relation between Markov measures and topological Markov chains.

Proposition 3.11. *If μ is the two-sided Markov measure associated to a stochastic pair (P, p), then* $\operatorname{supp} \mu = X_A$, *where A is the transition matrix defined by* (3.33).

Proof. For each $\omega = (\cdots i_0 \cdots) \in X_A$ and $m \in \mathbb{N}$, we have

$$\mu(C_{i_{-m}\cdots i_m}) = p_{i_{-m}} p_{i_{-m} i_{-m+1}} \cdots p_{i_{m-1} i_m} > 0.$$

Since the cylinder sets generate the topology, this shows that $\omega \in \operatorname{supp} \mu$.

Now we assume that $\omega = (\cdots i_0 \cdots) \notin X_A$. Then there exists $m \in \mathbb{Z}$ such that $a_{i_m i_{m+1}} = 0$. Therefore,

$$p_{i_m i_{m+1}} = 0 \quad \text{and} \quad \mu(C_{i_{-m-1}\cdots i_{m+1}}) = 0$$

when $m \geq 0$, and $\mu(C_{i_m \cdots i_{-m}}) = 0$ when $m < 0$. This shows that $\omega \notin \operatorname{supp} \mu$. $\qquad \square$

3.6   Topological Dynamics

We give in this section a brief introduction to the relation between topological dynamics and ergodic theory. In particular, we show that each ergodic property implies a corresponding topological property.

We start with an auxiliary statement concerning the support of a Borel probability measure μ in a metric space X. We recall that the *support* of μ is the set $\operatorname{supp} \mu \subset X$ formed by all points $x \in X$ such that $\mu(U) > 0$ for any open set U containing x.

Proposition 3.12. *If X is a metric space and μ is a Borel probability measure in X, then the following properties hold:*

1. $\operatorname{supp} \mu$ *is the complement of the largest open set of zero μ-measure.*
2. $\operatorname{supp} \mu$ *is a closed set.*
3. $\mu(X \setminus \operatorname{supp} \mu) = 0$.
4. *If A is a measurable set with $\mu(X \setminus A) = 0$, then $\overline{A} \supset \operatorname{supp} \mu$.*

Proof. Let $V \subset X$ be the largest open set of zero μ-measure. If $x \in X \setminus \operatorname{supp} \mu$, then there exists an open set $U \subset X$ containing x with $\mu(U) = 0$. Thus, $x \in U \subset V$, and we conclude that $X \setminus \operatorname{supp} \mu \subset V$. On the other hand, if $x \in V$, then clearly $x \notin \operatorname{supp} \mu$. This yields Property 1. Properties 2 and 3 follow immediately from Property 1.

For Property 4, we note that if $\operatorname{supp} \mu \setminus \overline{A} \neq \varnothing$, then there would exist an open set $U \subset X$ disjoint from $\overline{A}$ with $\mu(U) > 0$. On the other hand,

$$\mu(U) = \mu(U \cap \operatorname{supp} \mu) \leq \mu(\operatorname{supp} \mu \setminus \overline{A}) \leq \mu(X \setminus A) = 0.$$

This contradiction shows that $\overline{A} \supset \operatorname{supp} \mu$. $\qquad \square$

Now we recall some basic notions of topological dynamics. We continue to assume that X is a metric space.

Definition 3.12. Let $T\colon X \to X$ be a continuous transformation.

1. We define the *ω-limit set* of a point $x \in X$ by

$$\omega(x) = \bigcap_{n=1}^{\infty} \overline{\{T^k(x) : k \geq n\}}.$$

2. We say that x is a *recurrent* point of T if $x \in \omega(x)$.

Clearly, $\omega(x)$ is a closed set. One can easily verify that $y \in \omega(x)$ if and only if there exists an increasing sequence $(n_k)_{k \in \mathbb{N}} \subset \mathbb{N}$ such that $T^{n_k}(x) \to y$ when $k \to +\infty$. Therefore, x is a recurrent point of T if and only if for each open set U containing x, there exists $n \in \mathbb{N}$ such that $T^n(x) \in U$.

We also introduce some topological properties.

Definition 3.13. Let $T\colon X \to X$ be a continuous transformation.

1. We say that T is *topologically transitive* if for any nonempty open sets $U, V \subset X$, there exists $n \in \mathbb{N}$ such that $T^{-n}U \cap V \neq \varnothing$.
2. We say that T is *topologically mixing* if for any nonempty open sets $U, V \subset X$, there exists $m \in \mathbb{N}$ such that $T^{-n}U \cap V \neq \varnothing$ for every $n > m$.

Clearly if T is topologically mixing, then T is topologically transitive.

The following result makes explicit the relations between the ergodic properties of μ and the topological properties on the support of the measure:

Theorem 3.8. *Let $T\colon X \to X$ be a homeomorphism of a separable metric space and let μ be a T-invariant Borel probability measure in X. Then the following properties hold:*

1. $\operatorname{supp}\mu$ *is T-invariant.*
2. *If μ is ergodic, then $T \,|\, \operatorname{supp}\mu$ is topologically transitive.*
3. *If μ is mixing, then $T \,|\, \operatorname{supp}\mu$ is topologically mixing.*
4. *If X is compact and $T \,|\, \operatorname{supp}\mu$ is uniquely ergodic, then the two-sided orbit of any point in $\operatorname{supp}\mu$ is dense in $\operatorname{supp}\mu$.*

Proof. Let $x \in T^{-1}(\operatorname{supp}\mu)$. Given an open set $U \subset X$ containing x, the open set $T(U)$ contains the point $T(x) \in \operatorname{supp}\mu$. Therefore, $\mu(U) = \mu(T(U)) > 0$, and $x \in \operatorname{supp}\mu$. This shows that

$$T^{-1}(\operatorname{supp}\mu) \subset \operatorname{supp}\mu.$$

Now let $x \in \operatorname{supp}\mu$. Given an open set $U \subset X$ containing $T(x)$, the open set $T^{-1}(U)$ contains $x \in \operatorname{supp}\mu$. Therefore, $\mu(U) = \mu(T^{-1}(U)) > 0$ and $T(x) \in \operatorname{supp}\mu$. This shows that

$$\operatorname{supp}\mu \subset T^{-1}(\operatorname{supp}\mu).$$

Now we assume that μ is ergodic. Let U and V be nonempty open sets intersecting supp μ. By Proposition 3.4, we have

$$\lim_{n \to \infty} \frac{1}{n} \sum_{k=0}^{n-1} \mu(T^{-k}U \cap V) = \mu(U)\mu(V) > 0.$$

This implies that there exist infinitely many integers n with $T^{-n}U \cap V \neq \varnothing$, and in particular, $T \,|\, \text{supp} \,\mu$ is topologically transitive.

Now we assume that μ is mixing. Let again U and V be nonempty open sets intersecting supp μ. Then $\mu(U), \mu(V) > 0$, and by the mixing property, we obtain

$$\mu(T^{-n}U \cap V) \to \mu(U)\mu(V) > 0$$

when $n \to \infty$. Therefore, $T^{-n}U \cap V \neq \varnothing$ for all sufficiently large n.

Finally, in order to proceed by contradiction, we assume that there exists $x \in$ supp μ such that its two-sided orbit

$$A = \{T^n(x) : n \in \mathbb{Z}\}$$

is not dense in supp μ. Since supp μ is T-invariant, $\overline{A}$ is a proper subset of supp μ. On the other hand, since $\overline{A}$ is compact and $T|\overline{A}$ is continuous, by the Krylov–Bogolubov theorem (Theorem 3.2), there exists a T-invariant probability measure ν in X with supp $\nu \subset \overline{A}$. Since $\overline{A}$ is a proper subset of supp μ, we conclude that $\nu \neq \mu$. But this contradicts the unique ergodicity of $T \,|\, \text{supp} \,\mu$. $\qquad \square$

Example 3.10. Consider a Bernoulli measure μ in X_k^+. By Proposition 3.8, the measure μ is mixing. Since supp $\mu = X_k^+$, it follows from Theorem 3.8 that T is topologically mixing.

3.7 Exercises

Exercise 3.1. Let $T : X \to X$ be a continuous transformation of a compact metric space. Show that for each $x \in X$, the sequence of probability measures

$$\frac{1}{n} \sum_{k=0}^{n-1} \delta_{T^k(x)} \tag{3.38}$$

has at least one accumulation point in $\mathcal{M}(X)$ and that any of the accumulation points is a T-invariant probability measure. Hint: we have $T_*^k \delta_x = \delta_{T^k(x)}$ with T_* as in (3.6) and thus replacing μ by δ_x in (3.7) yields

$$T_* \left(\frac{1}{m_n} \sum_{k=0}^{m_n-1} \delta_{T^k(x)} \right) = \frac{1}{m_n} \sum_{k=0}^{m_n-1} \delta_{T^k(x)} + \frac{\delta_{T^{m_n}(x)}}{m_n} - \frac{\delta_x}{m_n}.$$

Exercise 3.2. Show that when x is an m-periodic point, the measure μ in (2.16) is the unique accumulation point of the sequence of measures in (3.38).

Exercise 3.3. Show that a uniquely ergodic transformation has at most one periodic orbit. Hint: consider the measure in (2.16).

Exercise 3.4. For each $n \in \mathbb{N}$, say whether there exists a uniquely ergodic toral automorphism of $\mathbb{T}^n$.

Exercise 3.5. Take $\alpha_1, \ldots, \alpha_n \in \mathbb{R}$ such that the set $\{1, \alpha_1, \ldots, \alpha_n\}$ is rationally independent. This means that if

$$k_0 + k_1\alpha_1 + \cdots + k_n\alpha_n = 0$$

for some integers $k_0, \ldots, k_n \in \mathbb{Z}$, then $k_0 = \cdots = k_n = 0$. Show that the toral translation $T_{\alpha_1,\ldots,\alpha_n} \colon \mathbb{T}^n \to \mathbb{T}^n$ in (2.62) is uniquely ergodic and that the unique $T_{\alpha_1,\ldots,\alpha_n}$-invariant probability measure is the Lebesgue measure.

Exercise 3.6. Given $p \in \mathbb{N}$, consider the sequence of the first digits in base-10 of the powers of p. Determine whether the limit in (3.18) exists, and if so, compute its value explicitly.

Exercise 3.7. Show that if $\alpha \notin \mathbb{Q}$, then the sequence of fractional parts of the polynomial $P(t) = \alpha t + \beta$ is uniformly distributed for every $\beta \in \mathbb{R}$.

Exercise 3.8. Given $a \in \mathbb{R}$, consider the set

$$K_a = \{\lfloor na \rfloor : n \in \mathbb{N}\},$$

where $\lfloor na \rfloor$ denotes the integer part of na.

1. Show that if $a, b \in (1, \infty)$ are such that $\{1, a^{-1}, b^{-1}\}$ is rationally independent (see Exercise 3.5), then

$$\lim_{n \to \infty} \frac{\operatorname{card}(K_a \cap K_b \cap [1, n])}{n} = \frac{1}{ab}.$$

2. Show that in general, the set $K_a \cap K_b$ may be empty. Hint: take $a > 1$ such that $a^{-1} + a^{-2} = 1$.

Exercise 3.9. Given $\alpha \in \mathbb{R} \setminus \mathbb{Q}$, we define a transformation $S \colon X \to X$ in the space $X = \mathbb{T} \times [0, 1]$ by

$$S(x, t) = (x + \alpha, t) \mod 1.$$

Show that for each $\varphi \in C(X)$, the sequence of functions in (3.8) converges uniformly, but that S is not uniquely ergodic. Hint: fix $t \in [0, 1]$ and observe that for the continuous function $\psi(x) = \varphi(x,t)$, the sequence of functions

$$\frac{1}{n}\sum_{k=0}^{n-1}\psi \circ T^k,$$

where $T(x) = x + \alpha \mod 1$, converges uniformly to $\int_0^1 \psi(x)\,dx$ when $n \to \infty$. This shows that the sequence of functions in (3.8) may converge uniformly even when T is not uniquely ergodic, although of course not to a constant.

Exercise 3.10. Let $\Phi = \{\varphi_t \colon X \to X\}_{t\in\mathbb{R}}$ be a group of measurable transformations, that is, a family of measurable transformations such that $\varphi_0(x) = x$ for every $x \in X$ and $\varphi_{t+s} = \varphi_t \circ \varphi_s$ for every $t, s \in \mathbb{R}$. We say that a measure μ in X is Φ-*invariant* if

$$\mu(\varphi_t(A)) = \mu(A)$$

for every $t \in \mathbb{R}$ and every measurable set $A \subset X$.

1. Show that if μ is a Φ-invariant finite measure in X and $g \in L^1(X,\mu)$, then for μ-almost every $x \in X$, the limit

$$\lim_{t\to+\infty}\frac{1}{t}\int_0^t g(\varphi_s(x))\,ds$$

   exists. Hint: consider the sequence

$$\frac{1}{n}\int_0^n g(\varphi_s(x))\,ds$$

   and use Birkhoff's ergodic theorem (Theorem 2.2).
2. Show that if μ is a Φ-invariant finite measure in X and $g, h \in L^1(X,\mu)$ with $\inf\{h(x) : x \in X\} > 0$, then for μ-almost every $x \in X$, the limit

$$\lim_{t\to+\infty}\frac{\int_0^t g(\varphi_s(x))\,ds}{\int_0^t h(\varphi_s(x))\,ds}$$

   exists.

Exercise 3.11. Show that if $\Phi = \{\varphi_t \colon \mathbb{T}^2 \to \mathbb{T}^2\}_{t\in\mathbb{R}}$ is the group of diffeomorphisms generated by the solutions of the differential equation $(x, y)' = (\alpha, \beta)$ in $\mathbb{T}^2$ for some $\alpha \neq 0$, then there exists a unique Φ-invariant probability measure in $\mathbb{T}^2$ (see Exercise 3.10) if and only if $\beta/\alpha \notin \mathbb{Q}$.

Exercise 3.12. Given $q \in \mathbb{Z}$ with $|q| > 1$, let $E_q \colon S^1 \to S^1$ be the expanding map of the circle given by $E_q(z) = z^q$. Show that the Lebesgue measure is mixing with respect to E_q.

Exercise 3.13. Prove or disprove the following statement: if (T, μ) is mixing, then (T^n, μ) is mixing for each $n \geq 2$.

Exercise 3.14. Equipping X_k^+ with the distance d_β in (3.32), show that:

1. The shift map $\sigma: X_k^+ \to X_k^+$ is continuous.
2. Each open ball is closed.
3. The cylinder sets are simultaneously open and closed, and they generate the topology of X_k^+.

Exercise 3.15. We recall that two distances d and d' are said to be *equivalent* if there exists a constant $c > 0$ such that

$$c^{-1} d(\omega, \omega') \leq d'(\omega, \omega') \leq cd(\omega, \omega')$$

for every $\omega, \omega' \in X_k^+$. Moreover, two distances d and d' are said to be *Hölder-equivalent* if there exist constants $c > 0$ and $\alpha \in (0, 1]$ such that

$$c^{-1} d(\omega, \omega')^{1/\alpha} \leq d'(\omega, \omega') \leq cd(\omega, \omega')^\alpha$$

for every $\omega, \omega' \in X_k^+$. Given distinct numbers $\beta, \beta' > 1$:

1. Show that the distances d_β and $d_{\beta'}$ are not equivalent.
2. Show that the distances d_β and $d_{\beta'}$ are Hölder-equivalent and conclude that they generate the same topology.

Exercise 3.16. Equipping $\{1, \ldots, k\}$ with the discrete topology, that is, the topology generated by all subsets of $\{1, \ldots, k\}$, show that the corresponding product topology in $X_k^+ = \{1, \ldots, k\}^{\mathbb{N}}$ coincides with the topology induced by each distance d_β. It follows from Tychonoff's theorem that the space X_k^+ is compact.

Exercise 3.17. A $k \times k$ matrix A is called *irreducible* if for each $i, j \in \{1, \ldots, k\}$, there exists $n \in \mathbb{N}$ (possibly depending on i and j) such that the entry (i, j) of A^n is positive. Show that the graph associated to the topological Markov chain $\sigma | X_A^+$ has an oriented path between any two vertices if and only if A is irreducible.

Exercise 3.18. Show that any two-sided Bernoulli measure is mixing.

Exercise 3.19. Let μ be a Markov measure with support X_A^+ for some matrix A. Show that if μ is mixing, then there exists $m \in \mathbb{N}$ such that all entries of the matrix A^m are positive.

Exercise 3.20. Equipping X_k with the distance d_β in (3.37), show that X_k is compact, and that $\sigma: X_k \to X_k$ is a homeomorphism.

Exercise 3.21. Consider the set of sequences $Y \subset \{1, 2\}^{\mathbb{Z}}$ in which 1 appears only in pairs. Since $\sigma(Y) \subset Y$, we can consider the restriction $\sigma | Y: Y \to Y$.

1. Show that $\sigma | Y$ is not a topological Markov chain.
2. Verify that the periodic entropy of $\sigma | Y$ is positive.

3. Show that the zeta function of $\sigma|Y$ is given by

$$\zeta(z) = \frac{1+z}{1-z-z^2}.$$

Exercise 3.22. For the shift map $\sigma: X_k^+ \to X_k^+$, find a recurrent point $\omega \in X_k^+$ which is not periodic.

Exercise 3.23. We say that two transformations $T_i: X_i \to X_i$ preserving a measure μ_i in X_i, for $i = 1, 2$, are *equivalent* if there is a measurable transformation $h: X_1 \to X_2$ such that:

1. h is bijective almost everywhere.
2. $h \circ T_1 = T_2 \circ h$ μ_1-almost everywhere in X_1.
3. $\mu_1(h^{-1}A) = \mu_2(A)$ for every measurable set $A \subset X_2$.

Show that if T_1 and T_2 are equivalent, then:

1. μ_1 is ergodic if and only if μ_2 is ergodic.
2. μ_1 is mixing if and only if μ_2 is mixing.

Exercise 3.24. Consider the expanding map of the circle $E_q: S^1 \to S^1$ together with the Lebesgue measure λ and the shift map $\sigma: X_k^+ \to X_k^+$ together with the Bernoulli measure ν with probability vector $(1/q, \ldots, 1/q)$. Show that:

1. The measurable transformation $h: X_k^+ \to S^1$ defined by

$$h(i_1 i_2 \cdots) = \exp\left(2\pi i \sum_{k=1}^{\infty} (i_k - 1) q^{-k}\right)$$

is bijective almost everywhere.
2. $E_q \circ h = h \circ \sigma$ everywhere in X_k^+.
3. $\nu(h^{-1}A) = \lambda(A)$ for any measurable set $A \subset S^1$.

This shows that the two transformations are equivalent (see Exercise 3.23).

Exercise 3.25. Consider the transformation in $T: \mathbb{R} \to \mathbb{R}$ given by

$$T(x) = \begin{cases} (x - 1/x)/2 & \text{if } x \neq 0, \\ 0 & \text{if } x = 0. \end{cases}$$

1. Show that T preserves the Borel probability measure

$$\mu(A) = \frac{1}{\pi} \int_A \frac{dx}{1 + x^2}.$$

Fig. 3.2 The transformation
in Exercise 3.28

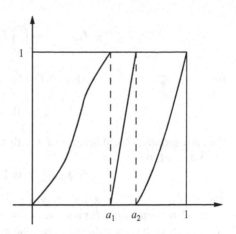

2. Show that the transformation T together with the measure μ is equivalent to the shift map $\sigma: X_2^+ \to X_2^+$ together with the Bernoulli measure with probability vector $(1/2, 1/2)$.

Exercise 3.26. For the expanding map of the circle E_2, find a point $z \in S^1$ such that $\omega(z) = S^1$.

Exercise 3.27. Under the assumptions of Theorem 3.8, show that if μ is ergodic, then for μ-almost every $x \in X$, the orbit of x is dense in $\operatorname{supp} \mu$. Hint: consider a countable base of open sets U_n for the induced topology in $\operatorname{supp} \mu$ (with $\mu(U_n) > 0$ for each n).

Exercise 3.28. For a transformation $T: [0, 1] \to [0, 1]$, we assume that there exist $p \in \mathbb{N}$ and $a_0, \dots, a_{p+1} \in \mathbb{R}$ with

$$0 = a_0 < a_1 < \cdots < a_p < a_{p+1} = 1$$

such that:

1. T is of class C^2 in $\bigcup_{i=0}^p (a_i, a_{i+1})$.
2. $T((a_i, a_{i+1})) = (0, 1)$ for $i = 0, \dots, p$.
3. There exists $\tau > 1$ such that $|T'(x)| > \tau$ for every $x \in \bigcup_{i=0}^p (a_i, a_{i+1})$.
4.
$$d_1 := \sup \left\{ \frac{|T''(x)|}{|T'(y)|} : x, y \in (a_i, a_{i+1}) \text{ and } i = 0, \dots, p \right\} < \infty.$$

See Fig. 3.2 for an example with $p = 2$. Other examples are given by the piecewise linear expanding maps (see Sect. 2.2.6). For each $n \in \mathbb{N}$, the transformation T^n is strictly monotone in each interval

$$I_{i_0\cdots i_{n-1}} = \bigcap_{k=0}^{n-1} T^{-k}(a_{i_k}, a_{i_k+1}),$$

for $i_0, \ldots, i_{n-1} \in \{0, \ldots, p\}$, and $T^n(I_{i_0\cdots i_{n-1}}) = (0, 1)$. We denote by

$$\psi_{i_0\cdots i_{n-1}} : (0, 1) \to I_{i_0\cdots i_{n-1}}$$

the corresponding local inverse of T^n, that is, the unique transformation of class C^2 in $(0, 1)$ such that

$$\psi_{i_0\cdots i_{n-1}}((0, 1)) = I_{i_0\cdots i_{n-1}}$$

and $T^n(\psi_{i_0\cdots i_{n-1}}(x)) = x$ for every $x \in (0, 1)$.

We also assume that T preserves a Borel probability measure μ that is absolutely continuous with respect to the Lebesgue measure m in $[0, 1]$. In addition, we assume that the Radon–Nikodym derivative $\rho = d\mu/dm$ is defined everywhere in $[0, 1]$, and that it satisfies

$$d_2 := \inf\{\rho(x) : x \in [0, 1]\} > 0 \quad \text{and} \quad d_3 := \sup\{\rho(x) : x \in [0, 1]\} < \infty.$$

Under these assumptions, show that:

1. For each $x, y \in (0, 1)$, $n \in \mathbb{N}$, and $i_0, \ldots, i_{n-1} \in \{0, \ldots, p\}$, we have

$$|\psi_{i_0\cdots i_{n-1}}(x) - \psi_{i_0\cdots i_{n-1}}(y)| \le \tau^{-n}|x - y|.$$

2. There exists $c_1 > 0$ such that if $x, y \in (0, 1)$, $n \in \mathbb{N}$, and $i_0, \ldots, i_{n-1} \in \{0, \ldots, p\}$, then

$$\frac{|\psi'_{i_0\cdots i_{n-1}}(x)|}{|\psi'_{i_0\cdots i_{n-1}}(y)|} \le c_1.$$

   Hint: note that

$$\frac{|(T^n)'(z)|}{|(T^n)'(w)|} = \prod_{k=0}^{n-1} \frac{|T'(T^k(z))|}{|T'(T^k(w))|}$$

$$\le \prod_{k=0}^{n-1} \left(1 + \frac{|T'(T^k(z)) - T'(T^k(w))|}{|T'(T^k(w))|}\right).$$

3. There exists $c_2 > 0$ such that if $n \in \mathbb{N}$, $i_0, \ldots, i_{n-1} \in \{0, \ldots, p\}$ and $x \in I_{i_0\cdots i_{n-1}}$, then

$$\frac{1}{c_2} \le \frac{\mu(I_{i_0\cdots i_{n-1}})}{|(T^n)'(x)|^{-1}} \le c_2.$$

4. There exists $c_3 > 0$ such that if $A \subset [0, 1]$ is measurable and $i_0, \ldots, i_{n-1} \in \{0, \ldots, p\}$, then

$$\mu(I_{i_0 \cdots i_{n-1}} \cap T^{-n} A) \geq c_3 \mu(I_{i_0 \cdots i_{n-1}})\mu(A).$$

Hint: note that

$$m(I_{i_0 \cdots i_{n-1}} \cap T^{-n} A) = \int_A |\psi'_{i_0 \cdots i_{n-1}}| \, dm.$$

5. The measure μ is ergodic.
6. For μ-almost every

$$x = \bigcap_{n=0}^{\infty} I_{i_0 \cdots i_{n-1}} \in [0, 1],$$

we have

$$\lim_{n \to \infty} -\frac{\log \mu(I_{i_0 \cdots i_{n-1}})}{n} = \int_0^1 \log|T'| \, d\mu.$$

Hint: show that the function $\log|T'|$ is μ-integrable.
7. The set of periodic points of T is countable and is dense in $[0, 1]$.
8. The periodic entropy of T is positive.
9. If $T'(x) > 0$ for $x \in \bigcup_{i=0}^{p}(a_i, a_{i+1})$ and $T(a_i) = 0$ for $i = 0, \ldots, p + 1$, then the zeta function of T (see Exercise 2.29) has a single pole.

Notes

For additional topics of ergodic theory, in addition to [26, 44, 57, 103], we recommend [23, 51, 66, 69, 77, 79, 82, 96, 97, 100]. The books [48, 55] are excellent sources for (abstract) symbolic dynamics. For the relation between symbolic dynamics and the theory of dynamical systems, see, for example, [20, 44]. Hadamard [37] laid the foundations of symbolic dynamics, later developed by Morse and Hedlund [63, 64]. Theorem 3.2 on the existence of invariant measures for a continuous transformation of a compact metric space is due to Krylov and Bogolubov [52]. Our proof of Theorem 3.7 is based on [24].

Part II
Entropy and Pressure

Chapter 4
Metric Entropy and Topological Entropy

This chapter is dedicated to the study of metric entropy, including its relation to topological entropy. After establishing some basic properties of metric entropy, we consider the notion of conditional entropy, and we show how generators can be used to compute metric entropy. We then establish the Shannon–McMillan–Breiman theorem, which can be seen as the fundamental theorem of entropy theory. In particular, it shows that metric entropy can be computed in terms of an invariant local quantity. We also introduce the notion of topological entropy for a continuous transformation of a compact metric space, and we establish the variational principle showing that the topological entropy is the supremum of the metric entropies over all invariant probability measures.

4.1 Introduction

The mathematical notion of entropy has its roots in thermodynamics with the work of Clausius, Maxwell, Boltzmann, and others. It appeared in the work of Shannon as the main concept of information theory. Let $p = (p_1, \ldots, p_n)$ be a vector with nonnegative entries such that $\sum_{i=1}^n p_i = 1$. The Shannon entropy of p is defined by

$$H(p) = -\sum_{i=1}^n p_i \log p_i,$$

with the convention that $0 \log 0 = 0$ (in fact with $\log_2$ instead of $\log$, but for simplicity, we shall take this form). Clearly,

$$H(p) = -\sum_{i=1}^n f(p_i), \quad \text{where} \quad f(x) = x \log x.$$

L. Barreira, *Ergodic Theory, Hyperbolic Dynamics and Dimension Theory*, Universitext, DOI 10.1007/978-3-642-28090-0_4, © Springer-Verlag Berlin Heidelberg 2012

One can show that up to a multiplicative constant, f is the unique function satisfying

$$\sum_{i=1}^{n}\sum_{j=1}^{m} f(p_i q_j) = \sum_{i=1}^{n} f(p_i) + \sum_{j=1}^{m} f(q_j) \tag{4.1}$$

for all $p_i, q_j \geq 0$ such that $\sum_{i=1}^{n} p_i = \sum_{j=1}^{m} q_j = 1$. Identity (4.1) can be written in the form

$$H(p * q) = H(p) + H(q),$$

where $p = (p_1, \ldots, p_n)$, $q = (q_1, \ldots, q_m)$ and $p * q \in \mathbb{R}^{n+m}$ is any vector with components $p_i q_j$. Now let μ be a probability measure in a space X and let $\xi = \{A_1, \ldots, A_n\}$ be a measurable partition of X, that is, a collection of pairwise disjoint measurable sets whose union is X. The information function I associated to ξ is defined by

$$I(x) = -\log \mu(A) \quad \text{for} \quad x \in A.$$

The entropy of the partition ξ is the mean value of the information, that is,

$$H_\mu(\xi) = \int_X I \, d\mu = H(\mu(A_1), \ldots, \mu(A_n)). \tag{4.2}$$

The notion of metric entropy is due to Kolmogorov and was extended to all dynamical systems by Sinai. It examines how a partition and its entropy vary under a dynamics. Namely, given a measurable transformation $T: X \to X$ preserving a probability measure μ in X, the entropy of T with respect to measurable partition ξ is defined by

$$h_\mu(T, \xi) = \inf_{n \in \mathbb{N}} \frac{1}{n} H_\mu(\xi_n),$$

where ξ_n is the partition of X formed by the sets

$$C_1 \cap T^{-1}C_2 \cap \cdots \cap T^{-(n-1)}C_n \tag{4.3}$$

with $C_1, \ldots, C_n \in \xi$. Roughly speaking, the speed with which the transformation T cuts smaller and smaller pieces under iteration is measured by the entropy.

In a similar manner to that in (4.2), the entropy $h_\mu(T, \xi)$ can be expressed as the integral of a local quantity. This is due to the Shannon–McMillan–Breiman theorem, which can be seen as the fundamental theorem of entropy theory. For ergodic measures, the theorem implies that most sets in (4.3) have approximately equal measure for large n (in a rigorous sense to be made precise later on).

There is a corresponding notion of entropy in topological dynamics, called topological entropy. Roughly speaking, it measures the exponential growth rate of the number of orbits that can be distinguished as time increases. For continuous transformations of a compact metric space, the so-called variational principle for the topological entropy says that the supremum of the metric entropies over all invariant probability measures is equal to the topological entropy. In fact, this identity could be used as a (working) definition for the notion of topological entropy.

4.2 Metric Entropy

We introduce in this section the notion of metric entropy, starting with the notion of entropy with respect to a measurable partition. We also establish some of its basic properties.

4.2.1 The Notion of Metric Entropy

We first introduce the notion of entropy with respect to a measurable partition. Let $(X, \mathcal{A}, \mu)$ be a measure space with $\mu(X) = 1$.

Definition 4.1. We say that a finite family $\xi \subset \mathcal{A}$ is a *pre-partition* of $(X, \mathcal{A}, \mu)$ or simply of X (with respect to μ) if:

1. $\mu(\bigcup_{C \in \xi} C) = 1$.
2. $\mu(C \cap D) = 0$ for any distinct $C, D \in \xi$.

We emphasize that we only consider finite pre-partitions. Now we introduce an equivalence relation. Given pre-partitions ξ and η of X, we write $\xi \sim \eta$ if:

1. For every $C \in \xi$, there exists $D \in \eta$ such that $\mu(C \setminus D) = 0$.
2. For every $D \in \eta$, there exists $C \in \xi$ such that $\mu(D \setminus C) = 0$.

In other words, two pre-partitions are equivalent if and only if their elements differ at most by sets of zero measure. It is easy to verify that this is indeed an equivalence relation.

Definition 4.2. Each equivalence class of this equivalence relation is called a *measurable partition* of X.

Since the elements of any pre-partition representing a given equivalence class are well defined up to sets of zero measure, in what follows, we shall never distinguish between any two pre-partitions representing the same measurable partition. In particular, we always consider pre-partitions without elements of zero measure.

Definition 4.3. The *entropy* of a measurable partition ξ of X (with respect to μ) is given by

$$H_\mu(\xi) = -\sum_{C \in \xi} \mu(C) \log \mu(C).$$

The number $H_\mu(\xi)$ can be interpreted as the amount of information obtained by cutting X into the pieces in ξ. We shall see that the smaller the pieces (in terms of measure), the largest amount of information we obtain.

Now we introduce the notion of refinement.

Definition 4.4. Given two measurable partitions ξ and η of X, we say that η is a *refinement* of ξ if for every $D \in \eta$ there exists $C \in \xi$ such that $\mu(D \setminus C) = 0$.

We note that if η is a refinement of ξ, then each element of ξ (or more precisely of any pre-partition representing ξ) is a union of elements of η (or more precisely of any pre-partition representing η) up to sets of zero measure.

Example 4.1. Consider the space $X = [0, 1]$ with the Lebesgue measure m. For each $k \in \mathbb{N}$, we consider the measurable partition

$$\xi_k = \{(j/k, (j+1)/k) : j = 0, \ldots, k-1\}$$

of X. We have

$$H_m(\xi_k) = -\sum_{j=0}^{k-1} \frac{1}{k} \log \frac{1}{k} = \log k.$$

In particular, if $k \mid \ell$, then ξ_ℓ is a refinement of ξ_k, and we have $H_m(\xi_k) \leq H_m(\xi_\ell)$, with equality if and only if $k = \ell$.

Now let $\psi : [0, 1] \to \mathbb{R}$ be the continuous function defined by

$$\psi(x) = \begin{cases} x \log x & \text{if } 0 < x \leq 1, \\ 0 & \text{if } x = 0. \end{cases} \tag{4.4}$$

We note that

$$H_\mu(\xi) = -\sum_{C \in \xi} \psi(\mu(C)).$$

Since $\psi''(x) = 1/x > 0$ for every $x > 0$, the function ψ is strictly convex, and hence,

$$\psi\left(\sum_{i=1}^{p} a_i x_i\right) \leq \sum_{i=1}^{p} a_i \psi(x_i) \tag{4.5}$$

for any numbers $x_1, \ldots, x_p, a_1, \ldots, a_p \in [0, 1]$ such that $\sum_{i=1}^{p} a_i = 1$. Moreover, (4.5) is an equality if and only if all numbers x_i corresponding to a nonzero a_i are equal. In particular, setting $k = \text{card}\,\xi$, we obtain

$$H_\mu(\xi) = -k \sum_{C \in \xi} \frac{1}{k} \psi(\mu(C))$$

$$\leq -k\psi\left(\sum_{C \in \xi} \frac{\mu(C)}{k}\right)$$

$$= -k\psi\left(\frac{1}{k}\right) = \log k,$$

that is,

$$H_\mu(\xi) \leq \log \text{card}\,\xi. \tag{4.6}$$

We also have the following property:

Proposition 4.1. *Let ξ and η be measurable partitions of X. If η is a refinement of ξ, then $H_\mu(\xi) \le H_\mu(\eta)$.*

Proof. We have

$$H_\mu(\eta) = -\sum_{D \in \eta} \psi(\mu(D))$$
$$= -\sum_{C \in \xi} \sum_{D \subset C} \psi(\mu(D)). \qquad (4.7)$$

For each $C \in \xi$, since $\sum_{D \subset C} \mu(D) = \mu(C)$, we obtain

$$\psi(\mu(C)) = \psi\left(\sum_{D \subset C} \mu(D)\right)$$
$$= \sum_{D \subset C} \mu(D) \log \sum_{D \subset C} \mu(D)$$
$$\ge \sum_{D \subset C} \mu(D) \log \mu(D) = \sum_{D \subset C} \psi(\mu(D)),$$

and it follows from (4.7) that

$$H_\mu(\eta) \le -\sum_{C \in \xi} \psi(\mu(C)) = H_\mu(\xi).$$

This yields the desired inequality. $\square$

Now we start discussing what happens to the entropy of a partition in the presence of a dynamics. Given two measurable partitions ξ and η of X, we define a new measurable partition

$$\xi \vee \eta = \{C \cap D : C \in \xi, D \in \eta\}.$$

As explained above, and without loss of generality, by discarding sets of zero measure, one can always assume that all elements of the partition $\xi \vee \eta$ have positive measure.

Proposition 4.2. *If $T : X \to X$ is a measurable transformation preserving a probability measure μ in X and ξ is a measurable partition of X, then the limit*

$$\lim_{n \to \infty} \frac{1}{n} H_\mu(\xi_n) = \inf_{n \in \mathbb{N}} \frac{1}{n} H_\mu(\xi_n) \qquad (4.8)$$

exists, where

$$\xi_n = \bigvee_{k=0}^{n-1} T^{-k}\xi \quad \text{for each} \quad n \in \mathbb{N}. \qquad (4.9)$$

Proof. We first prove some auxiliary lemmas.

Lemma 4.1. *If ξ and η are measurable partitions of X, then*

$$H_\mu(\xi \vee \eta) \le H_\mu(\xi) + H_\mu(\eta).$$

Proof of the lemma. We have

$$H_\mu(\xi \vee \eta) = - \sum_{C \in \xi, D \in \eta} \mu(C \cap D) \log \mu(C \cap D)$$

$$= - \sum_{C \in \xi, D \in \eta} \mu(C \cap D) \left[\log \frac{\mu(C \cap D)}{\mu(C)} + \log \mu(C) \right]$$

$$= - \sum_{D \in \eta} \sum_{C \in \xi} \mu(C) \psi \left(\frac{\mu(C \cap D)}{\mu(C)} \right)$$

$$- \sum_{C \in \xi, D \in \eta} \mu(C \cap D) \log \mu(C).$$

By (4.5), we obtain

$$H_\mu(\xi \vee \eta) \le - \sum_{D \in \eta} \psi \left(\sum_{C \in \xi} \mu(C) \frac{\mu(C \cap D)}{\mu(C)} \right) - \sum_{C \in \xi} \mu(C) \log \mu(C)$$

$$= - \sum_{D \in \eta} \psi(\mu(D)) + H_\mu(\xi)$$

$$= H_\mu(\eta) + H_\mu(\xi).$$

This completes the proof of the lemma. $\qquad\qquad\qquad\qquad\qquad\qquad\qquad$ $\square$

Now we consider the measurable partition

$$T^{-1}\xi = \{T^{-1}C : C \in \xi\}.$$

Lemma 4.2. *If ξ is a measurable partition of X, then $H_\mu(T^{-1}\xi) = H_\mu(\xi)$.*

Proof of the lemma. Since μ is T-invariant, we obtain

$$H_\mu(T^{-1}\xi) = - \sum_{C \in \xi} \mu(T^{-1}C) \log \mu(T^{-1}C)$$

$$= - \sum_{C \in \xi} \mu(C) \log \mu(C)$$

$$= H_\mu(\xi),$$

which yields the desired identity. $\qquad\qquad\qquad\qquad\qquad\qquad\qquad\qquad\qquad$ $\square$

Using these two lemmas, given $n, m \in \mathbb{N}$, we obtain

$$
\begin{aligned}
H_\mu(\xi_{n+m}) &= H_\mu(\xi_n \vee T^{-n}\xi_m) \\
&\leq H_\mu(\xi_n) + H_\mu(T^{-n}\xi_m) \\
&= H_\mu(\xi_n) + H_\mu(\xi_m).
\end{aligned}
\tag{4.10}
$$

We also need the following result:

Lemma 4.3. *If $(a_n)_{n\in\mathbb{N}}$ is a sequence of positive numbers such that $a_{n+m} \leq a_n + a_m$ for every $n, m \in \mathbb{N}$, then the limit*

$$
\lim_{n\to\infty} \frac{a_n}{n} = \inf\left\{\frac{a_k}{k} : k \in \mathbb{N}\right\} \geq 0
\tag{4.11}
$$

exists.

Proof of the lemma. Take $k \in \mathbb{N}$. Then $n = qk + r$ for some integers $q \in \mathbb{N}$ and $r \in \{0, \ldots, k-1\}$. We have

$$
\frac{a_n}{n} \leq \frac{a_{qk} + a_r}{qk + r} \leq \frac{qa_k + a_r}{qk + r}.
$$

Thus, letting $n \to \infty$ (and hence $q \to \infty$), we obtain

$$
\limsup_{n\to\infty} \frac{a_n}{n} \leq \frac{a_k}{k}.
$$

Therefore,

$$
\limsup_{n\to\infty} \frac{a_n}{n} \leq \inf\left\{\frac{a_k}{k} : k \in \mathbb{N}\right\} \leq \liminf_{n\to\infty} \frac{a_n}{n}.
$$

This establishes the existence of the limit in (4.11). □

By (4.10), it follows from Lemma 4.3 that the sequence $H_\mu(\xi_n)/n$ converges and that identity (4.8) holds. □

We use the limit in Proposition 4.2 to introduce the notion of entropy of a measure-preserving transformation.

Definition 4.5. Let $T\colon X \to X$ be a measurable transformation preserving a probability measure μ in X. We define the *(measure-theoretical* or *metric) entropy* of T with respect to μ and a measurable partition ξ of X by

$$
h_\mu(T, \xi) = \lim_{n\to\infty} \frac{1}{n} H_\mu(\xi_n),
\tag{4.12}
$$

with ξ_n as in (4.9). We also define the *(measure-theoretical* or *metric) entropy* of T with respect to μ by

$$h_\mu(T) = \sup\{h_\mu(T,\xi) : \xi \text{ is a measurable partition of } X\}. \qquad (4.13)$$

We give two examples of the computation of the entropy.

Example 4.2. Let $T: \mathbb{T} \to \mathbb{T}$ be the interval translation $T(x) = x + \alpha \bmod 1$ and let λ be the Lebesgue measure in $\mathbb{T}$. Given a measurable partition ξ of $\mathbb{T}$ by intervals, we have

$$\operatorname{card} \xi_n \le n \operatorname{card} \xi \quad \text{for each} \quad n \in \mathbb{N}.$$

Indeed, the number of boundary points of the intervals in $\bigcup_{i=0}^{n-1} T^{-i}\xi$, and thus the number of cuts made by these intervals in $\mathbb{T}$, is at most $n \operatorname{card} \xi$. Therefore, by (4.6), we have

$$h_\mu(T,\xi) = \lim_{n\to\infty} \frac{1}{n} H_\mu(\xi_n)$$

$$\le \lim_{n\to\infty} \frac{1}{n} \log(n \operatorname{card} \xi) = 0.$$

It follows from Theorem 4.2 below that $h_\mu(T) = 0$.

Example 4.3. Let $E_q(z) = z^q$ be an expanding map of the circle for some $q > 1$. For each $m \in \mathbb{N}$, we consider the measurable partition of S^1 given by

$$\xi^m = \left\{ h\left(\left(\frac{j}{q^m}, \frac{j+1}{q^m}\right)\right) : j = 0, \ldots, q^m - 1 \right\}, \qquad (4.14)$$

where h is the map in (2.11), with respect to the Lebesgue measure λ in S^1. It is easy to verify that

$$\bigvee_{i=0}^{m-1} E_q^{-i}\xi^1 = \xi^m.$$

Therefore,

$$h_\lambda(E_q,\xi^m) = \lim_{n\to\infty} \frac{1}{n} H_\lambda\left(\bigvee_{i=0}^{n-1} E_q^{-i}\xi^m\right)$$

$$= \lim_{n\to\infty} \frac{1}{n} H_\lambda(\xi^{m+n})$$

$$= \lim_{n\to\infty} \frac{1}{n} q^{m+n}\left(-\frac{1}{q^{m+n}} \log \frac{1}{q^{m+n}}\right)$$

$$= \log q.$$

It follows from Theorem 4.2 below that

$$h_\mu(E_q) = \log q.$$

Now we consider the powers of a transformation.

Proposition 4.3. *We have* $h_\mu(T^k) = k h_\mu(T)$ *for each* $k \in \mathbb{N}$.

Proof. We note that

$$\frac{1}{n} H_\mu \left(\bigvee_{\ell=0}^{n-1} T^{-k\ell} \xi_k \right) = \frac{1}{n} H_\mu \left(\bigvee_{i=0}^{nk-1} T^{-i} \xi \right).$$

Letting $n \to \infty$, we obtain

$$h_\mu(T^k, \xi_k) = k h_\mu(T, \xi),$$

and hence,

$$h_\mu(T^k) \geq \sup_\xi h_\mu(T^k, \xi_k) = k h_\mu(T).$$

On the other hand,

$$H_\mu \left(\bigvee_{i=0}^{n-1} T^{-ki} \xi \right) \leq H_\mu \left(\bigvee_{i=0}^{nk-1} T^{-i} \xi \right).$$

Therefore,

$$h_\mu(T^k, \xi) \leq \lim_{n \to \infty} \frac{1}{n} H_\mu \left(\bigvee_{i=0}^{nk-1} T^{-i} \xi \right) = k h_\mu(T, \xi).$$

Taking the supremum in ξ, this implies that $h_\mu(T^k) \leq k h_\mu(T)$. □

4.2.2 Conditional Entropy

We consider in this section the notion of conditional entropy. In particular, it allows us to give alternative characterizations of the notion of metric entropy.

Definition 4.6. Given measurable partitions ξ and ζ of X, we define the *conditional entropy* of ξ with respect to ζ by

$$H_\mu(\xi|\zeta) = - \sum_{C \in \xi, D \in \zeta} \mu(C \cap D) \log \frac{\mu(C \cap D)}{\mu(D)}.$$

Given a measurable partition ξ of X and a measurable set $D \subset X$, we denote by $\xi|D$ the measurable partition of D defined by

$$\xi|D = \{C \cap D : C \in \xi\}.$$

We also consider the conditional measure

$$\mu(C|D) = \frac{\mu(C \cap D)}{\mu(D)}.$$

In particular, for the probability measure $\mu_D = \mu/\mu(D)$ in D, we have

$$H_{\mu_D}(\xi|D) = -\sum_{C \in \xi} \mu_D(C) \log \mu_D(C)$$

$$= -\sum_{C \in \xi} \mu(C|D) \log \mu(C|D).$$

Therefore, given measurable partitions ξ and ζ of X, we can write

$$H_\mu(\xi|\zeta) = -\sum_{D \in \xi} \mu(D) \sum_{C \in \xi} \psi(\mu(C|D))$$

$$= \sum_{D \in \zeta} \mu(D) H_{\mu_D}(\xi|D).$$

Theorem 4.1. *If $T: X \to X$ is a measurable transformation preserving a probability measure μ in X and ξ is a measurable partition of X, then*

$$h_\mu(T, \xi) = \lim_{n \to \infty} H_\mu\left(\xi \Big| \bigvee_{i=1}^{n} T^{-i}\xi\right). \tag{4.15}$$

Proof. We first show that the limit in (4.15) exists. For that, we establish the following auxiliary result:

Lemma 4.4. *If ζ is a refinement of η, then*

$$H_\mu(\xi|\zeta) \le H_\mu(\xi|\eta).$$

Proof of the lemma. We have

$$H_\mu(\xi|\zeta) = -\sum_{C \in \xi, D \in \zeta} \mu(C \cap D) \log \frac{\mu(C \cap D)}{\mu(D)}$$

$$= -\sum_{C \in \xi, D \in \zeta, E \in \eta} \mu(D \cap E) \frac{\mu(C \cap D)}{\mu(D)} \log \frac{\mu(C \cap D)}{\mu(D)}$$

$$= -\sum_{C \in \xi, E \in \eta} \mu(E) \sum_{D \in \zeta} \frac{\mu(D \cap E)}{\mu(E)} \psi\left(\frac{\mu(C \cap D)}{\mu(D)}\right).$$

Thus, it follows from (4.5) that

$$H_\mu(\xi|\zeta) \le - \sum_{C \in \xi, E \in \eta} \mu(E)\psi\left(\sum_{D \in \zeta} \frac{\mu(D \cap E)}{\mu(E)} \frac{\mu(C \cap D)}{\mu(D)}\right).$$

But since ζ is a refinement of η, we have

$$\sum_{D \in \zeta} \frac{\mu(D \cap E)}{\mu(E)} \frac{\mu(C \cap D)}{\mu(D)} = \sum_{D \in \zeta, D \subset E} \frac{\mu(C \cap D)}{\mu(E)} = \frac{\mu(C \cap E)}{\mu(E)},$$

and hence,

$$H_\mu(\xi|\zeta) \le - \sum_{C \in \xi, E \in \eta} \mu(E)\psi\left(\frac{\mu(C \cap E)}{\mu(E)}\right)$$

$$= - \sum_{C \in \xi, E \in \eta} \mu(C \cap E) \log \frac{\mu(C \cap E)}{\mu(E)} = H_\mu(\xi|\eta),$$

which yields the desired inequality. $\qquad\square$

Since $\bigvee_{i=1}^{n+1} T^{-i}\xi$ is a refinement of $\bigvee_{i=1}^{n} T^{-i}\xi$, it follows from Lemma 4.4 that the sequence

$$H_\mu\left(\xi| \bigvee_{i=1}^{n} T^{-i}\xi\right)$$

is nonincreasing in n, and thus, there exists the limit in (4.15).

To complete the proof, we also need the following result:

Lemma 4.5. *If ξ, ζ, and η are measurable partitions of X, then*

$$H_\mu(\xi \vee \zeta|\eta) = H_\mu(\xi|\zeta \vee \eta) + H_\mu(\zeta|\eta).$$

Proof of the lemma. We have

$$H_\mu(\xi \vee \zeta|\eta) = - \sum_{C \in \xi, D \in \zeta, E \in \eta} \mu(C \cap D \cap E) \log \frac{\mu(C \cap D \cap E)}{\mu(D)}$$

$$= - \sum_{C \in \xi, D \in \zeta, E \in \eta} \mu(C \cap D \cap E) \log \frac{\mu(C \cap D \cap E)}{\mu(D \cap E)}$$

$$- \sum_{C \in \xi, D \in \zeta, E \in \eta} \mu(C \cap D \cap E) \log \frac{\mu(D \cap E)}{\mu(D)},$$

with the convention that $0 \log 0 = 0$. Therefore,

$$H_\mu(\xi \vee \zeta | \eta) = H_\mu(\xi | \zeta \vee \eta) - \sum_{D \in \zeta, E \in \eta} \mu(D \cap E) \log \frac{\mu(D \cap E)}{\mu(D)}$$

$$= H_\mu(\xi | \zeta \vee \eta) + H_\mu(\zeta | \eta),$$

which yields the desired inequality.   □

In particular, for the trivial partition $\eta = \{X\}$, it follows from Lemma 4.5 that

$$H_\mu(\xi \vee \zeta) = H_\mu(\xi | \zeta) + H_\mu(\zeta). \qquad (4.16)$$

Now we use induction to show that

$$H_\mu(\xi_n) = H_\mu(\xi) + \sum_{j=1}^{n-1} H_\mu\left(\xi \,\Big|\, \bigvee_{i=1}^{j} T^{-i}\xi\right) \qquad (4.17)$$

for every $n \in \mathbb{N}$. The identity is clear when $n = 1$. Now we assume that it holds for some n. Then, by Lemma 4.2 and (4.16), we have

$$H_\mu(\xi_{n+1}) = H_\mu\left(\bigvee_{i=1}^{n} T^{-i}\xi \vee \xi\right)$$

$$= H_\mu\left(\bigvee_{i=1}^{n} T^{-i}\xi\right) + H_\mu\left(\xi \,\Big|\, \bigvee_{i=1}^{n} T^{-i}\xi\right)$$

$$= H_\mu(\xi_n) + H_\mu\left(\xi \,\Big|\, \bigvee_{i=1}^{n} T^{-i}\xi\right)$$

$$= H_\mu(\xi) + \sum_{j=1}^{n} H_\mu\left(\xi \,\Big|\, \bigvee_{i=1}^{j} T^{-i}\xi\right),$$

using the induction hypothesis in the last line.

Since the sequence $H_\mu(\xi | \bigvee_{i=1}^{n} T^{-i}\xi)$ is convergent, it follows from (4.17) that

$$h_\mu(T, \xi) = \lim_{n \to \infty} \frac{1}{n} H_\mu(\xi_n)$$

$$= \lim_{n \to \infty} \frac{1}{n} \sum_{j=1}^{n-1} H_\mu\left(\xi \,\Big|\, \bigvee_{i=1}^{j} T^{-i}\xi\right)$$

$$= \lim_{n \to \infty} H_\mu\left(\xi \,\Big|\, \bigvee_{i=1}^{n} T^{-i}\xi\right).$$

This completes the proof of the theorem.   □

We also use the notion of conditional entropy to show that it is sufficient to consider increasing sequences of partitions generating the σ-algebra of X. Given a measurable partition ξ of X, we denote by $\mathcal{A}(\xi)$ the σ-algebra generated by ξ, that is, the smallest σ-algebra containing all the elements of ξ. Moreover, we denote by $\bigvee_{n \in I} \mathcal{A}(\xi_n)$ the σ-algebra generated by the measurable partitions ξ_n for $n \in I$. In particular, when I is a finite set, we have

$$\mathcal{A}\left(\bigvee_{n \in I} \xi_n\right) = \bigvee_{n \in I} \mathcal{A}(\xi_n).$$

Theorem 4.2. *Let $T: X \to X$ be a measurable transformation preserving a probability measure μ in X. If $(\alpha_n)_{n \in \mathbb{N}}$ is a sequence of measurable partitions of X with $\bigvee_{n=1}^{\infty} \mathcal{A}(\alpha_n) = \mathcal{A}$ such that α_{n+1} is a refinement of α_n for each $n \in \mathbb{N}$, then*

$$h_\mu(T) = \lim_{n \to \infty} h_\mu(T, \alpha_n) = \sup_{n \in \mathbb{N}} h_\mu(T, \alpha_n). \tag{4.18}$$

Proof. We first observe that the limit in (4.18) exists. Indeed, it follows from Proposition 4.1 that

$$H_\mu\left(\bigvee_{i=0}^{n-1} T^{-i} \alpha_m\right) \le H_\mu\left(\bigvee_{i=0}^{n-1} T^{-i} \alpha_{m+1}\right)$$

for any $m, n \in \mathbb{N}$, and hence,

$$h_\mu(T, \alpha_m) \le h_\mu(T, \alpha_{m+1}) \quad \text{for each} \quad m \in \mathbb{N}.$$

Now we establish an auxiliary result.

Lemma 4.6. *If η and ζ are measurable partitions of X, then*

$$h_\mu(T, \eta) \le h_\mu(T, \zeta) + H_\mu(\eta|\zeta). \tag{4.19}$$

Proof of the lemma. By Proposition 4.1 and (4.16), we have

$$H_\mu(\eta_n) \le H_\mu(\eta_n \vee \zeta_n) = H_\mu(\zeta_n) + H_\mu(\eta_n|\zeta_n), \tag{4.20}$$

where

$$\eta_n = \bigvee_{i=0}^{n-1} T^{-i}\eta \quad \text{and} \quad \zeta_n = \bigvee_{i=0}^{n-1} T^{-i}\zeta. \tag{4.21}$$

On the other hand, it follows from Lemmas 4.4 and 4.5 that

$$H_\mu(\alpha \vee \beta|\gamma) \le H_\mu(\alpha|\gamma) + H_\mu(\beta|\gamma)$$

for any measurable partitions α, β, and γ of X. Therefore, by (4.20), (4.21), and Lemma 4.4, we obtain

$$H_\mu(\eta_n) \leq H_\mu(\zeta_n) + \sum_{i=0}^{n-1} H_\mu(T^{-i}\eta|\zeta_n)$$

$$\leq H_\mu(\zeta_n) + \sum_{i=0}^{n-1} H_\mu(T^{-i}\eta|T^{-i}\zeta)$$

$$= H_\mu(\zeta_n) + n H_\mu(\eta|\zeta),$$

using the invariance of the measure μ in the last inequality. Thus, dividing by n and taking limits when $n \to \infty$ yields inequality (4.19). $\qquad\square$

To complete the proof, we use the following lemma:

Lemma 4.7. *For any measurable partition η of X, we have*

$$H_\mu(\eta|\alpha_n) \to 0 \quad when \quad n \to \infty.$$

Proof of the lemma. Write $\eta = \{C_1, \ldots, C_k\}$. For each $n \in \mathbb{N}$ and $i = 1, \ldots, k$, let $D_i^n \subset C_i$ be a set in $\mathcal{A}(\alpha_n)$ such that $\mu(C_i \setminus D_i^n) \to 0$ when $n \to \infty$. We set

$$\beta_n = \{D_0^n, D_1^n, \ldots, D_k^n\},$$

where $D_0^n = X \setminus \bigcup_{i=1}^k D_i^n$. Clearly, $\mu(C_i \cap D_0^n) \to 0$ when $n \to \infty$. Since

$$C_i \cap D_i^n = D_i^n \quad \text{and} \quad C_i \cap D_j^n = \emptyset$$

for $i = 1, \ldots, k$ and $j \neq i$, we obtain

$$
\begin{aligned}
H_\mu(\eta|\beta_n) = {}& -\sum_{i=1}^k \mu(C_i \cap D_i^n) \log \frac{\mu(C_i \cap D_i^n)}{\mu(D_i^n)} \\
& -\sum_{i=1}^k \mu(C_i \cap D_0^n) \log \frac{\mu(C_i \cap D_0^n)}{\mu(D_0^n)} \\
& -\sum_{j=1}^k \sum_{i \neq j} \mu(C_i \cap D_j^n) \log \frac{\mu(C_i \cap D_j^n)}{\mu(D_j^n)} \\
= {}& -\sum_{i=1}^k \mu(C_i \cap D_0^n) \log \frac{\mu(C_i \cap D_0^n)}{\mu(D_0^n)} \to 0
\end{aligned}
\tag{4.22}
$$

when $n \to \infty$. Since α_n is a refinement of β_n, it follows from Lemma 4.4 that

$$H_\mu(\eta|\alpha_n) \le H_\mu(\eta|\beta_n) \to 0 \quad \text{when} \quad n \to \infty,$$

which yields the desired statement.   □

Setting $\eta = \alpha_n$ in (4.19), it follows from Lemma 4.7 that

$$h_\mu(T, \eta) \le \lim_{n \to \infty} h_\mu(T, \alpha_n).$$

This completes the proof of the theorem.   □

4.2.3 Generators and Examples

Sometimes it is possible to compute the entropy of a measure-preserving transformation using a single partition, as we describe in this section.

Definition 4.7. Given a measure space $(X, \mathcal{A}, \mu)$, let $T: X \to X$ be a measurable transformation and let ξ be a measurable partition of X.

1. We say that ξ is a *one-sided generator* (with respect to T) if

$$\bigvee_{k=0}^{+\infty} \mathcal{A}(T^{-k}\xi) = \mathcal{A}.$$

2. We say that ξ is a *two-sided generator* (with respect to T) if

$$\bigvee_{k=-\infty}^{+\infty} \mathcal{A}(T^{-k}\xi) = \mathcal{A}.$$

Generators may be hard to find, but when they exist, the computation of the entropy reduces to the computation of the entropy with respect to the generator.

Theorem 4.3 (Kolmogorov–Sinai). *Let $T: X \to X$ be a measurable transformation preserving a probability measure μ in X. Then the following properties hold:*

1. *If ξ is a one-sided generator, then $h_\mu(T) = h_\mu(T, \xi)$.*
2. *If ξ is a two-sided generator and T is invertible almost everywhere, then $h_\mu(T) = h_\mu(T, \xi)$.*

Proof. We first assume that ξ is a one-sided generator. It is sufficient to show that $h_\mu(T, \eta) \le h_\mu(T, \xi)$ for any measurable partition η of X. Indeed, this yields

$$h_\mu(T, \xi) \le h_\mu(T) = \sup_\eta h_\mu(T, \eta) \le h_\mu(T, \xi).$$

We proceed with the proof. Setting $\zeta = \xi_{n+1}$ in (4.19), we obtain

$$h_\mu(T, \eta) \leq h_\mu(T, \xi_{n+1}) + H_\mu(\eta | \xi_{n+1}).$$

On the other hand, we have

$$
\begin{aligned}
h_\mu(T, \xi_{n+1}) &= \lim_{m \to \infty} \frac{1}{m} H_\mu\left(\bigvee_{k=0}^{m-1} T^{-k} \xi_{n+1}\right) \\
&= \lim_{m \to \infty} \frac{1}{m} H_\mu(\xi_{m+n}) \\
&= h_\mu(T, \xi),
\end{aligned}
\tag{4.23}
$$

and hence,

$$h_\mu(T, \eta) \leq h_\mu(T, \xi) + H_\mu(\eta | \xi_{n+1}). \tag{4.24}$$

Since ξ is a one-sided generator and

$$\mathcal{A}\left(\bigvee_{i=0}^{n} T^{-i} \xi\right) = \bigvee_{i=0}^{n} \mathcal{A}(T^{-i} \xi),$$

it follows from Lemma 4.7 that

$$H_\mu(\eta | \xi_{n+1}) \to 0 \quad \text{when} \quad n \to \infty.$$

Thus, we conclude from (4.24) that $h_\mu(T, \eta) \leq h_\mu(T, \xi)$.

Now we assume that ξ is a two-sided generator and that T is invertible almost everywhere. As in the case of one-sided generators, it is sufficient to show that $h_\mu(T, \eta) \leq h_\mu(T, \xi)$ for any measurable partition η of X. Setting $\zeta = \bigvee_{i=-n}^{n} T^{-i} \xi$ in (4.19), we obtain

$$h_\mu(T, \eta) \leq h_\mu\left(T, \bigvee_{i=-n}^{n} T^{-i} \xi\right) + H_\mu\left(\eta \middle| \bigvee_{i=-n}^{n} T^{-i} \xi\right).$$

We also have

$$h_\mu\left(T, \bigvee_{i=-n}^{n} T^{-i} \xi\right) = h_\mu(T, \xi_{2n+1}) = h_\mu(T, \xi)$$

by the invariance of the measure and (4.23). Therefore,

$$h_\mu(T, \eta) \leq h_\mu(T, \xi) + H_\mu\left(\eta \middle| \bigvee_{i=-n}^{n} T^{-i} \xi\right). \tag{4.25}$$

On the other hand, by Lemma 4.7, we obtain

$$H_\mu\left(\eta \Big| \bigvee_{i=-n}^{n} T^{-i}\xi\right) \to 0 \quad \text{when} \quad n \to \infty,$$

and it follows from (4.25) that $h_\mu(T, \eta) \le h_\mu(T, \xi)$. □

Now we give several examples of the computation of the entropy using generators.

Example 4.4. Let $E_q(z) = z^q$ be an expanding map of the circle, for some integer $q > 1$, and let λ be the Lebesgue measure in S^1. For each $n \in \mathbb{N}$, we consider again the measurable partitions ξ^m of S^1 in (4.14). Since

$$\xi_n^1 = \bigvee_{i=0}^{n-1} E_q^{-1}\xi^1 = \xi^n,$$

the partition ξ^1 is a one-sided generator (with respect to E_q). By Theorem 4.3 and Example 4.3, we thus obtain

$$h_\lambda(E_q) = h_\lambda(E_q, \xi^1) = \log q.$$

Example 4.5. Let $\sigma: X_k \to X_k$ be the shift map and let μ be the (two-sided) Bernoulli measure in X_k generated by the numbers $p_1, \dots, p_k$. The measurable partition into cylinder sets $\xi = \{C_1, \dots, C_k\}$ is a two-sided generator. Indeed,

$$\bigvee_{j=-n}^{n} \sigma^{-j}\xi = \{C_{i_{-n}\cdots i_n} : i_{-n}, \dots, i_n \in \{1, \dots, k\}\},$$

and the cylinder sets generate the σ-algebra of X_k. By Theorem 4.3, we obtain

$$h_\mu(\sigma) = \lim_{n\to\infty} \frac{1}{n} H_\mu\left(\bigvee_{\ell=0}^{n-1} \sigma^{-\ell}\xi\right)$$

$$= \lim_{n\to\infty} -\frac{1}{n} \sum_{i_0\cdots i_{n-1}} \mu(C_{i_0\cdots i_{n-1}}) \log \mu(C_{i_0\cdots i_{n-1}})$$

$$= \lim_{n\to\infty} -\frac{1}{n} \sum_{i_0\cdots i_{n-1}} p_{i_0}\cdots p_{i_{n-1}} \log(p_{i_0}\cdots p_{i_{n-1}})$$

$$= \lim_{n\to\infty} -\frac{1}{n} \sum_{i_0\cdots i_{n-1}} p_{i_0}\cdots p_{i_{n-1}}(\log p_{i_0} + \cdots + \log p_{i_{n-1}})$$

$$= \lim_{n\to\infty} -\frac{1}{n} n \sum_{i=1}^{k} p_i \log p_i = -\sum_{i=1}^{k} p_i \log p_i,$$

that is,

$$h_\mu(\sigma) = -\sum_{i=1}^{k} p_i \log p_i. \tag{4.26}$$

Since the function ψ in (4.4) is strictly convex, we obtain

$$h_\mu(\sigma) = -k \sum_{i=1}^{k} \frac{1}{k} \psi(p_i)$$

$$\leq -k\psi\left(\sum_{i=1}^{k} \frac{p_i}{k}\right)$$

$$= -k\psi\left(\frac{1}{k}\right) = \log k$$

for any Bernoulli measure μ, with equality if and only if $p_1 = \cdots = p_k = 1/k$. In particular,

$$\sup\{h_\mu(\sigma) : \mu \text{ is a Bernoulli measure}\} = \log k. \tag{4.27}$$

Example 4.6. We consider the shift map $\sigma_1 : \{1,2,3,4\}^{\mathbb{Z}} \to \{1,2,3,4\}^{\mathbb{Z}}$ with the Bernoulli measure μ_1 with the probabilities $\frac{1}{4}, \frac{1}{4}, \frac{1}{4}, \frac{1}{4}$. We also consider the shift map $\sigma_2 : \{1,2,3,4,5\}^{\mathbb{Z}} \to \{1,2,3,4,5\}^{\mathbb{Z}}$ with the Bernoulli measure μ_2 with probabilities $\frac{1}{2}, \frac{1}{8}, \frac{1}{8}, \frac{1}{8}, \frac{1}{8}$. It follows from (4.26) that

$$h_{\mu_1}(\sigma_1) = h_{\mu_2}(\sigma_2) = 2\log 2.$$

Now we compute the entropy of a Markov measure.

Example 4.7. Let $\sigma : X_k \to X_k$ be the shift map and let μ be the (two-sided) Markov measure associated to a stochastic pair (P, p). As in Example 4.5, we use the two-sided generator $\xi = \{C_1, \ldots, C_k\}$. We have

$$H_\mu\left(\bigvee_{\ell=0}^{n-1} \sigma^{-\ell}\xi\right) = -\sum_{i_0\cdots i_{n-1}} \mu(C_{i_0\cdots i_{n-1}}) \log \mu(C_{i_0\cdots i_{n-1}})$$

$$= -\sum_{i_0\cdots i_{n-1}} p_{i_0} p_{i_0 i_1} \cdots p_{i_{n-2} i_{n-1}} \log(p_{i_0} p_{i_0 i_1} \cdots p_{i_{n-2} i_{n-1}})$$

$$= -\sum_{i_0\cdots i_{n-1}} p_{i_0} p_{i_0 i_1} \cdots p_{i_{n-2} i_{n-1}} \log p_{i_0}$$

$$\quad - \sum_{i_0\cdots i_{n-1}} p_{i_0} p_{i_0 i_1} \cdots p_{i_{n-2} i_{n-1}} \sum_{j=0}^{n-2} \log p_{i_j i_{j+1}}$$

$$= -\sum_{i=1}^{k} p_i \log p_i - (n-1) \sum_{i=1}^{k} \sum_{j=1}^{k} p_i p_{ij} \log p_{ij},$$

and by Theorem 4.3, we obtain

$$h_\mu(\sigma) = \lim_{n\to\infty} \frac{1}{n} H_\mu\left(\bigvee_{\ell=0}^{n-1} \sigma^{-\ell}\xi\right) = -\sum_{i=1}^{k}\sum_{j=1}^{k} p_i\, p_{ij} \log p_{ij}.$$

Furthermore,

$$h_\mu(\sigma) = -\sum_{i=1}^{k}\sum_{j=1}^{k} p_i\psi(p_{ij})$$

$$= -\sum_{i=1}^{k} p_i\left(\sum_{j=1}^{k}\psi(p_{ij})\right)$$

$$\leq -\sum_{i=1}^{k} p_i k\psi\left(\sum_{j=1}^{k}\frac{p_{ij}}{k}\right)$$

$$= -\sum_{i=1}^{k} p_i k\psi\left(\frac{1}{k}\right) = \log k,$$

with equality if and only if $p_{ij} = 1/k$ for every i and j, that is, if and only if μ is a Bernoulli measure with $p_1 = \cdots = p_k = 1/k$. Therefore (compare with (4.27)),

$$\sup\{h_\mu(\sigma) : \mu \text{ is a Markov measure}\} = \log k.$$

4.3 Shannon–McMillan–Breiman Theorem

We describe in this section a different approach to the definition of metric entropy, based on the existence of a certain local quantity. The metric entropy is the integral of this local quantity.

We first note that if ξ is a measurable partition of X with respect to a measure μ, then for μ-almost every $x \in X$, there exists a single element

$$\xi_n(x) \in \xi_n = \bigvee_{k=0}^{n-1} T^{-k}\xi$$

such that $x \in \xi_n(x)$.

Theorem 4.4 (Shannon–McMillan–Breiman). *If* $T: X \to X$ *is a measurable transformation preserving a probability measure* μ *in* X *and* ξ *is a measurable partition of* X, *then the limit*

$$h_\mu(T, \xi, x) := \lim_{n \to \infty} -\frac{1}{n} \log \mu(\xi_n(x))$$

exists for μ-almost every $x \in X$. Moreover, the function $x \mapsto h_\mu(T, \xi, x)$ is T-invariant almost everywhere, is μ-integrable, and

$$h_\mu(T, \xi) = \int_X h_\mu(T, \xi, x) \, d\mu(x).$$

Proof. Given a measurable set $A \subset X$, we denote by $x \mapsto \mu(A|\xi)(x)$ the conditional expectation of χ_A with respect to the σ-algebra $\mathcal{B}(\xi)$ generated by ξ (see Sect. 2.5.2). This is a $\mathcal{B}(\xi)$-measurable function such that

$$\int_B \mu(A|\xi) \, d\mu = \mu(A \cap B)$$

for any set $B \in \mathcal{B}(\xi)$.

Lemma 4.8. *We have*

$$\log \mu(\xi_n(x)) = \log \mu(\xi(T^{n-1}(x))) + \sum_{j=1}^{n-1} \log \mu\left(\xi(T^{j-1}(x)) \Big| \bigvee_{k=1}^{n-j} T^{-k}\xi \right) \quad (4.28)$$

for every $n \in \mathbb{N}$ and μ-almost every $x \in X$.

Proof of the lemma. We first observe that if $A, B \subset X$ are measurable sets, then

$$\mu(A|\xi) = \sum_{B \in \xi} \frac{\mu(A \cap B)}{\mu(B)} \chi_B \quad (4.29)$$

μ-almost everywhere. Indeed, the right-hand side of (4.29) is $\mathcal{B}(\xi)$-measurable. Moreover, given $C \in \xi$, we have

$$\int_C \sum_{B \in \xi} \frac{\mu(A \cap B)}{\mu(B)} \chi_B \, d\mu = \mu(A \cap C) = \int_C \mu(A|\xi) \, d\mu.$$

This establishes (4.29).

Now we use induction to establish (4.28). When $n = 1$, identity (4.28) is clear. Now we assume that it holds for some $n \in \mathbb{N}$. Setting $\alpha_n = \bigvee_{k=1}^{n-1} T^{-k}\xi$ and using (4.29), we obtain

$$\mu(A|\alpha_n) = \sum_{B \in \alpha_n} \frac{\mu(A \cap B)}{\mu(B)} \chi_B,$$

and thus,

$$\log \mu(A|\alpha_n) = \sum_{B \in \alpha_n} \log \frac{\mu(A \cap B)}{\mu(B)} \chi_B \tag{4.30}$$

μ-almost everywhere for every $A \in \xi$. Therefore,

$$\log \mu(\xi_{n+1}(x)) = \sum_{A \in \xi, B \in \alpha_n} \log \mu(A \cap B) \chi_{A \cap B}(x)$$

$$= \sum_{A \in \xi, B \in \alpha_n} \log \mu(B) \chi_B(x) \chi_A(x) + \sum_{A \in \xi} \log \mu(A|\alpha_n) \chi_A(x)$$

$$= \sum_{B \in \alpha_n} \log \mu(B) \chi_B(x) + \sum_{A \in \xi} \log \mu(A|\alpha_n) \chi_A(x)$$

$$= \log \mu(\alpha_n(x)) + \log \mu(\xi(x)|\alpha_n),$$

and since $\alpha_n(x) = \xi_n(T(x))$, we obtain

$$\log \mu(\xi_{n+1}(x)) = \log \mu(\xi_n(T(x))) + \log \mu(\xi(x)|\alpha_n).$$

By the induction hypothesis, we conclude that

$$\log \mu(\xi_{n+1}(x)) = \log \mu(\xi(T^{n-1}(T(x)))) + \sum_{j=1}^{n-1} \log \mu\big(\xi(T^{j-1}(T(x))|\alpha_{n+1-j}\big)$$

$$+ \log \mu(\xi(x)|\alpha_n)$$

$$= \log \mu(\xi(T^n(x))) + \sum_{j=2}^{n-1} \log \mu\big(\xi(T^{j-1}(x)|\alpha_{n+1-j}\big)$$

$$+ \log \mu(\xi(x)|\alpha_n)$$

$$= \log \mu(\xi(T^n(x))) + \sum_{j=1}^{n-1} \log \mu\big(\xi(T^{j-1}(x)|\alpha_{n+1-j}\big).$$

This establishes identity (4.28).   □

We have thus shown that

$$\log \mu(\xi_n(x)) = \sum_{j=0}^{n-1} F_{n-j}(T^j(x)), \tag{4.31}$$

where

$$F_k(x) = \log \mu(\xi(x)|\alpha_k),$$

with the convention that $\alpha_1 = \{X\}$. Now we observe that the function $\varphi \colon X \to \mathbb{R}_0^-$ defined μ-almost everywhere by $\varphi(x) = \log \mu(\xi(x))$ is in $L^1(X, \mu)$, since

$$\int_X |\varphi| \, d\mu = -\sum_{C \in \xi} \mu(C) \log \mu(C) = H_\mu(\xi) < \infty.$$

Therefore, it follows from the increasing martingale theorem (Theorem A.7) that the sequence $(F_k)_{k \in \mathbb{N}}$ converges μ-almost everywhere and in $L^1(X, \mu)$ to the function

$$F(x) = \log \mu(\xi(x)|\alpha_\infty),$$

where $\alpha_\infty = \bigvee_{n=1}^\infty \mathcal{A}(\alpha_n)$ is the σ-algebra generated by the partitions α_n.
   On the other hand, it follows from (4.31) that

$$\frac{1}{n} \log \mu(\xi_n(x)) = \frac{1}{n} \sum_{j=0}^{n-1} F_{n-j}(T^j(x))$$

$$= \frac{1}{n} \sum_{j=0}^{n-1} F(T^j(x)) + \frac{1}{n} \sum_{j=0}^{n-1} (F_{n-j} - F)(T^j(x)). \tag{4.32}$$

By Birkhoff's ergodic theorem (Theorem 2.2), the limit

$$\psi(x) = \lim_{n \to \infty} \frac{1}{n} \sum_{j=0}^{n-1} F(T^j(x)) \tag{4.33}$$

exists for μ-almost every $x \in x$. Moreover, the function ψ is T-invariant μ-almost everywhere, is in $L^1(X, \mu)$, and

$$\int_X \psi \, d\mu = \int_X F \, d\mu.$$

We want to show that the second term in the right-hand sice of (4.32) converges to zero μ-almost everywhere. For this, we consider the function $F^* = \sup_{k \geq 1} |F_k|$.

Lemma 4.9. *The function F^* is in $L^1(X, \mu)$.*

Proof of the lemma. Given a measurable set $A \subset X$, $c > 0$, and $n \in \mathbb{N}$, we consider the set $B_n^A \subset X$ composed of the points $x \in X$ such that

$$-\log \mu(A|\alpha_k) \leq c \quad \text{for} \quad k = 1, \ldots, n-1$$

and $-\log \mu(A|\alpha_n) > c$. Clearly,

$$\mu(\{x \in A : F^*(x) > c\}) = \sum_{n=1}^\infty \mu(B_n^A \cap A).$$

Moreover, since $B_n^A \in \mathcal{B}(\alpha_n)$, we obtain

$$\mu(B_n^A \cap A) = \int_{B_n^A} X_A \, d\mu = \int_{B_n^A} \mu(A|\alpha_n) \, d\mu$$

$$\leq \int_{B_n^A} e^{-c} \, d\mu = e^{-c} \mu(B_n^A),$$

and thus,

$$\mu(\{x \in A : F^*(x) > c\}) = \sum_{n=1}^{\infty} \mu(B_n^A \cap A)$$

$$\leq e^{-c} \sum_{n=1}^{\infty} \mu(B_n^A) \leq e^{-c}$$

since $B_n^A \cap B_m^A = \varnothing$ for $n \neq m$. Therefore,

$$\int_X F^* \, d\mu = \sum_{A \in \xi} \int_A F^* \, d\mu$$

$$= \sum_{A \in \xi} \int_0^{\infty} \mu(\{x \in A : F^*(x) > c\}) \, dc$$

$$\leq \sum_{A \in \xi} \left(\int_0^{-\log \mu(A)} \mu(A) \, dc + \int_{-\log \mu(A)}^{\infty} e^{-c} \, dc \right)$$

$$= -\sum_{A \in \xi} (\mu(A) \log \mu(A) + \mu(A))$$

$$= H_\mu(\xi) + 1 < \infty,$$

which yields the desired result. □

Now we return to the last term in the right-hand side of (4.32). Given $k \in \mathbb{N}$, we set $G_k = |F_k - F|$ and $G_k^* = \sup_{n \geq k} G_n$. We note that

$$\frac{1}{n} \sum_{j=0}^{n-1} G_{n-j}(T^j(x)) = \frac{1}{n} \sum_{j=0}^{n-k} G_{n-j}(T^j(x)) + \frac{1}{n} \sum_{j=n-k+1}^{n-1} G_{n-j}(T^j(x))$$

$$\leq \frac{1}{n} \sum_{j=0}^{n-k} G_k^*(T^j(x)) + \frac{1}{n} \sum_{j=n-k+1}^{n-1} (F^* + F)(T^j(x)).$$

Letting $n \to \infty$, it follows from Lemma 4.9 and Birkhoff's ergodic theorem (see also Exercise 2.13) that

$$\limsup_{n\to\infty} \frac{1}{n} \sum_{j=0}^{n-1} G_{n-j}(T^j(x)) \le \lim_{n\to\infty} \frac{1}{n} \sum_{j=0}^{n-1} G_k^*(T^j(x)) =: \psi_k(x)$$

for μ-almost every $x \in X$, for some T-invariant almost everywhere function $\psi_k \in L^1(X, \mu)$ with

$$\int_X \psi_k \, d\mu = \int_X G_k^* \, d\mu.$$

Since $G_k^* \to 0$ μ-almost everywhere when $k \to \infty$, and $G_k^* \le F^* + F$ for $k \in \mathbb{N}$, with $F^* + F \in L^1(X, \mu)$, it follows from the dominated convergence theorem (Theorem A.3) that

$$\int_X \limsup_{n\to\infty} \frac{1}{n} \sum_{j=0}^{n-1} (G_{n-j} \circ T^j) \, d\mu \le \int_X G_k^* \, d\mu \to 0$$

when $k \to \infty$. Therefore,

$$\lim_{n\to\infty} \frac{1}{n} \sum_{j=0}^{n-1} G_{n-j}(T^j(x)) = 0$$

for μ-almost every $x \in X$, and by (4.32) and (4.33), we obtain

$$\lim_{n\to\infty} -\frac{1}{n} \log \mu(\xi_n(x)) = \psi(x)$$

for μ-almost every $x \in X$. This establishes the first statement in the theorem as well as the T-invariance μ-almost everywhere and the μ-integrability of the function $x \mapsto h_\mu(T, \xi, x)$. Moreover,

$$\int_X \lim_{n\to\infty} -\frac{1}{n} \log \mu(\xi_n(x)) \, d\mu(x) = -\int_X \psi \, d\mu = -\int_X F \, d\mu.$$

Since $F_n \to F$ in $L^1(X, \mu)$ when $n \to \infty$, it follows from (4.30) and Theorem 4.1 that

$$-\int_X F \, d\mu = -\lim_{n\to\infty} \int_X F_n \, d\mu = \lim_{n\to\infty} \int_X -\log \mu(\xi(x)|\alpha_n) \, d\mu(x)$$

$$= \lim_{n\to\infty} \sum_{A \in \xi} \int_A -\log \mu(A|\alpha_n) \, d\mu$$

$$= \lim_{n\to\infty} \sum_{A \in \xi} \int_A \sum_{B \in \alpha_n} -\log \frac{\mu(A \cap B)}{\mu(B)} \chi_B \, d\mu$$

$$= \lim_{n\to\infty} - \sum_{A\in\xi, B\in\alpha_n} \mu(A\cap B) \log \frac{\mu(A\cap B)}{\mu(B)}$$

$$= \lim_{n\to\infty} H_\mu(\xi|\alpha_n) = h_\mu(T,\xi).$$

This completes the proof of the theorem. $\qquad\square$

Now we give some examples.

Example 4.8. Let $E_q(z) = z^q$ be an expanding map of the circle, for some $q > 1$, and let λ be the Lebesgue measure in S^1. We consider again the measurable partitions ξ^n of S^1 in (4.14). For λ-almost every $x \in S^1$, we have $\xi_n^1(x) = \xi^n(x)$, and hence,

$$-\frac{1}{n} \log\mu(\xi_n(x)) = -\frac{1}{n} \log\frac{1}{q^n} = \log q.$$

It follows from Theorem 4.4 that $h_\mu(E_q, \xi^1) = \log q$.

Example 4.9. Let $\sigma: X_k^+ \to X_k^+$ be the shift map and let μ be the (one-sided) Bernoulli measure generated by the numbers $p_1, \ldots, p_k$. We also consider the measurable partition into cylinder sets $\xi = \{C_1, \ldots, C_k\}$. Given $\omega = (i_1 i_2\cdots) \in X_k^+$, we have $\xi_n(\omega) = C_{i_1\cdots i_n}$, and hence,

$$\log\mu(\xi_n(\omega)) = \log\mu(C_{i_1\cdots i_n})$$

$$= \sum_{j=1}^{n} \log p_{i_j} = \sum_{j=0}^{n-1} \varphi(\sigma^j(\omega)),$$

where $\varphi: X_k^+ \to \mathbb{R}$ is the continuous function given by $\varphi(i_1 i_2\cdots) = \log p_{i_1}$. Since μ is ergodic (see Proposition 3.8), for μ-almost every $\omega \in X_k^+$, we have

$$-\frac{1}{n} \log\mu(\xi_n(\omega)) \to -\int_{X_k^+} \varphi\,d\mu = -\sum_{j=1}^{k} \mu(C_j)\varphi|C_j$$

$$= -\sum_{j=1}^{k} \mu(C_j)\log\mu(C_j) = H_\mu(\xi).$$

By Theorem 4.4, we obtain

$$h_\mu(T,\xi) = \int_{X_k^+} \lim_{n\to\infty} -\frac{1}{n} \log\mu(\xi_n(\omega))\,d\mu(\omega) = H_\mu(\xi).$$

Moreover, since

$$\bigvee_{j=0}^{n-1} \sigma^{-j}\xi = \{C_{i_1\cdots i_n} : i_1, \ldots, i_n \in \{1, \ldots, k\}\},$$

the partition ξ is a one-sided generator, and it follows from Theorem 4.3 that

$$h_\mu(T) = h_\mu(T, \xi) = H_\mu(\xi) = -\sum_{i=1}^{k} p_i \log p_i.$$

Example 4.10. For a transformation $T: [0, 1] \to [0, 1]$ and a T-invariant probability measure μ as in Exercise 3.28, we consider the measurable partition

$$\xi = \{(a_i, a_{i+1}) : i = 0, \ldots, p\}.$$

Using the notation in that exercise, we know that

$$\lim_{n \to \infty} -\frac{1}{n} \log \mu(I_{i_0 \cdots i_{n-1}}) = \int_0^1 \log |T'| \, d\mu$$

for μ-almost every $x = \bigcap_{n=0}^{\infty} I_{i_0 \cdots i_{n-1}}$. Since ξ is a one-sided generator, it follows from Theorems 4.3 and 4.4 that

$$h_\mu(T) = h_\mu(T, \xi) = \int_0^1 \lim_{n \to \infty} -\frac{1}{n} \log \mu(I_{i_0 \cdots i_{n-1}}) \, d\mu(x) = \int_0^1 \log |T'| \, d\mu.$$

4.4 Topological Entropy

We introduce in this section the notion of topological entropy, and we establish some of its basic properties. Since the emphasis of this chapter is on the notion of metric entropy, we develop the material in a somewhat pragmatic manner. For completeness, some additional results correspond to exercises at the end of this chapter. For simplicity of the exposition, we only consider continuous transformations of a compact metric space, and not of an arbitrary compact topological space.

4.4.1 Basic Notions and Some Properties

Let $T: X \to X$ be a continuous transformation of a compact metric space (X, d). For each $n \in \mathbb{N}$, we introduce a new distance in X by

$$d_n(x, y) = d_{n,T}(x, y) = \max \{d(T^k(x), T^k(y)) : 0 \le k \le n - 1\}. \tag{4.34}$$

Moreover, given $\varepsilon > 0$, we denote by $N(d, \varepsilon)$ the maximum number of points in X at a d-distance at least ε.

Definition 4.8. We define the *topological entropy* of T by

$$h(T) = h^d(T) = \lim_{\varepsilon \to 0} \limsup_{n \to \infty} \frac{1}{n} \log N(d_n, \varepsilon). \tag{4.35}$$

We note that since the function

$$\varepsilon \mapsto \limsup_{n \to \infty} \frac{1}{n} \log N(d_n, \varepsilon)$$

is nondecreasing, the limit in (4.35) when $\varepsilon \to 0$ is well defined.

Now we give several examples.

Example 4.11. If T is an isometry, then $d(T^n(x), T^n(y)) = d(x, y)$, and hence,

$$d_n(x, y) = d(x, y) \quad \text{for every} \quad n \in \mathbb{N}.$$

Therefore, $N(d_n, \varepsilon) = N(d, \varepsilon)$, and

$$h(T) = \lim_{\varepsilon \to 0} \limsup_{n \to \infty} \frac{1}{n} \log N(d, \varepsilon) = 0.$$

Example 4.12. Let $T: \mathbb{T}^n \to \mathbb{T}^n$ be the toral translation

$$T(x_1, \ldots, x_n) = (x_1 + \alpha_1, \ldots, x_n + \alpha_n) \bmod 1.$$

Since T is an isometry, we have $h(T) = 0$.

Example 4.13. Now we consider the expanding map of the circle $E_q: S^1 \to S^1$, for some $q > 1$, with the distance d in S^1 obtained from normalizing the length. We note that if $d(x, y) < q^{-n}$, then

$$d_n(x, y) = d(E_q^{n-1}(x), E_q^{n-1}(y)) = q^{n-1} d(x, y). \tag{4.36}$$

Now given $k \in \mathbb{N}$, we consider the points $x_i = i/q^{n+k}$ for $i = 0, \ldots, q^{n+k} - 1$. By (4.36), we have $d_n(x_i, x_{i+1}) = q^{-(k+1)}$ for each i. Therefore, $d_n(x_i, x_j) \geq q^{-(k+1)}$ whenever $i \neq j$, and we conclude that

$$N(d_n, q^{-(k+1)}) \geq q^{n+k}.$$

On the other hand, given a set S in S^1 with at least $q^{n+k} + 1$ points, there exist $x, y \in S$ with $x \neq y$ such that $d(x, y) < q^{-(n+k)}$, and hence, also $d_n(x, y) < q^{-(k+1)}$. This shows that

$$N(d_n, q^{-(k+1)}) = q^{n+k} \quad \text{for} \quad n, k \in \mathbb{N},$$

and

$$h(E_q) = \lim_{k \to \infty} \limsup_{n \to \infty} \frac{1}{n} \log N(d_n, q^{-(k+1)})$$

$$= \lim_{k \to \infty} \limsup_{n \to \infty} \frac{n+k}{n} \log q$$

$$= \log q.$$

By (2.21) and Example 4.4, we have

$$h(E_q) = h_\lambda(E_q) = p(E_q),$$

where λ is the Lebesgue measure and where p is the periodic entropy.

Now we establish several properties of the topological entropy.

Proposition 4.4. *If $T: X \to X$ is a continuous transformation of a compact metric space, then the following properties hold:*

1. If $\Lambda \subset X$ is a closed T-invariant set, then $h(T|\Lambda) \le h(T)$.
2. If $X = \bigcup_{k=1}^{m} \Lambda_k$ where Λ_k is a closed T-invariant set for $k = 1, \ldots, m$, then

$$h(T) = \max \left\{ h(T|\Lambda_k) : 1 \le k \le m \right\}.$$

Proof. Let $\Lambda \subset X$ be a closed T-invariant set. Then $T|\Lambda: \Lambda \to \Lambda$ is also a continuous transformation of a compact metric space. Since $N(d_{n,T|\Lambda}, \varepsilon) \le N(d_{n,T}, \varepsilon)$, the first property is immediate.

For the second property, we first note that

$$N(d_{n,T}, \varepsilon) \le \sum_{k=1}^{m} N(d_{n,T|\Lambda_k}, \varepsilon).$$

Therefore, for each n and ε, there exists $k = k(n, \varepsilon) \in \mathbb{N}$ such that

$$N(d_{n,T|\Lambda_k}, \varepsilon) \ge \frac{1}{m} N(d_{n,T}, \varepsilon). \tag{4.37}$$

Now we take a sequence $(n_i)_{i \in \mathbb{N}} \subset \mathbb{N}$ such that

$$\lim_{i \to \infty} \frac{1}{n_i} \log N(d_{n_i, T}, \varepsilon) = \limsup_{n \to \infty} \frac{1}{n} \log N(d_{n,T}, \varepsilon).$$

Since $\{1, \ldots, m\}$ is a finite set, we can also assume that $k(n_i, \varepsilon)$ takes the same value, say j, for all $i \in \mathbb{N}$. By (4.37), we thus obtain

$$\limsup_{n \to \infty} \frac{1}{n} \log N(d_{n,T|\Lambda_j}, \varepsilon) \geq \limsup_{i \to \infty} \frac{1}{n_i} \log N(d_{n_i,T|\Lambda_j}, \varepsilon)$$

$$\geq \limsup_{i \to \infty} \frac{1}{n_i} \log N(d_{n_i,T}, \varepsilon)$$

$$= \limsup_{n \to \infty} \frac{1}{n} \log N(d_{n,T}, \varepsilon).$$

Therefore, $h(T|\Lambda_j) \geq h(T)$, and it follows from Property 1 that

$$h(T) \geq \max \{h(T|\Lambda_k) : 1 \leq k \leq m\} \geq h(T|\Lambda_j) \geq h(T).$$

This completes the proof of the proposition. □

We also give an equivalent description of the topological entropy. Given $\varepsilon > 0$, let $D(d, \varepsilon)$ be the minimum number of sets of d-diameter less than ε that are needed to cover X.

Theorem 4.5. *If $T: X \to X$ is a continuous transformation of a compact metric space, then*

$$h(T) = \lim_{\varepsilon \to 0} \lim_{n \to \infty} \frac{1}{n} \log D(d_n, \varepsilon). \tag{4.38}$$

Proof. It is easy to verify that

$$N(d_n, 2\varepsilon) \leq D(d_n, \varepsilon) \leq N(d_n, \varepsilon/2) \tag{4.39}$$

for every $n \in \mathbb{N}$ and $\varepsilon > 0$.

Lemma 4.10. *If $m, n \in \mathbb{N}$ and $\varepsilon > 0$, then*

$$D(d_{m+n}, \varepsilon) \leq D(d_m, \varepsilon) D(d_n, \varepsilon).$$

Proof of the lemma. Let $\{A_1, \ldots, A_k\}$ be a cover of X by sets of d_n-diameter less than ε, where $k = D(d_n, \varepsilon)$. Let also $\{B_1, \ldots, B_\ell\}$ be a cover of X by sets of d_m-diameter less than ε, where $\ell = D(d_m, \varepsilon)$. Then each set $A_i \cap T^{-n} B_j$ has d_{m+n}-diameter less than ε, and

$$\mathcal{U} = \{A_i \cap T^{-n} B_j : i = 1, \ldots, k \text{ and } j = 1, \ldots, \ell\}$$

is a cover of X. Therefore,

$$D(d_{m+n}, \varepsilon) \leq \operatorname{card} \mathcal{U} \leq \ell k = D(d_m, \varepsilon) D(d_n, \varepsilon),$$

which yields the desired inequality. □

It follows from Lemmas 4.3 and 4.10 that the limit

$$\lim_{n\to\infty} \frac{1}{n} \log D(d_n, \varepsilon)$$

exists. Hence, by (4.39), we conclude that identity (4.38) holds.   □

Finally, we compute the topological entropy of the powers of a given transformation.

Theorem 4.6. *If* $T: X \to X$ *is a continuous transformation of a compact metric space, then the following properties hold:*

1. $h(T^k) = k h(T)$ *for each* $k \in \mathbb{N}$.
2. *If* T *is a homeomorphism, then* $h(T^{-1}) = h(T)$.

Proof. We first note

$$d_{n,T^k}(x, y) = \max \left\{ d(T^{ik}(x), T^{ik}(y)) : 0 \le i \le n - 1 \right\}$$

$$\le \max \left\{ d(T^i(x), T^i(y)) : 0 \le i \le nk - 1 \right\}$$

$$= d_{nk,T}(x, y).$$

Therefore, $N(d_{n,T^k}, \varepsilon) \le N(d_{nk,T}, \varepsilon)$, and

$$h(T^k) \le \lim_{\varepsilon \to 0} \limsup_{n \to \infty} \frac{1}{n} \log N(d_{nk,T}, \varepsilon) \le k h(T).$$

For the converse inequality, we note that by the uniform continuity of T, given $\varepsilon > 0$ there exists $\delta(\varepsilon) \in (0, \varepsilon)$ such that $d_k(x, y) < \varepsilon$ whenever $d(x, y) < \delta(\varepsilon)$. Equivalently, $d(x, y) \ge \delta(\varepsilon)$ whenever $d_k(x, y) \ge \varepsilon$, and hence

$$d_{n,T^k}(x, y) \ge \delta(\varepsilon) \quad \text{whenever} \quad d_{kn}(x, y) \ge \varepsilon.$$

Therefore, $N(d_{kn}, \varepsilon) \le N(d_{n,T^k}, \delta(\varepsilon))$, and

$$\lim_{\varepsilon \to 0} \limsup_{n \to \infty} \frac{1}{n} \log N(d_{kn}, \varepsilon) \le \lim_{\varepsilon \to 0} \limsup_{n \to \infty} \frac{1}{n} \log N(d_{n,T^k}, \delta(\varepsilon)) = h(T^k), \quad (4.40)$$

where the last equality is due to the fact that $\delta(\varepsilon) \to 0$ when $\varepsilon \to 0$. Finally, it follows from (4.39) and (4.40) that

$$h(T^k) \ge \lim_{\varepsilon \to 0} \limsup_{n \to \infty} \frac{1}{n} \log N(d_{kn}, \varepsilon)$$

$$\ge \lim_{\varepsilon \to 0} \lim_{n \to \infty} \frac{1}{n} \log D(d_{kn}, 2\varepsilon)$$

$$= k \lim_{\varepsilon \to 0} \lim_{n \to \infty} \frac{1}{kn} \log D(d_{kn}, 2\varepsilon)$$

$$= k \lim_{\varepsilon \to 0} \lim_{n \to \infty} \frac{1}{n} \log D(d_n, 2\varepsilon) = kh(T),$$

which yields the first property.

For the second property, we observe that

$$d_{n,T}(x, y) = d_{n,T^{-1}}(T^{n-1}(x), T^{n-1}(y)).$$

This shows that if $x_1, \ldots, x_k$ are at a $d_{n,T}$-distance at least ε, then the points $T^{n-1}(x_1), \ldots, T^{n-1}(x_k)$ are at a $d_{n,T^{-1}}$-distance at least ε, and vice versa. Therefore,

$$N(d_{n,T^{-1}}, \varepsilon) = N(d_{n,T}, \varepsilon),$$

and hence, $h(T^{-1}) = h(T)$. $\qquad \square$

4.4.2 Topological Nature of the Entropy

The results described in this section highlight the topological nature of the topological entropy. We first show that $h(T)$ only depends on the topology induced by the distance d.

Proposition 4.5. *If the distances d and d' generate the same topology on X, then $h^d(T) = h^{d'}(T)$ for every continuous function $T : X \to X$.*

Proof. Given $\varepsilon > 0$, we consider the set

$$D_\varepsilon = \{(x, y) \in X \times X : d(x, y) \geq \varepsilon\}.$$

Since the function $d : X \times X \to \mathbb{R}$ is continuous, the set D_ε is compact. Moreover, since $d' : X \times X \to \mathbb{R}$ is also continuous, we have

$$\delta(\varepsilon) := \min\{d'(x, y) : (x, y) \in D_\varepsilon\} > 0,$$

and $\delta(\varepsilon) \to 0$ when $\varepsilon \to 0$. We note that

$$d'_n(x, y) \geq \delta(\varepsilon) \quad \text{whenever} \quad d_n(x, y) \geq \varepsilon.$$

Therefore, $N(d'_n, \delta(\varepsilon)) \geq N(d_n, \varepsilon)$, and

$$h^{d'}(T) = \lim_{\varepsilon \to 0} \limsup_{n \to \infty} \frac{1}{n} \log N(d'_n, \delta(\varepsilon))$$

$$\geq \lim_{\varepsilon \to 0} \limsup_{n \to \infty} \frac{1}{n} \log N(d_n, \varepsilon) = h^d(T).$$

Reversing the roles of d and d', we also obtain $h^{d'}(T) \leq h^d(T)$. $\qquad \square$

We also show that the topological entropy is a topological invariant, in the sense that topologically equivalent transformations have the same topological entropy. We recall that two continuous transformations $T: X \to X$ and $S: Y \to Y$ are *topologically equivalent* if there exists a homeomorphism $\chi: X \to Y$ such that $\chi \circ T = S \circ \chi$ in X. In other words, the diagram

$$
\begin{array}{ccc}
X & \xrightarrow{\ T\ } & X \\
\chi \downarrow & & \downarrow \chi \\
Y & \xrightarrow{\ S\ } & Y
\end{array}
\tag{4.41}
$$

is commutative.

Proposition 4.6. *If two continuous transformations $T: X \to X$ and $S: Y \to Y$ of compact metric spaces are topologically equivalent, then $h(T) = h(S)$.*

Proof. Let d be the distance in X. We introduce a distance in Y by

$$
d'(x, y) = d(\chi^{-1}(x), \chi^{-1}(y)),
$$

where χ is the homeomorphism in (4.41). We note that $\chi: (X, d) \to (Y, d')$ is an isometry, and hence,

$$
N(d_{n,T}, \varepsilon) = N(d'_{n,S}, \varepsilon).
\tag{4.42}
$$

Moreover, since χ is a homeomorphism, the distance d' generates the original topology of Y. Therefore, by Proposition 4.5 and (4.42), we obtain

$$
h(S) = h^{d'}(S) = h^d(T) = h(T),
$$

which yields the desired result.   □

4.5 Variational Principle

We end this chapter by establishing an important relation between the metric entropy and the topological entropy—the so-called *variational principle*. It says that the supremum of the metric entropies over all invariant probability measures is equal to the topological entropy.

Theorem 4.7 (Variational principle for the topological entropy). *If $T: X \to X$ is a continuous transformation of a compact metric space, then*

$$
h(T) = \sup \{h_\mu(T) : \mu \text{ is a } T\text{-invariant probability measure in } X\}.
\tag{4.43}
$$

Proof. Let $\eta = \{C_1, \ldots, C_k\}$ be a measurable partition of X. Given $\delta > 0$, for each $i = 1, \ldots, k$, let $D_i \subset C_i$ be a compact set such that $\mu(C_i \setminus D_i) < \delta$. Now let $\beta = \{D_0, D_1, \ldots, D_k\}$, where $D_0 = X \setminus \bigcup_{i=1}^{k} D_i$. Clearly, β is a measurable partition of X. Proceeding as in (4.22) with D_i^n replaced by D_i, we conclude that $H_\mu(\eta|\beta) < 1$ for any sufficiently small δ. Thus, it follows from (4.19) that

$$h_\mu(T^m, \eta) \leq h_\mu(T^m, \beta) + H_\mu(\eta|\beta) < h_\mu(T^m, \beta) + 1 \qquad (4.44)$$

for each $m \in \mathbb{N}$. Now we consider the open cover of X given by

$$\mathcal{U} = \{D_0 \cup D_1, \ldots, D_0 \cup D_k\},$$

as well as the open covers

$$\mathcal{U}_{mn} = \left\{ \bigcap_{i=0}^{n-1} T^{-im} U_i : U_0, \ldots, U_{n-1} \in \mathcal{U} \right\}$$

for each $m, n \in \mathbb{N}$. Since each element $D_0 \cup D_i$ of $\mathcal{U}$ intersects at most the elements D_0 and D_i of β, we have

$$\text{card } \beta_{mn} \leq 2^n \text{ card } \mathcal{U}_{mn},$$

where $\beta_{mn} = \bigvee_{i=0}^{n-1} T^{-im}\beta$. Moreover, by (4.6), we obtain

$$H_\mu(\beta_{mn}) \leq \log \text{card } \beta_{mn} \leq n \log 2 + \log \text{card } \mathcal{U}_{mn}. \qquad (4.45)$$

We observe that if ε is the Lebesgue number of $\mathcal{U}$ (i.e., if every ball of radius $r < \varepsilon$ is contained in some element of $\mathcal{U}$), then ε is the Lebesgue number of the cover $\mathcal{U}_{mn}$ with respect to the distance d_{n,T^m}. Moreover, by construction, each open set $U \in \mathcal{U}_{mn}$ contains at least a point x_U that is not in any other element of $\mathcal{U}_{mn}$. Therefore,

$$B_{d_{n,T^m}}(x_U, \varepsilon) \subset U \quad \text{for each} \quad U \in \mathcal{U}_{mn},$$

and thus, $d_{n,T^m}(x_U, x_V) \geq \varepsilon$ for any $V \in \mathcal{U}_{mn}$ distinct from U. This shows that

$$\text{card } \mathcal{U}_{mn} \leq N(d_{n,T^m}, \varepsilon),$$

and it follows from (4.45) that

$$h_\mu(T^m, \beta) = \lim_{n \to \infty} \frac{1}{n} H_\mu(\beta_{mn})$$

$$\leq \log 2 + \limsup_{n \to \infty} \frac{1}{n} \log N(d_{n,T^m}, \varepsilon).$$

By (4.44), letting $\varepsilon \to 0$ yields

$$h_\mu(T^m, \eta) \leq h(T^m) + \log 2 + 1,$$

and hence,
$$h_\mu(T^m) \le h(T^m) + \log 2 + 1.$$

Therefore, by Proposition 4.3 and Theorem 4.6,

$$h_\mu(T) = \frac{1}{m} h_\mu(T^m)$$

$$\le \frac{1}{m}\left(h(T^m) + \log 2 + 1\right)$$

$$= h(T) + \frac{1}{m}(\log 2 + 1)$$

for each $m \in \mathbb{N}$. Letting $m \to \infty$, we conclude that $h_\mu(T) \le h(T)$.

To complete the proof, given $\varepsilon > 0$, for each $n \in \mathbb{N}$, we consider a set E_n of points at a d-distance at least ε such that card $E_n = N(d_n, \varepsilon)$. We then define measures

$$\nu_n = \frac{1}{\text{card } E_n} \sum_{x \in E_n} \delta_x \quad \text{and} \quad \mu_n = \frac{1}{n} \sum_{i=0}^{n-1} T_*^i \nu_n,$$

with T_* as in (3.6). Given a sequence $(k_n)_{n \in \mathbb{N}} \subset \mathbb{N}$ such that

$$\lim_{n \to \infty} \frac{1}{k_n} \log \text{card } E_{k_n} = \limsup_{n \to \infty} \frac{1}{n} \log \text{card } E_n, \tag{4.46}$$

let μ be any accumulation point of the sequence of measures $(\mu_{k_n})_{n \in \mathbb{N}}$, which exists by Theorem 3.1. Moreover, if $(m_n)_{n \in \mathbb{N}}$ is a subsequence of $(k_n)_{n \in \mathbb{N}}$ such that $(\mu_{m_n})_{n \in \mathbb{N}}$ converges to μ, then it follows from (3.7) that μ is T-invariant. We note that in general, the measure μ may depend on ε.

Now we consider some particular measurable partitions. Let $\{B_1, \ldots, B_k\}$ be an open cover of X by balls of radius less than $\varepsilon/2$ such that $\mu(\partial B_i) = 0$ for $i = 1, \ldots, k$. This is always possible since for each $x \in X$, there are at most countably many values of $r > 0$ such that $\mu(\partial B(x, r)) > 0$. We define a partition $\xi = \{C_1, \ldots, C_k\}$ of X by

$$C_1 = \overline{B_1} \quad \text{and} \quad C_i = \overline{B_i} \setminus \bigcup_{j=1}^{i-1} \overline{B_j} \quad \text{for} \quad i = 2, \ldots, k.$$

Then diam $C_i < \varepsilon$ and $\mu(\partial C_i) = 0$ for each i since $\partial C_i \subset \bigcup_{j=1}^k \partial B_j$. Now we observe that since diam $C_i < \varepsilon$, each element of $\xi_n = \bigvee_{i=0}^{n-1} T^{-i}\xi$ contains at most one point in E_n. Therefore, there is exactly a number card E_n of elements of ξ_n with ν_n-measure equal to $1/\text{card } E_n$, and thus,

$$H_{\nu_n}(\xi_n) = \log \text{card } E_n. \tag{4.47}$$

Given $m, n \in \mathbb{N}$, we write $n = qm + r$, where $q \geq 0$ and $0 \leq r < m$. We have

$$\xi_n = \xi_{qm+r} = \bigvee_{j=0}^{q-1} T^{-jm}\xi_m \vee \bigvee_{j=qm}^{qm+r-1} T^{-j}\xi,$$

and thus, for each $i = 0, \ldots, m-1$, the measurable partition

$$\eta = \bigvee_{j=0}^{q-1} T^{-jm-i}\xi_m \vee \left(\bigvee_{j=qm}^{qm+r-1} T^{-j}\xi \vee \xi_i \right)$$

is a refinement of ξ_n. Since

$$\operatorname{card}\left(\bigvee_{j=qm}^{qm+r-1} T^{-j}\xi \vee \xi_i \right) \leq (\operatorname{card}\xi)^{2m},$$

it follows from Lemma 4.1 and (4.6) that

$$H_{\nu_n}(\xi_n) \leq H_{\nu_n}(\eta) \leq \sum_{j=0}^{q-1} H_{\nu_n}(T^{-jm-i}\xi_m) + 2m \log \operatorname{card}\xi. \tag{4.48}$$

Now we observe that since the function ψ in (4.4) is convex, we have

$$\frac{1}{n} \sum_{i=0}^{m-1} \sum_{j=0}^{q-1} H_{\nu_n}(T^{-jm-i}\xi_m) \leq \frac{1}{n} \sum_{\ell=0}^{n-1} H_{\nu_n}(T^{-\ell}\xi_m)$$

$$= -\sum_{A \in \xi_m} \sum_{\ell=0}^{n-1} \frac{1}{n} \psi\left(\nu_n(T^{-\ell}A)\right)$$

$$\leq -\sum_{A \in \xi_m} \psi\left(\sum_{\ell=0}^{n-1} \frac{1}{n}\nu_n(T^{-\ell}A) \right) \tag{4.49}$$

$$= -\sum_{A \in \xi_m} \psi(\mu_n(A)) = H_{\mu_n}(\xi_m),$$

and thus, by (4.48),

$$\frac{m}{n} H_{\nu_n}(\xi_n) \leq H_{\mu_n}(\xi_m) + \frac{2m^2}{n} \log \operatorname{card}\xi.$$

It follows from (4.47) with n replaced by m_n that

$$\frac{1}{m_n} \log \operatorname{card} E_{m_n} \leq \frac{1}{m} H_{\mu_{m_n}} (\xi_m) + \frac{2m}{m_n} \log \operatorname{card} \xi.$$

On the other hand, by (4.46), letting $n \to \infty$ yields

$$\begin{aligned}
\limsup_{n \to \infty} \frac{1}{n} \log N(d_n, \varepsilon) &= \limsup_{n \to \infty} \frac{1}{n} \log \operatorname{card} E_n \\
&\leq \frac{1}{m} \lim_{n \to \infty} H_{\mu_{m_n}} (\xi_m) = \frac{1}{m} H_\mu(\xi_m).
\end{aligned} \tag{4.50}$$

Indeed, let $A \subset X$ be a measurable set with $\mu(\partial A) = 0$. Since $(\mu_{m_n})_{n \in \mathbb{N}}$ converges to μ, if $\varphi_k \colon X \to \mathbb{R}_0^+$ is a sequence of continuous functions decreasing to $\chi_{\overline{A}}$ when $k \to \infty$, then

$$\begin{aligned}
\limsup_{n \to \infty} \mu_{m_n}(\overline{A}) &\leq \limsup_{n \to \infty} \int_X \varphi_k \, d\mu_{m_n} \\
&= \int_X \varphi_k \, d\mu \to \mu(\overline{A})
\end{aligned}$$

when $k \to \infty$. Therefore, since $\mu(\partial A) = 0$, we have

$$\limsup_{n \to \infty} \mu_{m_n}(A) \leq \limsup_{n \to \infty} \mu_{m_n}(\overline{A}) \leq \mu(\overline{A}) = \mu(A). \tag{4.51}$$

Similarly, since $\partial(X \setminus A) = \partial A$, we also have

$$\limsup_{n \to \infty} \mu_{m_n}(X \setminus A) \leq \mu(X \setminus A),$$

which yields

$$\liminf_{n \to \infty} \mu_{m_n}(A) \geq \mu(A).$$

Together with (4.51), this implies that

$$\lim_{n \to \infty} \mu_{m_n}(A) = \mu(A),$$

and hence,

$$\lim_{n \to \infty} H_{\mu_{m_n}} (\xi_m) = H_\mu(\xi_m).$$

Letting $m \to \infty$ in (4.50), we obtain

$$\limsup_{n \to \infty} \frac{1}{n} \log N(d_n, \varepsilon) \leq h_\mu(T, \xi) \leq h_\mu(T) \leq \sup_\nu h_\nu(T),$$

where the supremum is taken over all T-invariant probability measures ν in X. Letting $\varepsilon \to 0$ yields

$$h(T) \leq \sup_{\nu} h_{\nu}(T).$$

This completes the proof of the theorem. □

We note that identity (4.43) could be used as an alternative definition of topological entropy.

4.6 Exercises

Exercise 4.1. Let $T: X \to X$ be a measurable transformation preserving a probability measure μ in X. Show that if T is invertible almost everywhere and T^{-1} is measurable, then

$$h_{\mu}(T^{-1}) = h_{\mu}(T).$$

Exercise 4.2. Let $T: \mathbb{R}^m \to \mathbb{R}^m$ be a diffeomorphism preserving a probability measure μ in $\mathbb{R}^m$ such that

$$h_{\mu}(T^n) = \int_{\mathbb{R}^m} \log \|d_x T^n\| \, d\mu$$

for each $n \in \mathbb{N}$. Show that

$$h_{\mu}(T) \leq \lim_{n \to \infty} \frac{1}{n} \log \sup_{x \in \mathbb{R}^m} \|d_x T^n\|.$$

Exercise 4.3. Consider the transformation $T: \{0, 1\}^{\mathbb{N}} \to \{0, 1\}^{\mathbb{N}}$ defined by

$$(T(x))_n = \begin{cases} 1 - x_n & \text{if } n = 1 \text{ or if } x_m = 1 \text{ for every } m < n, \\ x_n & \text{otherwise.} \end{cases}$$

It is called an *infinite adding machine* since it can be written in the form $T(x) = x + (10 \cdots)$, where each element is in base 2 with the addition effected from the left to the right. Show that:

1. T preserves the Bernoulli measure μ with probability vector $(1/2, 1/2)$.
2. $h_{\mu}(T) = 0$.

Exercise 4.4. For $i = 1, 2$, let $T_i: X_i \to X_i$ be a measurable transformation preserving a probability measure μ_i in X_i. Show that if (T_1, μ_1) and (T_2, μ_2) are equivalent (see Exercise 3.23), then $h_{\mu_1}(T_1) = h_{\mu_2}(T_2)$.

Exercise 4.5. Find a function $h: \{1, 2, 3, 4\}^{\mathbb{Z}} \to \{1, 2, 3, 4, 5\}^{\mathbb{Z}}$ as in Exercise 3.23 for the shift maps σ_1 and σ_2 in Example 4.6.

Exercise 4.6. Say if there exist transformations such that:

1. (T, μ) and (T^2, μ) are ergodic, but (T^3, μ) is not ergodic.
2. (T, μ) and (T^3, μ) are ergodic, but (T^2, μ) is not ergodic.
3. (T, μ) is ergodic, but (T^2, μ) and (T^3, μ) are not ergodic.

Exercise 4.7. Show that $H_\mu(\xi|\zeta) = 0$ if and only if ζ is a refinement of ξ. Hint: if $H_\mu(\xi|\zeta) = 0$, then

$$\mu(C \cap D) \log \frac{\mu(C \cap D)}{\mu(D)} = 0 \quad \text{for} \quad C \in \xi, \ D \in \zeta,$$

and hence, either $\mu(C \cap D) = 0$ or $\mu(C \cap D) = \mu(D)$ for each $C \in \xi$ and $D \in \zeta$.

Exercise 4.8. Show that

$$d(\xi, \eta) = H_\mu(\xi|\eta) + H_\mu(\eta|\xi) \tag{4.52}$$

is a distance (*Rohklin's distance*) in the space of measurable partitions of X. Hint: by Proposition 4.1 and Lemmas 4.4 and 4.5, we have

$$H_\mu(\xi|\eta) \le H_\mu(\xi \vee \zeta|\eta)$$
$$= H_\mu(\xi|\zeta \vee \eta) + H_\mu(\zeta|\eta)$$
$$\le H_\mu(\xi|\zeta) + H_\mu(\zeta|\eta).$$

Exercise 4.9. Show that if $T: X \to X$ is a measurable transformation and ξ and η are measurable partitions of X, then

$$|h_\mu(T, \xi) - h_\mu(T, \zeta)| \le d(\xi, \eta),$$

where d is the distance in (4.52). Hint: use inequality (4.19).

Exercise 4.10. Compute $h_\mu(T)$ for the map T and the measure μ in Exercise 3.28.

Exercise 4.11. Compute the entropy of a one-sided Markov measure.

Exercise 4.12. Let $T: X \to X$ be a measurable transformation preserving a probability measure μ in X. Show that if ξ is a one-sided generator and T is invertible almost everywhere, then $h_\mu(T) = 0$.

Exercise 4.13. Let $T: X \to X$ be a measurable transformation preserving a probability measure μ in X and let ξ be a finite (one-sided or two-sided) generator.

1. Show that $h_\mu(T) \le \log \operatorname{card} \xi$.
2. When $h_\mu(T) = \log \operatorname{card} \xi$, show that for each $n \in \mathbb{N}$, the measurable partition $\xi_n = \bigvee_{k=0}^{n-1} T^{-k}\xi$ has exactly $(\operatorname{card} \xi)^n$ elements all with the same measure. Hint: the function ψ defined by (4.4) is strictly convex.

3. Show that if
$$h_\mu(T^n) \le \log n + an$$

for every $n \in \mathbb{N}$, then $h_\mu(T) \le a$.

Exercise 4.14. For the shift map $\sigma: X_k^+ \to X_k^+$, show that $h(\sigma) = \log k$.

Exercise 4.15. For the one-sided topological Markov chain $\sigma|X_A^+: X_A^+ \to X_A^+$, show that $h(\sigma) = \log \rho(A)$, where $\rho(A)$ is the spectral radius of the matrix A.

Exercise 4.16. For a continuous transformation $T: X \to X$ of a compact metric space, show that
$$h(T) = \lim_{\varepsilon \to 0} \liminf_{n \to \infty} \frac{1}{n} \log N(d_n, \varepsilon).$$

Hint: see the proof of Theorem 4.5.

Exercise 4.17. Let $T: X \to X$ be a continuous transformation of a compact metric space. Given $\varepsilon > 0$, we denote by $M(d, \varepsilon)$ the least number of points $p_1, \ldots, p_m \in X$ such that any $x \in X$ satisfies $d(x, p_i) < \varepsilon$ for some i. Show that:

1. For each $\varepsilon > 0$,

$$D(d, 2\varepsilon) \le M(d, \varepsilon) \le N(d, \varepsilon) \le M(d, \varepsilon/2) \le D(d, \varepsilon/2).$$

2.

$$h(T) = \lim_{\varepsilon \to 0} \limsup_{n \to \infty} \frac{1}{n} \log M(d_n, \varepsilon)$$

$$= \lim_{\varepsilon \to 0} \liminf_{n \to \infty} \frac{1}{n} \log M(d_n, \varepsilon).$$

Exercise 4.18. Given a matrix $A \in S(2, \mathbb{Z})$ without eigenvalues in S^1, compute the topological entropy of the toral automorphism $T_A: \mathbb{T}^2 \to \mathbb{T}^2$. Hint: use Exercise 4.17.

Exercise 4.19. Let $T: X \to X$ be an invertible measurable transformation with measurable inverse, preserving a probability measure μ in X. Given a function $\varphi: X \to \mathbb{R}$, we define a new function $U_T(\varphi): X \to \mathbb{R}$ by $U_T(\varphi) = \varphi \circ T$. Show that:

1. $U_T: L^2(X, \mu) \to L^2(X, \mu)$ is a linear transformation with norm $\|U_T\| = 1$.
2. T is ergodic if and only if 1 is an eigenvalue of U_T.
3. For a function $\varphi \in L^2(X, \mu)$ such that

$$\lim_{n \to \infty} \int_X \varphi U_T^n(\varphi) \, d\mu = \left(\int_X \varphi \, d\mu \right)^2, \tag{4.53}$$

if S is the smallest closed subspace of $L^2(X, \mu)$ containing the functions 1 and $U_T^n(\varphi)$ for every $n \in \mathbb{N}$, then both S and $S^\perp$ (the orthogonal complement of S) are contained in

$$\left\{ \psi \in L^2(X, \mu) : \lim_{n \to \infty} \int_X \psi U_T^n(\varphi)\, d\mu = \int_X \psi\, d\mu \int_X \varphi\, d\mu \right\}.$$

4. μ is mixing if and only if (4.53) holds for every function $\varphi \in L^2(X, \mu)$.

Exercise 4.20. Let X be a compact topological space and let $T: X \to X$ be a continuous transformation. Show that for each finite open cover $\mathcal{U}$ of X, the limit

$$\lim_{n \to \infty} \frac{1}{n} \log \mathrm{card} \bigvee_{k=0}^{n-1} T^{-k}\mathcal{U}$$

exists, where

$$\bigvee_{k=0}^{n-1} T^{-k}\mathcal{U} = \left\{ \bigcap_{k=0}^{n-1} T^{-k} U_k : U_0, \dots, U_{n-1} \in \mathcal{U} \right\}.$$

Exercise 4.21. Let X be a compact metric space and let $T: X \to X$ be a continuous transformation. Show that

$$h(T) = \sup_{\mathcal{U}} \lim_{n \to \infty} \frac{1}{n} \log \mathrm{card} \bigvee_{k=0}^{n-1} T^{-k}\mathcal{U},$$

where the supremum is taken over all finite open covers $\mathcal{U}$ of X.

Notes

The notion of metric entropy is due to Kolmogorov [49, 50]. It was extended to all dynamical systems by Sinai [93] in the form (4.12)–(4.13). Theorem 4.4 was proven successively in more general forms by several authors. Shannon [87] considered Markov measures, although the statement was only derived rigorously by Khinchin [46] (see also [47]). McMillan [60] obtained the L^1 convergence, and Breiman [22] obtained the convergence almost everywhere. Our proof of Theorem 4.4 is based on [77]. The original definition of topological entropy is due to Adler, Konheim and McAndrew [1] (in the form described in Exercises 4.20 and 4.21). The definition in (4.35) was introduced independently by Bowen [19] and Dinaburg [27]. The variational principle for the topological entropy in Theorem 4.7 is a combination of work of Goodwyn [36] (showing that the topological entropy bounds the metric entropy), Dinaburg [27] (for a finite-dimensional space X), and Goodman [34] (for an arbitrary space). Our proof of Theorem 4.7 is based on [77], which follows the simpler proof of Misiurewicz [61].

Chapter 5
Thermodynamic Formalism

This chapter is an introduction to the thermodynamic formalism. We first introduce the notion of topological pressure, which includes topological entropy as a special case. In particular, we establish a somewhat explicit formula for the topological pressure in the case of symbolic dynamics. This formula is particularly useful in dimension theory of hyperbolic dynamics. We also establish the variational principle for the topological pressure. Finally, we show that there exist equilibrium measures for any expansive transformation. These are invariant probability measures attaining the supremum in the variational principle.

5.1 Introduction

The notion of topological pressure, which is the most basic notion of the thermo-dynamic formalism, was introduced by Ruelle for expansive transformations and by Walters in the general case. The thermodynamic formalism (following Ruelle's original expression) can be described as a rigorous study of certain mathematical structures inspired in thermodynamics. For a continuous transformation $T: X \to X$ of a compact metric space, the topological pressure of a continuous function $\varphi: X \to \mathbb{R}$ is defined by

$$P(\varphi) = \lim_{\varepsilon \to 0} \limsup_{n \to \infty} \frac{1}{n} \log \sup_E \sum_{x \in E} \exp \sum_{k=0}^{n-1} \varphi(T^k(x)),$$

where the supremum is taken over all (n, ε)-separated sets $E \subset X$ (see Sect. 5.2 for the definition). For example, taking $\varphi = 0$, we recover the notion of topological entropy

$$h(T) = \lim_{\varepsilon \to 0} \limsup_{n \to \infty} \frac{1}{n} \log N_{n,\varepsilon},$$

where $N_{n,\varepsilon}$ is the maximal number of elements of an (n, ε)-separated set $E \subset X$.

L. Barreira, *Ergodic Theory, Hyperbolic Dynamics and Dimension Theory*, Universitext,   147
DOI 10.1007/978-3-642-28090-0_5, © Springer-Verlag Berlin Heidelberg 2012

The variational principle relating topological pressure to Kolmogorov–Sinai entropy was established by Ruelle for expansive transformations and by Walters in the general case. It says that

$$P(\varphi) = \sup_{\mu} \left(h_{\mu}(T) + \int_X \varphi \, d\mu \right),$$

where the supremum is taken over all T-invariant probability measures μ in X. The theory also includes a discussion of the existence and uniqueness of equilibrium and Gibbs measures. In particular, a T-invariant probability measure μ is called an equilibrium measure for φ if

$$P(\varphi) = h_{\mu}(T) + \int_X \varphi \, d\mu.$$

As we already mentioned earlier, the possibility of coding repellers and hyperbolic sets via symbolic dynamics often allows one to give simpler proofs. Thus, it is of interest to have explicit formulas for the topological pressure with respect to the shift map. One can show that the topological pressure of a continuous function $\varphi \colon X_k^+ \to \mathbb{R}$ is given by

$$P(\varphi) = \lim_{n \to \infty} \frac{1}{n} \log \sum_{i_1 \cdots i_n} \exp \sup_{C_{i_1 \cdots i_n}} \sum_{k=0}^{n-1} \varphi \circ \sigma^k,$$

where $C_{i_1 \cdots i_n}$ are the cylinder sets. For example, for the function

$$\varphi(i_1 i_2 \cdots) = \log \lambda_{i_1},$$

which occurs later on in dimension theory of hyperbolic dynamics, we have

$$P(\varphi) = \log \sum_{j=1}^{k} \lambda_j.$$

5.2 Topological Pressure

We first introduce the notion of topological pressure. Let $T \colon X \to X$ be a continuous transformation of a compact metric space (X, d). For each $n \in \mathbb{N}$, we consider the distance d_n in X defined by (4.34).

Definition 5.1. Given $\varepsilon > 0$, a set $E \subset X$ is called (n, ε)-*separated* if $d_n(x, y) > \varepsilon$ for every $x, y \in E$ with $x \neq y$.

We note that since X is compact, each (n, ε)-separated set E is finite. The notion of topological pressure can now be introduced as follows:

Definition 5.2. The *topological pressure* of a continuous function $\varphi \colon X \to \mathbb{R}$ (with respect to T) is defined by

$$P_T(\varphi) = \lim_{\varepsilon \to 0} \limsup_{n \to \infty} \frac{1}{n} \log \sup_E \sum_{x \in E} \exp \sum_{k=0}^{n-1} \varphi(T^k(x)), \tag{5.1}$$

where the supremum is taken over all (n, ε)-separated sets $E \subset X$.

Since the function

$$\varepsilon \mapsto \limsup_{n \to \infty} \frac{1}{n} \log \sup_E \sum_{x \in E} \exp \sum_{k=0}^{n-1} \varphi(T^k(x))$$

is nondecreasing, the limit in (5.1) when $\varepsilon \to 0$ is well defined.

The following example shows that the topological entropy is a particular case of the topological pressure:

Example 5.1. For the constant function $\varphi = c$, and any (n, ε)-separated set E, we have

$$\sum_{x \in E} \exp \sum_{k=0}^{n-1} \varphi(T^k(x)) = \sum_{x \in E} e^{nc} = e^{nc} \operatorname{card} E.$$

Therefore,

$$\sup_E \sum_{x \in E} \exp \sum_{k=0}^{n-1} \varphi(T^k(x)) = e^{nc} N(d_n, \varepsilon),$$

with $N(d_n, \varepsilon)$ as in Sect. 4.4.1. By (4.35), we obtain

$$P_T(c) = \lim_{\varepsilon \to 0} \limsup_{n \to \infty} \frac{1}{n} \log \left(e^{nc} N(d_n, \varepsilon) \right) = c + h(T),$$

and in particular,

$$P_T(0) = \lim_{\varepsilon \to 0} \limsup_{n \to \infty} \frac{1}{n} \log N(d_n, \varepsilon) = h(T). \tag{5.2}$$

5.3 Symbolic Dynamics

We consider in this section the particular case of symbolic dynamics.

Let $\sigma \colon X_k^+ \to X_k^+$ be the (one-sided) shift map. We recall that X_k^+ is a compact metric space with the distance $d = d_\beta$ in (3.32) for each $\beta > 1$. Given $\varepsilon > 0$, let $m = m(\varepsilon) \in \mathbb{N}$ be the largest integer such that

$$m < -\log \varepsilon / \log \beta, \quad \text{that is,} \quad \beta^{-m} > \varepsilon.$$

Then the (n, ε)-separated sets are exactly the collections of sequences

$$\omega_{i_1 \cdots i_{m+n-1}} \in C_{i_1 \cdots i_{m+n-1}} \quad \text{for} \quad i_1, \ldots, i_{m+n-1} \in \{1, \ldots, k\}.$$

Now we give a formula for the topological pressure with respect to the shift map.

Theorem 5.1. *For each continuous function $\varphi \colon X_k^+ \to \mathbb{R}$, we have*

$$P_\sigma(\varphi) = \lim_{n \to \infty} \frac{1}{n} \log \sum_{i_1 \cdots i_n} \exp \sup_{C_{i_1 \cdots i_n}} \sum_{k=0}^{n-1} \varphi \circ \sigma^k. \tag{5.3}$$

Proof. Since X_k^+ is compact, the function φ is uniformly continuous, and thus, for each $\delta > 0$, there exists $n \in \mathbb{N}$ such that

$$\sup_{C_{i_1 \cdots i_n}} \varphi - \inf_{C_{i_1 \cdots i_n}} \varphi < \delta$$

for every $i_1, \ldots, i_n \in \{1, \ldots, k\}$. Writing for simplicity $D_n = C_{i_1 \cdots i_n}$, we thus have

$$\delta_n := \max_{i_1 \cdots i_n} \left(\sup_{D_n} \varphi - \inf_{D_n} \varphi \right) \to 0$$

when $n \to \infty$. This implies that

$$0 \le \log \sum_{i_1 \cdots i_{m+n-1}} \exp \sup_{D_{m+n-1}} \varphi_n - \log \sum_{i_1 \cdots i_{m+n-1}} \exp \inf_{D_{m+n-1}} \varphi_n \le n\delta_m$$

for each $m, n \in \mathbb{N}$, where

$$\varphi_n = \sum_{k=0}^{n-1} \varphi \circ \sigma^k.$$

On the other hand, given $m, n \in \mathbb{N}$ and a (n, ε)-separated set $E \subset X_k^+$, we have

$$\sum_{i_1 \cdots i_{m+n-1}} \exp \inf_{D_{m+n-1}} \varphi_n \le \sum_{x \in E} \exp \varphi_n(x) \le \sum_{i_1 \cdots i_{m+n-1}} \exp \sup_{D_{m+n-1}} \varphi_n,$$

and thus,

$$-\delta_m + \limsup_{n\to\infty} \frac{1}{n} \log \sum_{i_1\cdots i_{m+n-1}} \exp \sup_{D_{m+n-1}} \varphi_n$$

$$\leq \limsup_{n\to\infty} \frac{1}{n} \log \sum_{x\in E} \exp \varphi_n(x)$$

$$\leq \limsup_{n\to\infty} \frac{1}{n} \log \sum_{i_1\cdots i_{m+n-1}} \exp \sup_{D_{m+n-1}} \varphi_n.$$

Letting $m \to \infty$, we obtain

$$P_\sigma(\varphi) = \lim_{m\to\infty} \limsup_{n\to\infty} \frac{1}{n} \log \sum_{i_1\cdots i_{m+n-1}} \exp \sup_{D_{m+n-1}} \varphi_n$$

$$= \lim_{m\to\infty} \limsup_{n\to\infty} \frac{1}{n} \log \sum_{i_1\cdots i_{m+n-1}} \exp \inf_{D_{m+n-1}} \varphi_n.$$

Moreover, writing

$$S_n = \sum_{i_1\cdots i_n} \exp \sup_{C_{i_1\cdots i_n}} \varphi_n,$$

we have

$$\sum_{i_1\cdots i_{m+n-1}} \exp \sup_{D_{m+n-1}} \varphi_n \leq k^{m-1} S_n,$$

and

$$\sum_{i_1\cdots i_{m+n-1}} \exp \sup_{D_{m+n-1}} \varphi_n \geq e^{-(m-1)\|\varphi\|_\infty} S_{m+n-1},$$

where

$$\|\varphi\|_\infty = \sup \{|\varphi(\omega)| : \omega \in \Sigma_k^+\}.$$

This implies that

$$\lim_{m\to\infty} \limsup_{n\to\infty} \frac{1}{n} \log \sum_{i_1\cdots i_{m+n-1}} \exp \sup_{D_{m+n-1}} \varphi_n \leq \limsup_{n\to\infty} \frac{1}{n} \log S_n,$$

and

$$\lim_{m\to\infty} \limsup_{n\to\infty} \frac{1}{n} \log \sum_{i_1\cdots i_{m+n-1}} \exp \sup_{D_{m+n-1}} \varphi_n \geq \lim_{m\to\infty} \limsup_{n\to\infty} \frac{1}{n} \log S_{m+n-1}$$

$$= \limsup_{m\to\infty} \frac{1}{n} \log S_n,$$

that is,

$$\lim_{m\to\infty} \limsup_{n\to\infty} \frac{1}{n} \log \sum_{i_1\cdots i_{m+n-1}} \exp \sup_{D_{m+n-1}} \varphi_n = \limsup_{n\to\infty} \frac{1}{n} \log S_n. \qquad (5.4)$$

Finally, since

$$\sup_{C_{i_1\cdots i_{m+n}}} \varphi_{m+n} \leq \sup_{C_{i_1\cdots i_{m+n}}} \varphi_m + \sup_{C_{i_1\cdots i_{m+n}}} \varphi_n \circ T^m$$

$$\leq \sup_{C_{i_1\cdots i_m}} \varphi_m + \sup_{C_{i_{m+1}\cdots i_{m+n}}} \varphi_n,$$

we have

$$S_{m+n} \leq \sum_{i_1\cdots i_{m+n}} \exp\left(\sup_{C_{i_1\cdots i_m}} \varphi_m + \sup_{C_{i_{m+1}\cdots i_{m+n}}} \varphi_n \right)$$

$$= \sum_{i_1\cdots i_m} \exp \sup_{C_{i_1\cdots i_m}} \varphi_m \sum_{i_{m+1}\cdots i_{m+n}} \exp \sup_{C_{i_{m+1}\cdots i_{m+n}}} \varphi_n = S_m S_n.$$

The formula for the topological pressure $P_\sigma(\varphi)$ in (5.3) follows now readily from Lemma 4.3 together with identity (5.4). □

The following are applications of Theorem 5.1:

Example 5.2. Given numbers $\lambda_1, \ldots, \lambda_k > 0$, we consider the continuous function $\varphi: X_k^+ \to \mathbb{R}$ defined by

$$\varphi(i_1 i_2 \cdots) = \log \lambda_{i_1}. \qquad (5.5)$$

It follows from (5.3) that

$$P_\sigma(\varphi) = \lim_{n\to\infty} \frac{1}{n} \log \sum_{i_1\cdots i_n} \exp \sum_{k=1}^{n} \log \lambda_{i_k}$$

$$= \lim_{n\to\infty} \frac{1}{n} \log \sum_{i_1\cdots i_n} \prod_{k=1}^{n} \lambda_{i_k}$$

$$= \lim_{n\to\infty} \frac{1}{n} \log \left(\sum_{j=1}^{k} \lambda_j \right)^n \qquad (5.6)$$

$$= \log \sum_{j=1}^{k} \lambda_j.$$

We also consider functions depending on the first two symbols of each sequence.

Example 5.3. Given numbers $\lambda_{ij} > 0$ for $i, j = 1, \ldots, k$, we consider the continuous function $\varphi: X_k^+ \to \mathbb{R}$ defined by

$$\varphi(i_1 i_2 \cdots) = \log \lambda_{i_1 i_2}.$$

It follows again from (5.3) that

$$P_\sigma(\varphi) = \lim_{n\to\infty} \frac{1}{n} \log \sum_{i_1\cdots i_n} \exp \max_{i_{n+1}} \sum_{k=1}^{n} \log \lambda_{i_k i_{k+1}}$$

$$= \lim_{n\to\infty} \frac{1}{n} \log \sum_{i_1\cdots i_n} \max_{i_{n+1}} \prod_{k=1}^{n} \lambda_{i_k i_{k+1}}. \tag{5.7}$$

Since

$$\left(\prod_{k=1}^{n-1} \lambda_{i_k i_{k+1}}\right) \min_{i,j} \lambda_{ij} \le \max_{i_{n+1}} \prod_{k=1}^{n} \lambda_{i_k i_{k+1}} \le \left(\prod_{k=1}^{n-1} \lambda_{i_k i_{k+1}}\right) \max_{i,j} \lambda_{ij},$$

considering the matrix $B = (\lambda_{ij})$, it follows from (5.7) that

$$P_\sigma(\varphi) = \lim_{n\to\infty} \frac{1}{n} \log \sum_{i_1\cdots i_n} \prod_{k=1}^{n-1} \lambda_{i_k i_{k+1}}$$

$$= \lim_{n\to\infty} \frac{1}{n} \log \operatorname{tr}(B^{n-1})$$

$$= \log \lim_{n\to\infty} \sqrt[n]{\operatorname{tr}(B^n)}$$

$$= \log \rho(B),$$

where $\rho(B)$ is the spectral radius of B.

5.4   Variational Principle

We establish in this section the variational principle for the topological pressure that includes as a particular case the variational principle for the topological entropy in Theorem 4.7.

Theorem 5.2 (Variational principle for the topological pressure). *Given a continuous transformation $T: X \to X$ of a compact metric space, if $\varphi: X \to \mathbb{R}$ is a continuous function, then*

$$P_T(\varphi) = \sup_{\mu} \left\{ h_\mu(T) + \int_X \varphi \, d\mu \right\}, \tag{5.8}$$

where the supremum is taken over all T-invariant probability measures μ in X.

Proof. The argument is an elaboration of the proof of Theorem 4.7. We consider again a measurable partition $\eta = \{C_1, \ldots, C_k\}$ of X. Given $\delta > 0$, for each $i = 1, \ldots, k$, let $D_i \subset C_i$ be a compact set such that $\mu(C_i \setminus D_i) < \delta$. We also consider the measurable partition $\beta = \{D_0, D_1, \ldots, D_k\}$, where $D_0 = X \setminus \bigcup_{i=1}^{k} D_i$. As in the proof of Theorem 4.7, for any sufficiently small δ, we have

$$h_\mu(T, \eta) < h_\mu(T, \beta) + 1.$$

Now let

$$\Delta = \inf\{d(x, y) : x \in D_i, y \in D_j, i \neq j\}.$$

Clearly, $\Delta > 0$. Moreover, take $\varepsilon \in (0, \Delta/2)$ such that

$$|\varphi(x) - \varphi(y)| < 1 \quad \text{whenever} \quad d(x, y) < \varepsilon. \tag{5.9}$$

For each $n \in \mathbb{N}$ and $C \in \beta_n := \bigvee_{j=0}^{n-1} T^{-j}\beta$, there exists $x_C \in \overline{C}$ such that

$$\varphi_n(x_C) = \sup\{\varphi_n(x) : x \in C\},$$

where

$$\varphi_n = \sum_{j=0}^{n-1} \varphi \circ T^j.$$

If E is a $(n, \varepsilon/2)$-separated set with card $E = N(d_n, \varepsilon/2)$, then for each C, there exists $p_C \in E$ such that $d_n(x_C, p_C) < \varepsilon$. By (5.9), we thus obtain

$$\varphi_n(x_C) \leq \varphi_n(p_C) + n. \tag{5.10}$$

On the other hand, since $\varepsilon < \Delta/2$, for each $x \in E$ and $j \in \{0, \ldots, n-1\}$, the point $T^j(x)$ can be at most in two elements of β. Therefore,

$$\text{card}\{C \in \beta_n : p_C = x\} \leq 2^n. \tag{5.11}$$

To proceed with the proof, we need the following auxiliary result:

Lemma 5.1. *For any $p_i \geq 0$ with $\sum_{i=1}^{k} p_i = 1$, and any $c_i \in \mathbb{R}$, we have*

$$\sum_{i=1}^{k} p_i(-\log p_i + c_i) \leq \log \sum_{i=1}^{k} e^{c_i}, \tag{5.12}$$

with equality if and only if $p_i = e^{c_i} / \sum_{i=1}^{k} e^{c_i}$ for $i = 1, \ldots, k$.

Proof of the lemma. Setting

$$a_i = \frac{e^{c_i}}{\sum_{i=1}^{k} e^{c_i}} \quad \text{and} \quad x_i = \frac{p_i}{e^{c_i}} \sum_{i=1}^{k} e^{c_i}$$

for each i, we obtain

$$\sum_{i=1}^{k} a_i x_i = \sum_{i=1}^{k} p_1 = 1.$$

Since the function ψ in (4.4) is convex, we obtain

$$0 = \psi\left(\sum_{i=1}^{k} a_i x_i\right) \le \sum_{i=1}^{k} a_i \psi(x_i)$$

$$= \sum_{i=1}^{k} \frac{e^{c_i}}{\sum_{i=1}^{k} e^{c_i}} \cdot \frac{p_i}{e^{c_i}} \sum_{i=1}^{k} e^{c_i} \log\left(\frac{p_i}{e^{c_i}} \sum_{i=1}^{k} e^{c_i}\right)$$

$$= \sum_{i=1}^{k} p_i \left(\log p_i - c_i + \log \sum_{i=1}^{k} e^{c_i}\right)$$

$$= \log \sum_{i=1}^{k} e^{c_i} - \sum_{i=1}^{k} p_i(-\log p_i + c_i).$$

This establishes inequality (5.12). Moreover, by the strict convexity of the function ψ, this is an identity if and only if $x_1 = \cdots = x_k = d$ for some $d \ge 0$, that is, if and only if

$$p_i = \frac{d e^{c_i}}{\sum_{i=1}^{k} e^{c_i}} \quad \text{for} \quad i = 1,\dots,k.$$

Summing over i yields $d = 1$, and hence, inequality (5.12) is an identity if and only if $p_i = e^{c_i}/\sum_{i=1}^{k} e^{c_i}$ for $i = 1,\dots,k$.   □

By Lemma 5.1 together with (5.10) and (5.11), we obtain

$$H_\mu(\beta_n) + \int_X \varphi_n \, d\mu \le \sum_{C \in \beta_n} \mu(C)(-\log \mu(C) + \varphi_n(x_C))$$

$$\le \log \sum_{C \in \beta_n} e^{\varphi_n(x_C)}$$

$$\le \log \sum_{C \in \beta_n} e^{\varphi_n(p_C)+n}$$

$$\le n + \log\left(2^n \sum_{x \in E} e^{\varphi_n(x)}\right),$$

and hence,

$$\frac{1}{n} H_\mu(\beta_n) + \int_X \varphi \, d\mu = \frac{1}{n} H_\mu(\beta_n) + \frac{1}{n} \int_X \varphi_n \, d\mu$$

$$\leq 1 + \log 2 + \frac{1}{n} \log \sup_E \sum_{x \in E} e^{\varphi_n(x)}.$$

This implies that

$$h_\mu(T, \eta) + \int_X \varphi \, d\mu < h_\mu(T, \beta) + 1 + \int_X \varphi \, d\mu$$

$$\leq 2 + \log 2 + \limsup_{n \to \infty} \frac{1}{n} \log \sup_E \sum_{x \in E} e^{\varphi_n(x)},$$

and letting $\varepsilon \to 0$ yields

$$h_\mu(T) + \int_X \varphi \, d\mu = \sup_\eta \left(h_\mu(T, \eta) + \int_X \varphi \, d\mu \right) \tag{5.13}$$

$$\leq 2 + \log 2 + P_T(\varphi).$$

Since $P_{T^m}(\varphi_m) = m P_T(\varphi)$, replacing T by T^m and φ by φ_m in (5.13), we obtain

$$h_\mu(T) + \int_X \varphi \, d\mu = \frac{1}{m} \left(h_\mu(T^m) + \int_X \varphi_m \, d\mu \right)$$

$$\leq \frac{1}{m} \left(2 + \log 2 + P_{T^m}(\varphi_m) \right)$$

$$= \frac{2 + \log 2}{m} + P_T(\varphi).$$

Finally, letting $m \to \infty$ yields

$$h_\mu(T) + \int_X \varphi \, d\mu \leq P_T(\varphi).$$

To complete the proof, given $\varepsilon > 0$, for each $n \in \mathbb{N}$, we consider a set E_n of points at a d_n-distance at least ε such that

$$\log \sum_{x \in E_n} e^{\varphi_n(x)} > \log \sup_E \sum_{x \in E} e^{\varphi_n(x)} - 1, \tag{5.14}$$

where the supremum is taken over all (n, ε)-separated sets. We then define probability measures

$$\nu_n = \frac{\sum_{x \in E_n} e^{\varphi_n(x)} \delta_x}{\sum_{x \in E_n} e^{\varphi_n(x)}}$$

and

$$\mu_n = \frac{1}{n} \sum_{i=0}^{n-1} T_*^i \nu_n,$$

with T_* as in (3.6). Given a sequence $(k_n)_{n \in \mathbb{N}} \subset \mathbb{N}$ such that

$$\lim_{n \to \infty} \frac{1}{k_n} \log \sum_{x \in E_{k_n}} e^{\varphi_{k_n}(x)} = \limsup_{n \to \infty} \frac{1}{n} \log \sum_{x \in E_n} e^{\varphi_n(x)}, \qquad (5.15)$$

let μ be any accumulation point of the sequence of measure $(\mu_{k_n})_{n \in \mathbb{N}}$. Then μ is a T-invariant measure. As in the proof of Theorem 4.7, we consider a partition ξ of X such that $\mathrm{diam}\, C < \varepsilon$ and $\mu(\partial C) = 0$ for each $C \in \xi$. Write $E_n = \{x_1, \ldots, x_k\}$, $p_i = \nu_n(\{x_i\})$, and $c_i = \varphi_n(x_i)$ for $i = 1, \ldots, k$. We note that

$$p_i = \frac{e^{\varphi_n(x_i)}}{\sum_{x \in E_n} e^{\varphi_n(x)}} = \frac{e^{c_i}}{\sum_{i=1}^{N} e^{c_i}}.$$

By Lemma 5.1, this ensures that (5.12) is an identity, and hence,

$$H_{\nu_n}(\xi_n) + n \int_X \varphi \, d\mu_n = H_{\nu_n}(\xi_n) + \int_X \varphi_n \, d\nu_n$$

$$= \sum_{x \in E_n} \nu_n(\{x\}) \big(-\log \nu_n(\{x\}) + \varphi_n(x) \big)$$

$$= \log \sum_{x \in E_n} e^{\varphi_n(x)}.$$

Now given $m, n \in \mathbb{N}$, we write $n = qm + r$, where $q \geq 0$ and $0 \leq r < m$. Then by (4.48) and (4.49), we have

$$\frac{m}{n} \log \sum_{x \in E_n} e^{\varphi_n(x)} = \frac{m}{n} H_{\nu_n}(\xi_n) + m \int_X \varphi \, d\mu_n$$

$$= \frac{1}{n} \sum_{i=0}^{m-1} H_{\nu_n}(\xi_n) + m \int_X \varphi \, d\mu_n$$

$$\leq \frac{1}{n} \sum_{i=0}^{m-1} \sum_{j=0}^{q-1} H_{\nu_n}(T^{-jm-i}\xi_m) + \frac{2m^2}{n} \log \mathrm{card}\, \xi + m \int_X \varphi \, d\mu_n$$

$$\leq H_{\mu_n}(\xi_m) + \frac{2m^2}{n} \log \mathrm{card}\, \xi + m \int_X \varphi \, d\mu_n.$$

Therefore, by (5.15), in a similar manner to that in the proof of Theorem 4.7, we obtain

$$\lim_{n\to\infty} \frac{1}{k_n} \log \sum_{x\in E_{k_n}} e^{\varphi_{k_n}(x)} \le \frac{1}{m} H_\mu(\xi_m) + \int_X \varphi\, d\mu.$$

Letting $m \to \infty$ yields

$$\lim_{n\to\infty} \frac{1}{k_n} \log \sum_{x\in E_{k_n}} e^{\varphi_{k_n}(x)} \le h_\mu(T,\xi) + \int_X \varphi\, d\mu$$

$$\le h_\mu(T) + \int_X \varphi\, d\mu,$$

and hence,

$$\limsup_{n\to\infty} \frac{1}{n} \log \sum_{x\in E_n} e^{\varphi_n(x)} \le \sup_\nu \left(h_\nu(T) + \int_X \varphi\, d\nu \right),$$

with the supremum taken over all T-invariant probability measures ν in X. Finally, by (5.14), letting $\varepsilon \to 0$ yields

$$P_T(\varphi) \le \sup_\nu \left(h_\nu(T) + \int_X \varphi\, d\nu \right).$$

This completes the proof of the theorem.   □

We note that identity (5.8) can be used as an alternative definition of the topological pressure. Setting $\varphi = 0$, one recovers the variational principle for the topological entropy in Theorem 4.7.

5.5 Equilibrium Measures

We consider in this section the class of measures at which the supremum in (5.8) is attained. Let $T: X \to X$ be a continuous transformation of a compact metric space.

Definition 5.3. Given a continuous function $\varphi: X \to \mathbb{R}$, a T-invariant probability measure μ in X is called an *equilibrium measure* for φ (with respect to T) if

$$P_T(\varphi) = h_\mu(T) + \int_X \varphi\, d\mu.$$

We give an example in the particular case of symbolic dynamics.

Example 5.4. Given numbers $\lambda_1, \ldots, \lambda_k > 0$, we consider the continuous function $\varphi: X_k^+ \to \mathbb{R}$ defined by (5.5). By (5.6), its topological pressure is given by

$$P_\sigma(\varphi) = \log \sum_{j=1}^{k} \lambda_j.$$

Thus, it follows from the variational principle in Theorem 5.2 that

$$\log \sum_{j=1}^{k} \lambda_j = \sup_\mu \left\{ h_\mu(\sigma) + \int_{X_k^+} \varphi \, d\mu \right\}$$

$$= \sup_\mu \left\{ h_\mu(\sigma) + \sum_{i=1}^{k} \mu(C_i) \log \lambda_i \right\}, \tag{5.16}$$

where the supremum is taken over all σ-invariant probability measures μ in X_k^+ and where $C_1, \ldots, C_k$ are cylinder sets.

Now let μ be the (one-sided) Bernoulli measure generated by the numbers $p_1, \ldots, p_k$. By (4.26), we have

$$h_\mu(\sigma) = - \sum_{i=1}^{k} p_i \log p_i.$$

Hence, by Lemma 5.1 (or again by Theorem 5.2), we obtain

$$h_\mu(\sigma) + \int_{X_k^+} \varphi \, d\mu = \sum_{i-1}^{k} p_i (- \log p_i + \log \lambda_i)$$

$$\leq \log \sum_{i=1}^{k} \lambda_i = P_\sigma(\varphi). \tag{5.17}$$

It also follows from Lemma 5.1 that we have an equality in (5.17) if and only if $p_i = \lambda_i / \sum_{i=1}^{k} \lambda_i$ for $i = 1, \ldots, k$. In particular, this shows that the suprema in (5.16) are attained at the Bernoulli measure μ generated by these numbers. In other words, this is an equilibrium measure for the function φ, and no other Bernoulli measure is an equilibrium measure.

Now we consider a particular class of transformations.

Definition 5.4. A transformation $T: X \to X$ is said to be *one-sided expansive* if there exists $\varepsilon > 0$ such that if

$$d(T^n(x), T^n(y)) < \varepsilon \quad \text{for every} \quad n \in \mathbb{N} \cup \{0\}, \tag{5.18}$$

then $x = y$.

Example 5.5. No isometry T is one-sided expansive, since it satisfies

$$d(T^n(x), T^n(y)) = d(x, y) \quad \text{for every} \quad n \in \mathbb{N}.$$

Example 5.6. For the shift map $\sigma: X_k^+ \to X_k^+$, we note that if

$$d(\sigma^n(\omega), \sigma^n(\omega')) < 1 \quad \text{for every} \quad n \in \mathbb{N} \cup \{0\},$$

then it follows readily from (3.32) that $\omega = \omega'$. Therefore, the shift map in X_k^+ is one-sided expansive.

Example 5.7. Let us consider the expanding map $E_q: S^1 \to S^1$. If $d(z, w) < 1/q^2$ with $z \neq w$ (where d is the distance in S^1), then there exists $n \in \mathbb{N}$ such that

$$d(E_q^n(z), E_q^n(w)) = q^n d(z, w) \geq \frac{1}{q^2}.$$

This implies that if

$$d(E_q^n(z), E_q^n(w)) < \frac{1}{q^2} \quad \text{for any} \quad n \geq 0,$$

then $z = w$, and thus, the expanding map E_q is expansive.

The following statement establishes the existence of equilibrium measures for any one-sided expansive transformation:

Theorem 5.3. *If $T: X \to X$ is a one-sided expansive continuous transformation of a compact metric space, then any continuous function $\varphi: X \to \mathbb{R}$ has at least one equilibrium measure.*

Proof. For ε as in Definition 5.4, let η be a measurable partition of X with diameter diam $\eta < \varepsilon$. We show that the partitions $\eta_n = \bigvee_{k=0}^{n-1} T^{-k} \eta$ satisfy

$$\text{diam } \eta_n \to 0 \quad \text{when} \quad n \to \infty. \tag{5.19}$$

Otherwise, there would exist $\delta > 0$, an increasing sequence $(n_p)_{p \in \mathbb{N}} \subset \mathbb{N}$, and points x_p and y_p for each $p \in \mathbb{N}$ such that

$$d(x_p, y_p) \geq \delta \quad \text{and} \quad x_p, y_p \in \bigcap_{k=0}^{n_p-1} T^{-k} C_{pk}$$

for some sets $C_{pk} \in \eta$. Since X is compact, we can also assume that $x_p \to x$ and $y_p \to y$ when $p \to \infty$ for some points $x, y \in X$. Clearly, $d(x, y) \geq \delta$. Since η is finite, for each k infinitely many sets C_{pk} coincide, say with some $D_k \in \eta$. Therefore, $x_p, y_p \in T^{-k} D_k$ for infinitely many integers p, and hence, $x, y \in T^{-k} \overline{D_k}$. This shows that (5.18) holds, and since T is one-sided expansive, we conclude that $x = y$. But this contradicts the inequality $d(x, y) \geq \delta$. We have thus established (5.19). This implies that η is a one-sided generator (see Definition 4.7), and hence, it follows from Theorem 4.3 that

$$h_\mu(T) = h_\mu(T, \eta). \tag{5.20}$$

Now we show that the transformation $\mu \mapsto h_\mu(T)$ is upper semicontinuous in the set $\mathcal{M}(X)$ of all Borel probability measures in X. This means that given a T-invariant measure $\mu \in \mathcal{M}(X)$ and $\delta > 0$, we have $h_\nu(T) < h_\mu(T) + \delta$ for any T-invariant measure $\nu \in \mathcal{M}(X)$ in some open neighborhood of μ, with respect to the distance d in (3.2). Since the transformation $\mu \mapsto \int_X \varphi\, d\mu$ is continuous for each given continuous function $\varphi \colon X \to \mathbb{R}$, this shows that

$$\mu \mapsto h_\mu(T) + \int_X \varphi\, d\mu$$

is upper semicontinuous. Since an upper semicontinuous function in a compact space has a maximum, it follows from Theorem 5.2 that each continuous function φ has an equilibrium measure.

We proceed with the proof of the upper semicontinuity. Let μ be a T-invariant measure in X. Let also $\xi = \{C_1, \ldots, C_k\}$ be a measurable partition of X with diam $\xi < \varepsilon$. Given $\delta > 0$, take $n \in \mathbb{N}$ such that

$$\frac{1}{n} H_\mu(\xi_n) < h_\mu(T) + \delta, \tag{5.21}$$

where $\xi_n = \bigvee_{j=0}^{n-1} T^{-j}\xi$. Given $\alpha > 0$, for each $i_1, \ldots, i_n \in \{1, \ldots, k\}$, let

$$K_{i_1 \cdots i_n} \subset \bigcap_{j=0}^{n-1} T^{-j} C_{i_{j+1}}$$

be a compact set with

$$\mu\left(\bigcap_{j=0}^{n-1} T^{-j} C_{i_{j+1}} \setminus K_{i_1 \cdots i_n}\right) < \alpha. \tag{5.22}$$

Now we consider the sets

$$E_i := \bigcup_{j=0}^{n-1} \bigcup_{i_j = i} T^j(K_{i_1 \cdots i_n}) \subset C_i,$$

for $i = 1, \ldots, k$. Since these are pairwise disjoint compact sets, there exists a measurable partition $\eta = \{D_1, \ldots, D_k\}$ of X with diam $\eta < \varepsilon$ such that $E_i \subset \operatorname{int} D_i$ for $i = 1, \ldots, k$. Clearly,

$$K_{i_1 \cdots i_n} \subset \operatorname{int} \bigcap_{j=0}^{n-1} T^{-j} D_{i_{j+1}}.$$

By Urysohn's lemma, for each $i_1, \ldots, i_n \in \{1, \ldots, k\}$, there exists a continuous function $\varphi_{i_1 \cdots i_n} : X \to [0, 1]$ that is 0 on $X \setminus \mathrm{int} \bigcap_{j=0}^{n-1} T^{-j} D_{i_{j+1}}$ and 1 on $K_{i_1 \cdots i_n}$. Now we consider the set $V_{i_1 \cdots i_n}$ of all T-invariant measures $v \in \mathcal{M}(X)$ such that

$$\left| \int_X \varphi_{i_1 \cdots i_n} \, dv - \int_X \varphi_{i_1 \cdots i_n} \, d\mu \right| < \alpha.$$

We note that $V_{i_1 \cdots i_n}$ is an open neighborhood of μ_i. Then

$$v \left(\bigcap_{j=0}^{n-1} T^{-j} D_{i_{j+1}} \right) \geq \int_X \varphi_{i_1 \cdots i_n} \, dv$$

$$> \int_X \varphi_{i_1 \cdots i_n} \, d\mu - \alpha$$

$$\geq \mu(K_{i_1 \cdots i_n}) - \alpha.$$

By (5.22), this implies that

$$\mu \left(\bigcap_{j=0}^{n-1} T^{-j} C_{i_{j+1}} \right) - v \left(\bigcap_{j=0}^{n-1} T^{-j} D_{i_{j+1}} \right) < 2\alpha. \qquad (5.23)$$

Now let $V = \bigcap_{i_1 \cdots i_n} V_{i_1 \cdots i_n}$. For each $v \in U$ and $i_1, \ldots, i_n \in \{1, \ldots, k\}$, since

$$\sum_{\ell_1 \cdots \ell_n} v \left(\bigcap_{j=0}^{n-1} T^{-j} D_{\ell_{j+1}} \right) = \sum_{\ell_1 \cdots \ell_n} \mu \left(\bigcap_{j=0}^{n-1} T^{-j} C_{\ell_{j+1}} \right) = 1,$$

we have

$$v \left(\bigcap_{j=0}^{n-1} T^{-j} D_{i_{j+1}} \right) - \mu \left(\bigcap_{j=0}^{n-1} T^{-j} C_{i_{j+1}} \right)$$

$$= \sum_{(\ell_1 \cdots \ell_n) \neq (i_1 \cdots i_n)} \left[\mu \left(\bigcap_{j=0}^{n-1} T^{-j} C_{\ell_{j+1}} \right) - v \left(\bigcap_{j=0}^{n-1} T^{-j} D_{\ell_{j+1}} \right) \right] \leq 2\alpha k^n.$$

Together with (5.23), this implies that

$$\left| v \left(\bigcap_{j=0}^{n-1} T^{-j} D_{i_{j+1}} \right) - \mu \left(\bigcap_{j=0}^{n-1} T^{-j} C_{i_{j+1}} \right) \right| \leq 2\alpha k^n.$$

Therefore, provided that α is sufficiently small, we obtain

$$\frac{1}{n}H_\nu(\eta_n) \le \frac{1}{n}H_\mu(\xi_n) + \delta,$$

and by (4.8) and (5.21), we conclude that

$$h_\nu(T) = h_\nu(T, \eta) \le \frac{1}{n}H_\nu(\eta_n)$$

$$\le \frac{1}{n}H_\mu(\xi_n) + \delta \le h_\mu(T) + 2\delta.$$

This completes the proof of the theorem. □

The following is an immediate consequence of Theorem 5.3:

Theorem 5.4. *If $T: X \to X$ is a one-sided expansive continuous transformation of a compact metric space, then there exists a T-invariant probability measure μ in X with $h_\mu(T) = h(T)$.*

Proof. Setting $\varphi = 0$ in Theorem 5.3 yields the existence of a T-invariant probability measure μ in X such that

$$P_T(0) = h_\nu(T) + \int_X 0 \, d\mu = h_\mu(T).$$

The desired result follows now from (5.2). □

In the case of homeomorphisms, we consider a weaker notion of expansivity.

Definition 5.5. An invertible transformation $T: X \to X$ is said to be *two-sided expansive* if there exists $\varepsilon > 0$ such that if

$$d(T^n(x), T^n(y)) < \varepsilon \quad \text{for every} \quad n \in \mathbb{Z},$$

then $x = y$.

Example 5.8. The shift map $\sigma: X_k \to X_k$ is invertible. We note that if

$$d(\sigma^n(\omega), \sigma^n(\omega')) < 1 \quad \text{for every} \quad n \in \mathbb{Z},$$

then it follows readily from (3.37) that $\omega = \omega'$. Therefore, the shift map in X_k is two-sided expansive.

The following statement establishes the existence of equilibrium measures for any two-sided expansive homeomorphism:

Theorem 5.5. *If $T: X \to X$ is a two-sided expansive homeomorphism of a compact metric space, then any continuous function $\varphi: X \to \mathbb{R}$ has at least one equilibrium measure.*

Proof. For ε as in Definition 5.5, let η be a finite measurable partition of X with diam $\eta < \varepsilon$. We can show in a similar manner to that in the proof of Theorem 5.3 that the partitions $\eta'_n = \bigvee_{k=-n}^{n} T^{-k}\eta$ satisfy diam $\eta'_n \to 0$ when $n \to \infty$. This implies that η is a two-sided generator (see Definition 4.7), and thus, it follows from Theorem 4.3 that (5.20) holds. This allows us to repeat arguments in the proof of Theorem 5.3 to show that in this new situation, the transformation $\mu \mapsto h_\mu(T)$ is also upper semicontinuous. Therefore, for each continuous function $\varphi: X \to \mathbb{R}$, the transformation $\mu \mapsto h_\mu(T) + \int_X \varphi \, d\mu$ is upper semicontinuous, and hence, φ has at least one equilibrium measure. $\square$

Similarly, the following is an immediate consequence of Theorem 5.5:

Theorem 5.6. *If $T: X \to X$ is a two-sided expansive homeomorphism of a compact metric space, then there exists a T-invariant probability measure μ in X with $h_\mu(T) = h(T)$.*

5.6   Exercises

Exercise 5.1. Let $T: X \to X$ be a continuous transformation of a compact metric space, and let $\varphi, \psi: X \to \mathbb{R}$ be continuous functions. Show that:

1. If $\varphi \le \psi$, then $P_T(\varphi) \le P_T(\psi)$.
2. If $P_T(\varphi)$ and $P_T(\psi)$ are finite, then

$$|P_T(\varphi) - P_T(\psi)| \le \|\varphi - \psi\|_\infty.$$

3. $P_T(\varphi + \psi \circ T - \psi) = P_T(\varphi)$.

Exercise 5.2. Let $T: X \to X$ be a continuous transformation of a compact metric space and let $\varphi: X \to \mathbb{R}$ be a continuous function. Show that $P_{T^n}(\varphi_n) - nP_T(\varphi)$ for every $n \in \mathbb{N}$, where $\varphi_n = \sum_{k=0}^{n-1} \varphi \circ T^k$.

Exercise 5.3. Let $T: X \to X$ be a homeomorphism of a compact metric space. Show that $P_T(\varphi) = P_{T^{-1}}(\varphi)$ for every continuous function $\varphi: X \to \mathbb{R}$.

Exercise 5.4. Given a continuous transformation $T: X \to X$ of a compact metric space and a continuous function $\varphi: X \to \mathbb{R}$, for each $n \in \mathbb{N}$ and $\varepsilon > 0$, let

$$R_n(\varphi, \varepsilon) = \inf_{\mathcal{V}} \sum_{V \in \mathcal{V}} \exp \inf_V \sum_{k=0}^{n-1} \varphi \circ T^k$$

and

$$S_n(\varphi, \varepsilon) = \inf_{\mathcal{V}} \sum_{V \in \mathcal{V}} \exp \sup_V \sum_{k=0}^{n-1} \varphi \circ T^k,$$

where the infimum in $\mathcal{V}$ is taken over all finite open covers $\mathcal{V}$ of X by d_n-balls of radius ε. Show that

$$P_T(\varphi) = \lim_{\varepsilon \to 0} \limsup_{n \to \infty} \frac{1}{n} \log R_n(\varphi, \varepsilon)$$

$$= \lim_{\varepsilon \to 0} \limsup_{n \to \infty} \frac{1}{n} \log S_n(\varphi, \varepsilon).$$

Exercise 5.5. Show that if $\varphi: X_k^+ \to \mathbb{R}$ is a continuous function, then

$$P_\sigma(\varphi) = \lim_{n \to \infty} \frac{1}{n} \log \sum_{i_1 \cdots i_n} \exp \inf_{C_{i_1 \cdots i_n}} \varphi_n,$$

where $\varphi_n = \sum_{k=0}^{n-1} \varphi \circ \sigma^k$. Hint: note that

$$\sum_{i_1 \cdots i_{m+n-1}} \inf_{C_{i_1 \cdots i_{m+n-1}}} \varphi_n \geq \sum_{i_1 \cdots i_n} \exp \inf_{C_{i_1 \cdots i_n}} \varphi_n,$$

and

$$\sum_{i_1 \cdots i_{m+n-1}} \inf_{C_{i_1 \cdots i_{m+n-1}}} \varphi_n \leq e^{(m-1)\|\varphi\|_\infty} \sum_{i_1 \cdots i_{m+n-1}} \exp \inf_{C_{i_1 \cdots i_{m+n-1}}} \varphi_{m+n-1}.$$

Exercise 5.6. Show that if $\varphi: X_k^+ \to \mathbb{R}$ is a continuous function, then

$$P_\sigma(\varphi) = \lim_{n \to \infty} \frac{1}{n} \log \sum_{\omega \in P_n} \exp \sum_{k=0}^{n-1} \varphi(\sigma^k(\omega)),$$

where

$$P_n = \{\omega \in \Sigma_A^+ : \sigma^n(\omega) = \omega\}$$

is the set of n-periodic points.

Exercise 5.7. Show that if $\varphi: X_k \to \mathbb{R}$ is a continuous function, then

$$P_\sigma(\varphi) = \lim_{n \to \infty} \frac{1}{n} \log \sum_{i_1 \cdots i_n} \exp \sup_{C_{i_1 \cdots i_n}} \sum_{k=0}^{n-1} \varphi \circ \sigma^k.$$

Exercise 5.8. Given a $k \times k$ matrix A with entries in $\{0, 1\}$, show that:

1. If $\varphi: X_A^+ \to \mathbb{R}$ is a continuous function, then

$$P_{\sigma|X_A^+}(\varphi) = \lim_{n \to \infty} \frac{1}{n} \log \sum_{i_1 \cdots i_n} \exp \sum_{k=0}^{n-1} \varphi \circ \sigma^k,$$

where the supremum is taken over all finite sequences $i_1 \cdots i_n$ that are the first n elements of some sequence in X_A^+.

2. If $\varphi \colon X_A \to \mathbb{R}$ is a continuous function, then

$$P_{\sigma|X_A}(\varphi) = \lim_{n \to \infty} \frac{1}{n} \log \sum_{i_1 \cdots i_n} \exp \sum_{k=0}^{n-1} \varphi \circ \sigma^k,$$

where the supremum is taken over all finite sequences $i_1 \cdots i_n$ that are the elements $i_1(\omega) \cdots i_n(\omega)$ of some sequence $\omega \in X_A$.

Exercise 5.9. Show that if $\varphi_n \colon X_k^+ \to \mathbb{R}$ are continuous functions such that

$$\varphi_{n+m} \le \varphi_n + \varphi_m \circ \sigma^n$$

for every $m, n \in \mathbb{N}$, then there exists the limit

$$p = \lim_{n \to \infty} \frac{1}{n} \log \sum_{i_1 \cdots i_n} \exp \sup_{C_{i_1 \cdots i_n}} \varphi_n.$$

Exercise 5.10. Given numbers $\lambda_1, \ldots, \lambda_k > 0$, show that the continuous function $\varphi \colon X_A^+ \to \mathbb{R}$ defined by (5.5) has topological pressure

$$P_{\sigma|X_A^+}(\varphi) = \log \rho(AB),$$

where B is the $k \times k$ diagonal matrix with entries $\lambda_1, \ldots, \lambda_k$ in the diagonal.

Exercise 5.11. Show that if a continuous transformation $T \colon X \to X$ is uniquely ergodic and μ is the unique T-invariant probability measure in X, then

$$P_T(\varphi) = h_\mu(T) + \int_X \varphi \, d\mu$$

for every continuous function $\varphi \colon X \to \mathbb{R}$.

Exercise 5.12. Let $T \colon X \to X$ be a continuous transformation of a compact metric space. Show that if $h(T) = \infty$, then there is a T-invariant probability measure μ in X with $h_\mu(T) = \infty$. Hint: take T-invariant probability measures μ_n with $h_{\mu_n}(T) > 2^n$ and consider the measure

$$\mu = \sum_{n=1}^{\infty} \frac{1}{2^n} \mu_n.$$

Exercise 5.13. Show that any toral automorphism induced by a matrix without eigenvalues in S^1 is two-sided expansive.

Exercise 5.14. Prove or disprove the following statement: any toral automorphism as in Exercise 5.13 is one-sided expansive.

Exercise 5.15. Let $T: X \to X$ be a continuous transformation of a compact metric space. Show that if T is one-sided expansive, then its topological entropy is given by

$$h(T) = \lim_{n \to \infty} \frac{1}{n} \log N(d_n, \delta)$$

$$= \lim_{n \to \infty} \frac{1}{n} \log M(d_n, \delta)$$

$$= \lim_{n \to \infty} \frac{1}{n} \log D(d_n, \delta)$$

for any sufficiently small $\delta > 0$, with $D(d_n, \delta)$ as in (4.39) and $M(d_n, \delta)$ as in Exercise 4.17. Hint: show that given positive numbers $\delta < \alpha < \varepsilon$, with ε as in Definition 5.4, there exists $m = m(\delta, \alpha) \in \mathbb{N}$ such that if $d(x, y) \geq \delta$, then

$$d(f^i(x), f^i(y)) > \alpha \quad \text{for some} \quad i \in \{0, \dots, m\}$$

and conclude that $N(d_n, \delta) \leq N(d_{n+2m}, \alpha)$.

Exercise 5.16. Show that if $T: X \to X$ is a two-sided expansive homeomorphism of a compact metric space, with

$$a_n := \text{card}\left\{x \in X : T^n(x) = x\right\} < \infty$$

for every $n \in \mathbb{N}$, then

$$h(T) \geq \limsup_{n \to \infty} \frac{1}{n} \log a_n.$$

Exercise 5.17. Given a continuous transformation $T: X \to X$ of a compact metric space and continuous functions $\varphi, \psi: X \to \mathbb{R}$, show that

$$\liminf_{t \to 0} \frac{P_T(\varphi + t\psi) - P_T(\varphi)}{t} \geq \int_X \psi \, d\mu,$$

where μ is any equilibrium measure for φ.

Exercise 5.18. Let $T: X \to X$ be a continuous transformation of a compact metric space with $h(T) < \infty$. Show that if μ is a T-invariant probability measure in X with

$$h_\mu(T) = \inf_\varphi \left\{ P_T(\varphi) - \int_X \varphi \, d\mu \right\},$$

where the infimum is taken over all continuous functions $\varphi: X \to \mathbb{R}$, then the map $\nu \mapsto h_\nu(T)$ is upper semicontinuous at $\nu = \mu$.

Notes

The notion of topological pressure was introduced by Ruelle [83] for expansive transformations and by Walters [102] in the general case. They also established corresponding versions of the variational principle for the topological pressure (Theorem 5.2). Our proof of Theorem 5.2 is based on [103], which follows the simpler proof of Misiurewicz [61]. Theorem 5.3 is due to Ruelle [83] (for $\varphi = 0$, the statement was first established by Goodman [35]). The argument for the upper semicontinuity of the entropy in the proof of Theorem 5.3 is based on [103]. Ruelle's book [84] contains a detailed discussion of the relation between the thermodynamic formalism and the theory of dynamical systems. For further developments, we refer to [44, 45, 70, 103].

Part III
Hyperbolic Dynamics

Chapter 6
Basic Notions and Examples

We introduce in this chapter the basic notions of hyperbolic dynamics, starting with the concept of hyperbolicity. We also establish several basic properties of hyperbolic sets, including the continuous dependence of the stable and unstable subspaces on the base point. In addition, we discuss several examples of hyperbolic sets. These include hyperbolic fixed points, the Smale horseshoe, and hyperbolic automorphisms of the 2-torus. We also construct coding maps via symbolic dynamics. Finally, we consider noninvertible transformations and their repellers, and we construct corresponding Markov partitions. We refer to the following chapter for the more elaborate construction of Markov partitions for hyperbolic sets.

6.1 Introduction

The study of hyperbolicity goes back to seminal work of Hadamard on the geodesic flow in the unit tangent bundle of a surface with negative curvature, in particular revealing its instability with respect to initial conditions. In the case of constant negative curvature, the geodesic flow can be described as follows. Consider the upper-half plane

$$H = \{z \in \mathbb{C} : \Im z > 0\},$$

with the inner product in the tangent space $T_z H = \mathbb{C}$ given by

$$\langle v, w \rangle_z = \frac{\langle v, w \rangle}{(\Im z)^2},$$

where $\langle v, w \rangle$ is the standard inner product in $\mathbb{R}^2$. Consider also the group G of matrices $A = \left(\begin{smallmatrix} a & b \\ c & d \end{smallmatrix} \right)$ with real entries and determinant 1 or -1 and define Möbius transformations T_A in H by

L. Barreira, *Ergodic Theory, Hyperbolic Dynamics and Dimension Theory*, Universitext, DOI 10.1007/978-3-642-28090-0_6, © Springer-Verlag Berlin Heidelberg 2012

$$T_A(z) = \frac{az + b}{cz + d} \quad \text{or} \quad T_A(z) = \frac{a\bar{z} + b}{c\bar{z} + d},$$

respectively, when det A is 1 or -1. Then $G/\{\mathrm{Id}, -\mathrm{Id}\}$ is the group of isometries of H. Now take

$$(z, v) \in SH = \{(z, v) \in H \times \mathbb{C} : |v|_z = 1\}.$$

One can show that there exists a Möbius transformation T such that $T(z) = i$ and $T'(z)v = i$, thus taking the geodesic passing through z with tangent v onto the geodesic ie^t traversing the positive part of the imaginary axis. The geodesic flow $\varphi_t : SH \to SH$ is given by

$$\varphi_t(z, v) = (\gamma(t), \gamma'(t)),$$

where $\gamma(t) = T^{-1}(ie^t)$. One can show that the geodesic flow preserves volume and thus it also exhibits a nontrivial recurrence. The case of nonconstant negative curvature is more elaborate, but the geodesic flow still preserves volume (in fact, any geodesic flow is a Hamiltonian flow). A considerable activity took place during the 1920s and 1930s in particular with the important contributions of Hedlund and Hopf who established several topological and ergodic properties of geodesic flows.

Moreover, the geodesic flow on a surface with negative curvature is hyperbolic. Hyperbolicity corresponds to the existence of complementary transverse subspaces, called stable and unstable, exhibiting, respectively, expansion and contraction. More precisely, the expansion and contraction is required for the linear maps approximating the dynamics. Hyperbolicity gives rise to a very rich structure and in particular to the existence of stable and unstable manifolds: it follows from a substantial generalization of the Hadamard–Perron theorem that for every point x in a hyperbolic set of a C^1 diffeomorphism f and any sufficiently small $\varepsilon > 0$, the sets

$$V^s(x) = \{y \in B(x, \varepsilon) : d(f^n(x), f^n(y)) < \varepsilon \text{ for every } n \geq 0\}$$

and

$$V^u(x) = \{y \in B(x, \varepsilon) : d(f^n(x), f^n(y)) < \varepsilon \text{ for every } n \leq 0\}$$

are invariant manifolds tangent, respectively, to the stable and unstable subspaces.

There is a corresponding theory for noninvertible transformations, in which case the notion of hyperbolic set is replaced by the notion of repeller, with the dynamics exhibiting only expansion. While there are some important differences between the two theories, it is sometimes simpler to first consider some notions for repellers and then consider appropriate elaborations for hyperbolic sets. This is the case for example with the construction of Markov partitions for repellers. The corresponding construction for hyperbolic sets is an elaboration of this approach.

6.2 Hyperbolic Sets

We introduce in this section the notion of hyperbolicity. We also discuss some of the basic properties of hyperbolic sets.

6.2.1 The Notion of Hyperbolicity

Let $f : M \to M$ be a C^1 diffeomorphism of a smooth manifold M. For each point $x \in M$, we consider the inner product $\langle \cdot, \cdot \rangle_x$ and the corresponding norm $\|\cdot\|_x$ in the tangent space $T_x M$. Whenever there is no danger of confusion, we simply write $\langle \cdot, \cdot \rangle$ and $\|\cdot\|$, without making explicit the dependence on x.

Definition 6.1. A compact f-invariant set $\Lambda \subset M$ is said to be a *hyperbolic set* for f if there exist $\tau \in (0, 1)$, $c > 0$ and a decomposition

$$T_x M = E^s(x) \oplus E^u(x) \tag{6.1}$$

for each $x \in \Lambda$ such that

$$d_x f E^s(x) = E^s(f(x)), \quad d_x f E^u(x) = E^u(f(x)), \tag{6.2}$$

$$\|d_x f^n v\| \le c\tau^n \|v\| \quad \text{whenever} \quad v \in E^s(x),$$

and

$$\|d_x f^{-n} v\| \le c\tau^n \|v\| \quad \text{whenever} \quad v \in E^u(x)$$

for every $x \in \Lambda$ and $n \in \mathbb{N}$. We then call $E^s(x)$ and $E^u(x)$, respectively, the *stable* and *unstable subspaces* at x.

As first examples, we consider fixed points and periodic points.

Definition 6.2. Let f be a diffeomorphism.

1. A fixed point $x = f(x)$ of f is said to be *hyperbolic* if $\{x\}$ is a hyperbolic set.
2. An m-periodic point $x = f^m(x)$ of f is said to be *hyperbolic* if its orbit

$$\mathcal{O}_f(x) = \{ f^k(x) : k = 0, \ldots, m - 1 \}$$

is a hyperbolic set.

We denote by $\mathrm{Sp}(A)$ the *spectrum* of a square matrix A, that is, the set of its eigenvalues. Note that when x is a fixed point of f, we have $d_x f(T_x M) \subset T_x M$, and thus, the symbol $\mathrm{Sp}(d_x f)$ is well defined. The following is a characterization of the hyperbolicity of a fixed point x in terms of the spectrum of $d_x f$:

Proposition 6.1. *A fixed point x of a diffeomorphism f is hyperbolic if and only if* $\mathrm{Sp}(d_x f) \cap S^1 = \varnothing$.

Proof. We consider the complexification

$$T_x M^{\mathbb{C}} = \{u + iv : u, v \in T_x M\}$$

of $T_x M$, equipped with the norm

$$\|u + iv\| = \sqrt{\|u\|^2 + \|v\|^2}, \quad u, v \in T_x M,$$

and the linear operator $A : T_x M^{\mathbb{C}} \to T_x M^{\mathbb{C}}$ defined by

$$A(u + iv) = d_x f u + i d_x f v \quad \text{for each} \quad u, v \in T_x M.$$

We first assume that $\{x\}$ is a hyperbolic set. If $v \in T_x M^{\mathbb{C}}$ and $\rho \in S^1$ are such that $Av = \rho v$, then

$$\|A^n v\| = |\rho|^n \|v\| = \|v\|. \tag{6.3}$$

Writing $v = v^s + v^u$, where $v^s \in E^s(x)^{\mathbb{C}}$ and $v^u \in E^u(x)^{\mathbb{C}}$, we obtain

$$\begin{aligned}
\|v\| &\geq \|A^n v^u\| - \|A^n v^s\| \\
&\geq c^{-1} \tau^{-n} \|v^u\| - c \tau^n \|v^s\|.
\end{aligned} \tag{6.4}$$

Letting $n \to \infty$ in (6.4) yields $v^u = 0$, and thus, $v \in E^s(x)^{\mathbb{C}}$. Then it follows from (6.3) that $v = 0$. This shows that ρ is not an eigenvalue of $d_x f$, and thus, $\mathrm{Sp}(d_x f) \cap S^1 = \varnothing$.

Now we assume that $d_x f$ has no eigenvalues in S^1. Let $F^s \subset T_x M^{\mathbb{C}}$ be the linear subspace generated by all vectors $v \in T_x M^{\mathbb{C}}$ satisfying

$$(A - \tau \, \mathrm{Id})^k v = 0 \tag{6.5}$$

for some $k \in \mathbb{N}$ and some $\tau \in \mathrm{Sp}(d_x f)$ with $|\tau| < 1$. Similarly, let $F^u \subset T_x M^{\mathbb{C}}$ be the linear subspace generated by all vectors $v \in T_x M^{\mathbb{C}}$ satisfying (6.5) for some $k \in \mathbb{N}$ and some $\tau \in \mathrm{Sp}(d_x f)$ with $|\tau| > 1$. We can easily verify that

$$A F^s \subset F^s \quad \text{and} \quad A F^u \subset F^u. \tag{6.6}$$

Indeed, it is sufficient to note that

$$(A - \tau \, \mathrm{Id})^k A v = A (A - \tau \, \mathrm{Id})^k v.$$

Moreover, since $d_x f$ has no eigenvalues in S^1, we have $T_x M^{\mathbb{C}} = F^s \oplus F^u$. Therefore, setting

$$E^s(x) = F^s \cap T_x M \quad \text{and} \quad E^u(x) = F^u \cap T_x M,$$

we obtain

$$T_x M = T_x M^{\mathbb{C}} \cap T_x M = E^s(x) \oplus E^u(x),$$

and it follows from (6.6) that

$$d_x f E^s(x) = E^s(x) \quad \text{and} \quad d_x f E^u(x) = E^u(x).$$

Hence, property (6.2) holds. Now we take $\rho \in (0, 1)$ such that

$$(\rho, 1/\rho) \cap \{|\tau| : \tau \in \mathrm{Sp}(d_x f)\} = \varnothing.$$

Using Jordan's canonical form, it is easy to verify that there is a constant $c > 0$ such that

$$\|d_x f^n v\| \le c\rho^n \|v\| \quad \text{for} \quad v \in E^s(x), \, n \in \mathbb{N},$$

and

$$\|d_x f^{-n} v\| \le c\rho^n \|v\| \quad \text{for} \quad v \in E^u(x), \, n \in \mathbb{N}.$$

Therefore, $\{x\}$ is a hyperbolic set. $\qquad\square$

Now we describe an example where the hyperbolic set is the whole manifold.

Example 6.1. Let $T: \mathbb{T}^2 \to \mathbb{T}^2$ be the toral automorphism induced by the matrix $B = \left(\begin{smallmatrix} 2 & 1 \\ 1 & 1 \end{smallmatrix}\right)$, which has eigenvalues

$$\tau = (3 + \sqrt{5})/2 > 1 \quad \text{and} \quad \tau^{-1} = (3 - \sqrt{5})/2 < 1,$$

both outside S^1. We denote respectively by F^u and F^s the eigenspaces of B corresponding to the eigenvalues τ and τ^{-1}. Since $d_x T = B$ for every $x \in \mathbb{T}^2$, we obtain

$$\|d_x T^n v\| = \tau^n \|v\| \quad \text{for} \quad v \in F^u, \, n \in \mathbb{N},$$

and

$$\|d_x T^n v\| = \tau^{-n} \|v\| \quad \text{for} \quad v \in F^s, \, n \in \mathbb{N}.$$

Furthermore, $F^u \oplus F^s = \mathbb{R}^2 = T_x \mathbb{T}^2$,

$$d_x T F^s = F^s, \quad \text{and} \quad d_x T F^u = F^u.$$

Therefore, $\mathbb{T}^2$ is a hyperbolic set for T.

This example can be readily generalized to the class of toral automorphisms induced by matrices without eigenvalues in S^1.

Definition 6.3. We say that a square matrix B is *hyperbolic* if $\mathrm{Sp}(B) \cap S^1 = \varnothing$, and we say that a toral automorphism of $\mathbb{T}^n$ is *hyperbolic* if $\mathbb{T}^n$ is a hyperbolic set for the automorphism.

6.2.2   Some Properties of Hyperbolic Sets

We study in this section some elementary properties of hyperbolic sets. We start with the study of the dependence of the stable and unstable subspaces on the base point. We denote by $\angle(E, F)$ the angle between two subspaces E and F.

Proposition 6.2. *If Λ is a hyperbolic set, then the following properties hold:*

1. *The spaces $E^s(x)$ and $E^u(x)$ vary continuously with $x \in \Lambda$.*
2. $\inf\{\angle(E^s(x), E^u(x)) : x \in \Lambda\} > 0$.

Proof. Let $(x_p)_{p \in \mathbb{N}} \subset \Lambda$ be a sequence converging to $x \in \Lambda$ when $p \to \infty$. We consider a subsequence $(y_p)_{p \in \mathbb{N}}$ such that the numbers $\dim E^s(y_p)$ and $\dim E^u(y_p)$ are independent of p (this is always possible because the dimension of the subspaces can only take finitely many values). Now let $v_{1p}, \ldots, v_{kp} \in E^s(y_p)$ be an orthonormal basis of $E^s(y_p)$, where $k = \dim E^s(y_p)$. By the compactness of the unit tangle bundle $S_\Lambda M$, we can also assume that for each $i = 1, \ldots, k$, there exists $v_i \in S_x M$ such that $v_{ip} \to v_i$ when $p \to \infty$. On the other hand, for $i = 1, \ldots, k$ and $n \in \mathbb{N}$, we have

$$\|d_{y_p} f^n v_{ip}\| \le c\tau^n \|v_{ip}\|.$$

Letting $p \to \infty$, we conclude that

$$\|d_x f^n v_i\| \le c\tau^n \|v_i\|,$$

which implies that $v_1, \ldots, v_k \in E^s(x)$, and hence, $\dim E^s(x) \ge k$. One can show in a similar manner that $\dim E^u(x) \ge \dim M - k$. It follows from (6.1) that

$$\dim E^s(x) = k \quad \text{and} \quad \dim E^u(x) = \dim M - k.$$

Therefore, for any sufficiently large p, the numbers $\dim E^s(x_p)$ and $\dim E^u(x_p)$ are independent of p. This completes the proof of Property 1.

For the second property, we note that the function

$$x \mapsto \angle(E^s(x), E^u(x)) \in (0, \pi/2]$$

is the composition of the continuous functions

$$(E, F) \mapsto \angle(E, F) \quad \text{and} \quad x \mapsto E^s(x) \oplus E^u(x).$$

Therefore, Property 2 follows readily from the compactness of Λ. $\square$

Now we show that it is always possible to redefine the inner product so that one can take $c = 1$ in Definition 6.1.

Proposition 6.3. *If Λ is a hyperbolic set for a diffeomorphism f, then there exists a inner product $\langle \cdot, \cdot \rangle'$ in TM with respect to which the following properties hold:*

1. $\angle'(E^s(x), E^u(x)) = \pi/2$ *for every* $x \in \Lambda$.
2. *There exists $\rho \in (\tau, 1)$ such that for each $x \in \Lambda$, we have*

$$\|d_x f v\|' \le \rho \|v\|' \quad \text{whenever} \quad v \in E^s(x),$$

and

$$\|d_x f^{-1} v\|' \le \rho \|v\|' \quad \text{whenever} \quad v \in E^u(x).$$

Proof. Fix $\varepsilon > 0$. Given $v, w \in E^s(x)$, we define

$$\langle v, w \rangle'_x = \sum_{n=0}^{\infty} \langle d_x f^n v, d_x f^n w \rangle \tau^{-2n} e^{-2\varepsilon n}. \tag{6.7}$$

To show that the series converges, it is sufficient to observe that

$$\sum_{n=0}^{\infty} |\langle d_x f^n v, d_x f^n w \rangle| \tau^{-2n} e^{-2\varepsilon n} \le \sum_{n=0}^{\infty} c^2 e^{-2\varepsilon n} \|v\| \cdot \|w\| < \infty.$$

For each $v \in E^s(x)$, we have

$$\left(\|d_x f v\|'_{f(x)} \right)^2 = \sum_{n=0}^{\infty} \|d_x f^{n+1} v\|^2 \tau^{-2n} e^{-2\varepsilon n}$$

$$\le \sum_{n=0}^{\infty} \|d_x f^n v\|^2 \tau^{-2(n-1)} e^{-2\varepsilon(n-1)}$$

$$= \tau^2 e^{2\varepsilon} (\|v\|'_x)^2,$$

and hence,

$$\|d_x f v\|'_{f(x)} \le \tau e^{\varepsilon} \|v\|'_x. \tag{6.8}$$

Similarly, given $v, w \in E^u(x)$, we define

$$\langle v, w \rangle'_x = \sum_{n=0}^{\infty} \langle d_x f^{-n} v, d_x f^{-n} w \rangle \tau^{-2n} e^{-2\varepsilon n}, \tag{6.9}$$

and for each $v \in E^u(x)$, we have

$$\|d_x f^{-1} v\|'_{f^{-1}(x)} \le \tau e^{\varepsilon} \|v\|'_x. \tag{6.10}$$

Now we extend $\langle \cdot, \cdot \rangle'_x$ to the whole tangent space $T_x M$. Namely, given $v = v^s + v^u$ and $w = w^s + w^u$ in $T_x M$ with $v^s, w^s \in E^s(x)$ and $v^u, w^u \in E^u(x)$, we define

$$\langle v, w \rangle'_x = \langle v^s, w^s \rangle'_x + \langle v^u, w^u \rangle'_x, \tag{6.11}$$

where $\langle v^s, w^s \rangle'_x$ and $\langle v^u, w^u \rangle'_x$ are given, respectively, by (6.7) and (6.9). If $v \in E^s(x)$ and $w \in E^u(x)$, then $v^u = w^s = 0$, and it follows from (6.11) that $\langle v, w \rangle'_x = 0$. This establishes the first property. Setting $\rho = \tau e^\varepsilon$ for some $\varepsilon > 0$ such that $\rho < 1$, the second property follows readily from (6.8) and (6.10). $\square$

In fact, to prove Proposition 6.3, one can also use finite sums, instead of the series in (6.7) and (6.9) (see Exercise 6.4).

6.3 Smale Horseshoe

We consider in this section a more elaborate example of hyperbolic set. It will also help us motivating the use of symbolic dynamics.

6.3.1 Construction of the Horseshoe

We consider a diffeomorphism f in an open neighborhood of the square $Q = [0, 1]^2$ with the behavior indicated in Fig. 6.1. To define f more explicitly, we consider the horizontal strips (see Fig. 6.2)

$$H_1 = [0, 1] \times \left[\frac{1}{9}, \frac{4}{9} \right] \quad \text{and} \quad H_2 = [0, 1] \times \left[\frac{5}{9}, \frac{8}{9} \right],$$

and the vertical strips (see Fig. 6.2)

$$V_1 = \left[\frac{1}{9}, \frac{4}{9} \right] \times [0, 1] \quad \text{and} \quad V_2 = \left[\frac{5}{9}, \frac{8}{9} \right] \times [0, 1].$$

Now we define a transformation $g \colon H_1 \cup H_2 \to \mathbb{R}^2$ by

$$g(x, y) = \begin{cases} (x/3, 3y) + (1/9, -1/3) & \text{if } (x, y) \in H_1, \\ (-x/3, -3y) + (8/9, 8/3) & \text{if } (x, y) \in H_2, \end{cases} \tag{6.12}$$

and we consider any diffeomorphism f with the behavior indicated in Fig. 6.1 such that

$$f|(H_1 \cup H_2) = g.$$

We note that f transforms horizontal strips into vertical strips, that is,

$$f(H_1) = V_1 \quad \text{and} \quad f(H_2) = V_2.$$

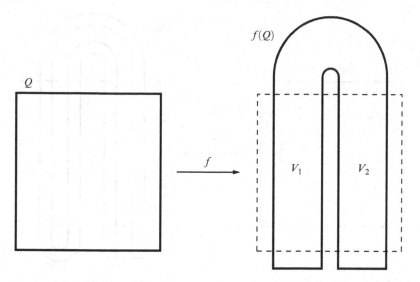

Fig. 6.1 The diffeomorphism f

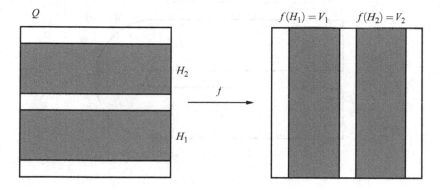

Fig. 6.2 Horizontal and vertical strips

Applying f a second time, we obtain the set $f^2(Q)$ in Fig. 6.3. We note that the intersection $Q \cap f^2(Q)$ is composed of four vertical strips of width $1/9$. Similarly, applying successively f^{-1}, we obtain the sets in Fig. 6.4.

Definition 6.4. The set $\Lambda \subset Q$ defined by

$$\Lambda = \bigcap_{n \in \mathbb{Z}} f^n(Q) \tag{6.13}$$

is called a *Smale horseshoe*.

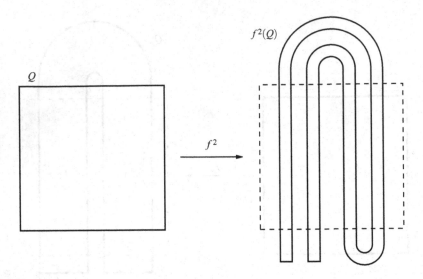

Fig. 6.3 Second iterate of the diffeomorphism

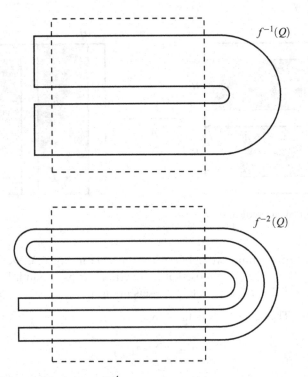

Fig. 6.4 First iterates of the inverse f^{-1}

Fig. 6.5 The sets Q, Q_1, and Q_2

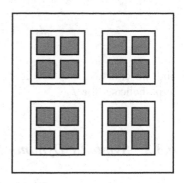

We note that Λ is the intersection of the decreasing sequence of closed sets

$$Q_n = \bigcap_{k=-n}^{n} f^k(Q), \quad n \in \mathbb{N},$$

that is,

$$\Lambda = \bigcap_{n \in \mathbb{N}} Q_n.$$

This implies that Λ is compact and nonempty. Each set Q_n is the union of 4^n disjoint squares with sides of length 3^{-n}. In Fig. 6.5, we represent Q_1 and Q_2, respectively, with white and gray backgrounds.

It follows from (6.13) that Λ is f-invariant, that is, $f^{-1}(\Lambda) = \Lambda$. Since f is a diffeomorphism, this is equivalent to $f(\Lambda) = \Lambda$. In particular, one can consider the restriction $f|\Lambda \colon \Lambda \to \Lambda$.

Proposition 6.4. *The Smale horseshoe Λ is a hyperbolic set for the diffeomorphism f.*

Proof. By (6.12), we have

$$d_x f = \begin{cases} B & \text{if } x \in H_1, \\ -B & \text{if } x \in H_2, \end{cases} \tag{6.14}$$

where

$$B = \begin{pmatrix} 3^{-1} & 0 \\ 0 & 3 \end{pmatrix}.$$

Now we consider the decomposition

$$T_x \mathbb{R}^2 = E^s(x) \oplus E^u(x),$$

where $E^s(x)$ and $E^u(x)$ are, respectively, the horizontal and vertical axes. Since the matrix B is diagonal, this decomposition satisfies (6.2). Furthermore, by (6.14),

$$\|d_x f v\| = \begin{cases} 3^{-1}\|v\| & \text{if } v \in E^s(x), \\ 3\|v\| & \text{if } v \in E^u(x), \end{cases}$$

and we can take $\tau = 3^{-1}$ and $c = 1$ in Definition 6.1. This shows that Λ is a hyperbolic set for f. $\square$

6.3.2 Symbolic Dynamics

We consider again the Smale horseshoe Λ and the diffeomorphism f introduced in Sect. 6.3.1. We show that $f|\Lambda$ has a certain symbolic representation. This is one of the important properties of the Smale horseshoe. In particular, it can be used to obtain information that otherwise would be difficult to obtain.

We first recall the relevant material from Sect. 3.5.4. Let $X = X_2 = \{1, 2\}^{\mathbb{Z}}$ be the space of two-sided sequences

$$\omega = (\cdots i_{-1}(\omega) i_0(\omega) i_1(\omega) \cdots)$$

of numbers in $\{1, 2\}$. Given $\beta > 1$, we equip X with the distance d_β in (3.37). Let also $\sigma \colon X \to X$ be the shift map defined by $\sigma(\omega) = \omega'$, where $i_n(\omega') = i_{n+1}(\omega)$ for each $n \in \mathbb{Z}$. Since

$$\beta^{-1} d_\beta(\omega, \omega') \le d_\beta(\sigma(\omega), \sigma(\omega')) \le \beta d_\beta(\omega, \omega')$$

for every $\omega, \omega' \in X$, the shift map is a homeomorphism.

For each $\omega \in X$ and $n \in \mathbb{N}$, we define the set

$$Q_n(\omega) = \bigcap_{k=-n}^{n-1} f^{-k}(V_{i_k(\omega)}). \tag{6.15}$$

We note that $Q_n(\omega)$ is a square with sides of length 3^{-n} (see Fig. 6.5). Furthermore, let

$$H(\omega) = \bigcap_{n \in \mathbb{Z}} f^{-n}(V_{i_n(\omega)}) = \bigcap_{n \in \mathbb{N}} Q_n(\omega).$$

It follows from (6.13) that $H(\omega) \subset \Lambda$. Moreover:

1. card $H(\omega) \ge 1$ since $Q_n(\omega)$ is a decreasing sequence of closed sets.
2. card $H(\omega) \le 1$ since diam $Q_n(\omega) \to 0$ when $n \to \infty$.

Therefore, card $H(\omega) = 1$ for every $\omega \in X$. Denoting the single element of $H(\omega)$ by $h(\omega)$, we can define a transformation $h \colon X \to \Lambda$ by

$$h(\omega) = \bigcap_{n \in \mathbb{Z}} f^{-n}(V_{i_n(\omega)}).$$ (6.16)

Proposition 6.5. *The following properties hold:*

1. The transformation h is a homeomorphism.
2. The inverse of h is given by $h^{-1}(x) = (\cdots i_{-1} i_0 i_1 \cdots)$, where

$$i_n = \begin{cases} 1 & \text{if } f^n(x) \in V_1, \\ 2 & \text{if } f^n(x) \in V_2 \end{cases}$$ (6.17)

for each $n \in \mathbb{Z}$.
3. $h \circ \sigma = f \circ h$, that is, the diagram

$$
\begin{array}{ccc}
X & \xrightarrow{\;\sigma\;} & X \\
{\scriptstyle h}\Big\downarrow & & \Big\downarrow{\scriptstyle h} \\
\Lambda & \xrightarrow{\;f\;} & \Lambda
\end{array}
$$

is commutative.

Proof. It follows from the definition of the Smale horseshoe that

$$\Lambda = \bigcap_{n \in \mathbb{Z}} f^n(V_1 \cup V_2)$$

$$= \bigcup_{\omega \in X} \bigcap_{n \in \mathbb{Z}} f^n(V_{i_n(\omega)}) = h(X),$$

and hence, h is onto. To show that h is injective, we take sequences $\omega, \omega' \in X$ with $\omega \neq \omega'$. Then $i_m(\omega) \neq i_m(\omega')$ or equivalently $V_{i_m(\omega)} \cap V_{i_m(\omega')} = \varnothing$, for some $m \in \mathbb{Z}$. Therefore,

$$H(\omega) \cap H(\omega') = \left(\bigcap_{n \in \mathbb{Z}} f^{-n}(V_{i_n(\omega)}) \right) \cap \left(\bigcap_{n \in \mathbb{Z}} f^{-n}(V_{i_n(\omega')}) \right) = \varnothing,$$

and $h(\omega) \neq h(\omega')$, which shows that h is injective. To show that h is a homeomorphism, let

$$Q'_n(\omega) = \bigcap_{k=-n}^{n} f^{-k}(V_{i_k(\omega)}).$$

Then $h^{-1}(Q'_n(\omega))$ is a cylinder set (see (3.36)). Since it is open and since the sets $Q'_n(\omega)$ generate the topology of Λ, we conclude that h is continuous. Moreover,

$$h(C_{i_{-n}(\omega) \cdots i_n(\omega)}) = \Lambda \cap \text{int } Q'_n(\omega)$$

is open in Λ, and since the cylinder sets generate the topology of X, the transformation h^{-1} is continuous.

Identity (6.17) follows readily from (6.16). Finally, for the third property, we note that

$$h(\sigma(\omega)) = \bigcap_{n\in\mathbb{Z}} f^{-n}(V_{i_{n+1}(\omega)})$$

$$= \bigcap_{n\in\mathbb{Z}} f^{1-n}(V_{i_n(\omega)}) = f(h(\omega)).$$

This completes the proof of the proposition. $\square$

The transformation h is called a *coding map* (of the Smale horseshoe) and acts as a dictionary between the dynamics $f|\Lambda$ on the Smale horseshoe and the symbolic dynamics $\sigma|X$ given by the shift map. In many situations, a coding map allows one to study in a somewhat easier manner some aspects of the dynamics. For example, by Proposition 6.5, we have

$$h \circ \sigma^m = f^m \circ h$$

for each $m \in \mathbb{N}$, and hence,

$$\sigma^m(\omega) = \omega \quad \text{if and only if} \quad f^m(h(\omega)) = h(\omega). \tag{6.18}$$

In other words, the study of the periodic points of $f|\Lambda$ can be reduced to the study of the periodic points of σ. We first consider the shift map.

Proposition 6.6. *The following properties hold:*

1. For each $m \in \mathbb{N}$,
$$\operatorname{card}\{\omega \in X : \sigma^m(\omega) = \omega\} = 2^m.$$

2. The set of periodic points of σ is dense in X.

Proof. The transition matrix of $\sigma|X$ is $A = \left(\begin{smallmatrix} 1 & 1 \\ 1 & 1 \end{smallmatrix}\right)$. Therefore, $\operatorname{tr}(A^m) = 2^m$ for each $m \in \mathbb{N}$, and the first property follows readily from (3.34).

For the second property, since the cylinder sets generate the topology of X, it is sufficient to show that each set $C_{i_{-n}\cdots i_n}$ contains a periodic point. One can easily verify that the sequence $\omega \in C_{i_{-n}\cdots i_n}$ obtained from repeating the finite sequence $(i_{-n}\cdots i_n)$ satisfies $\sigma^{2n+1}(\omega) = \omega$. Therefore, ω is a periodic point. $\square$

Now we study the periodic points of $f|\Lambda$.

Proposition 6.7. *The following properties hold:*

1. For each $m \in \mathbb{N}$,
$$\operatorname{card}\{x \in \Lambda : f^m(x) = x\} = 2^m.$$

2. The set of periodic points of $f|\Lambda$ is dense in Λ.

Proof. It follows from (6.18) and Property 1 in Proposition 6.6 that

$$\text{card}\{x \in \Lambda : f^m(x) = x\} = \text{card}\{\omega \in X : \sigma^m(\omega) = \omega\} = 2^m$$

for each $m \in \mathbb{N}$. Since h is a homeomorphism, the second property follows readily from Proposition 6.6. $\qquad\square$

By Propositions 6.6 and 6.7, we have

$$p(f|\Lambda) = p(\sigma) = \log 2 > 0,$$

where p is the periodic entropy introduced in Definition 2.6.

6.4   Hyperbolic Automorphisms of $\mathbb{T}^2$

We show in this section how to introduce a coding map for a hyperbolic toral automorphism of $\mathbb{T}^2$. In order to make the construction more explicit, we consider a particular automorphism.

6.4.1   Construction of a Coding Map

Let $T : \mathbb{T}^2 \to \mathbb{T}^2$ be the hyperbolic automorphism of $\mathbb{T}^2$ induced by the matrix $B = \left(\begin{smallmatrix} 2 & 1 \\ 1 & 1 \end{smallmatrix}\right)$. This matrix has eigenvalues

$$\tau = \frac{3 + \sqrt{5}}{2} \quad \text{and} \quad \tau^{-1} = \frac{3 - \sqrt{5}}{2}$$

with orthogonal eigendirections defined by the vectors

$$\left(\frac{1 + \sqrt{5}}{2}, 1\right) \quad \text{and} \quad \left(\frac{1 - \sqrt{5}}{2}, 1\right).$$

We consider a partition of $\mathbb{T}^2$ into closed rectangles S_1 and S_2. This means that

$$\mathbb{T}^2 = S_1 \cup S_2 \quad \text{and} \quad S_1 \cap S_2 = \partial S_1 \cup \partial S_2.$$

Namely, we consider the partition indicated in Fig. 6.6, with the sides of the rectangles S_1 and S_2 parallel to the stable and unstable directions or more precisely to the eigendirections of the matrix B. Applying the automorphism T to the

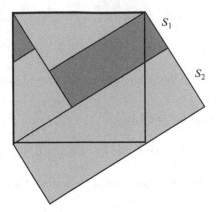

Fig. 6.6 Partition of the torus into rectangles

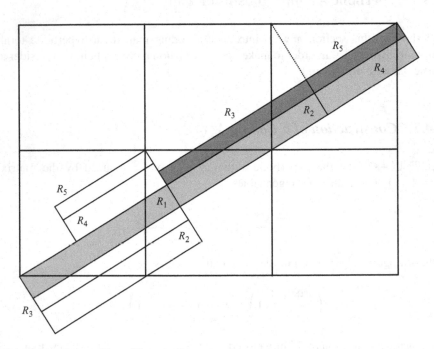

Fig. 6.7 Another partition of the torus into rectangles

rectangles S_1 and S_2, we obtain a new partition of $\mathbb{T}^2$ into five closed rectangles R_1, R_2, R_3, R_4, and R_5 with disjoint interiors such that (see Fig. 6.7)

$$T(S_1) = R_3 \cup R_5 \quad \text{and} \quad T(S_2) = R_1 \cup R_2 \cup R_4.$$

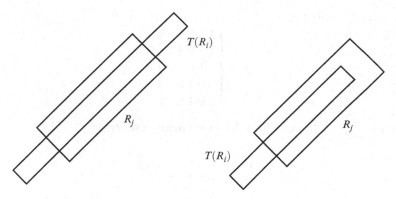

Fig. 6.8 Two types of intersections of rectangles

One can easily verify that:

1. If int $T(R_i) \cap$ int $R_j \neq \varnothing$, then $T(R_i)$ intersects R_j along the whole unstable direction.
2. If int $T^{-1}(R_i) \cap$ int $R_j \neq \varnothing$, then $T^{-1}(R_i)$ intersects R_j along the whole stable direction.

This is called the *Markov property* of the partition (with respect to T). See Fig. 6.8 for two possible types of intersections. The first type satisfies the Markov property, while the second one does not. We shall see that the Markov property is crucial in order to obtain a coding map for the automorphism T.

The following is a particular case of a more general notion introduced in Sect. 7.6:

Definition 6.5. Given a hyperbolic automorphism T of $\mathbb{T}^2$, let $R_1, \ldots, R_k \subset \mathbb{T}^2$ be closed rectangles with sides parallel to the stable and unstable directions such that

$$\mathbb{T}^2 = \bigcup_{i=1}^{k} R_i \quad \text{and} \quad R_i \cap R_j = \partial R_i \cap \partial R_j$$

whenever $i \neq j$. If the Markov property holds, then $\{R_i : i = 1, \ldots, k\}$ is called a *Markov partition* of $\mathbb{T}^2$ (with respect to T).

Now we return to the hyperbolic automorphism $T = T_B$. We define a 5×5 transition matrix A with entries

$$a_{ij} = \begin{cases} 1 & \text{if int } T(R_i) \cap \text{int } R_j \neq \varnothing, \\ 0 & \text{if int } T(R_i) \cap \text{int } R_j = \varnothing. \end{cases} \tag{6.19}$$

One can easily verify that

$$A = \begin{pmatrix} 1 & 1 & 0 & 1 & 0 \\ 1 & 1 & 0 & 1 & 0 \\ 1 & 1 & 0 & 1 & 0 \\ 0 & 0 & 1 & 0 & 1 \\ 0 & 0 & 1 & 0 & 1 \end{pmatrix}. \tag{6.20}$$

For each $\omega \in X_A \subset X_5$ (see Sect. 3.5.4), we define a set $H(\omega) \subset \mathbb{T}^2$ by

$$H(\omega) = \bigcap_{n \in \mathbb{Z}} T^{-n} R_{i_n(\omega)}.$$

Since the sides of the rectangles are parallel to the stable and unstable directions, there is a constant $c > 0$ such that for each $n \in \mathbb{N}$, the closed set

$$Q_n(\omega) = \bigcap_{k=-n}^{n} T^{-k} R_{i_k(\omega)} \tag{6.21}$$

has diameter at most $c\tau^{-n}$. This ensures that card $H(\omega) \leq 1$. We use the Markov property to show that in fact card $H(\omega) = 1$ for every $\omega \in X_A$. Let R_i, R_j, and R_k be rectangles such that

$$\text{int } T(R_i) \cap \text{int } R_j \neq \emptyset \quad \text{and} \quad \text{int } T(R_j) \cap \text{int } R_k \neq \emptyset.$$

By the Markov property, int $T^2(R_i)$ intersects int $T(R_j)$ along the whole unstable direction of R_k. This implies that

$$\text{int } T^2(R_i) \cap \text{int } T(R_j) \cap \text{int } R_k \neq \emptyset.$$

One can use induction to show that if

$$\text{int } T(R_{i_k}) \cap \text{int } R_{i_{k+1}} \neq \emptyset \quad \text{for} \quad k = -n, \dots, n-1,$$

then

$$\text{int } T^{2n}(R_{i_{-n}}) \cap \text{int } T^{2n-1}(R_{i_{-n+1}}) \cap \cdots \cap \text{int } R_{i_n} \neq \emptyset.$$

Therefore, the closed set $Q_n(\omega)$ in (6.21) is nonempty for each $n \in \mathbb{N}$. This implies that card $H(\omega) = 1$ for every $\omega \in X_A$, and hence, we can define a *coding map* $h: X_A \to \mathbb{T}^2$ by

$$h(\omega) = \bigcap_{n \in \mathbb{Z}} T^{-n} R_{i_n}. \tag{6.22}$$

Incidentally, one can use a similar construction to obtain a Markov partition and thus also a coding map for any hyperbolic automorphism of $\mathbb{T}^2$.

Repeating arguments in the proof of Proposition 6.5, we obtain the following statement:

Proposition 6.8. *The transformation* $h: X_A \to \mathbb{T}^2$ *is continuous and onto, and it satisfies* $T \circ h = h \circ \sigma$, *that is, the diagram*

$$
\begin{array}{ccc}
X_A & \xrightarrow{\ \sigma\ } & X_A \\
h\downarrow & & \downarrow h \\
\mathbb{T}^2 & \xrightarrow{\ T\ } & \mathbb{T}^2
\end{array}
$$

is commutative.

Contrarily to what happens in the case of the Smale horseshoe (see Sect. 6.3.2), the coding map h in (6.22) is not injective. This can be seen as follows. The fixed points of $\sigma|X_A$ are the constant sequences. By direct inspection of the matrix A in (6.20), the fixed points are thus

$$(\cdots 000 \cdots), \ (\cdots 111 \cdots), \text{ and } (\cdots 444 \cdots).$$

However, one can show that 0 is the unique fixed point of T (see also (6.26) below), and thus,

$$h(\cdots 000 \cdots) = h(\cdots 111 \cdots) = h(\cdots 444 \cdots) = 0. \tag{6.23}$$

We also consider the periodic points of the hyperbolic automorphism $T = T_B$. We first observe that the matrix A in (6.20) has eigenvalues (counted with their multiplicities)

$$\frac{3 + \sqrt{5}}{2}, \frac{3 - \sqrt{5}}{2}, 0, 0, 0.$$

By (3.34), we have

$$
\begin{aligned}
\text{card}\,\{\omega \in X_A : \sigma^m(\omega) = \omega\} &= \text{tr}(A^m) \\
&= \left(\frac{3 + \sqrt{5}}{2}\right)^m + \left(\frac{3 - \sqrt{5}}{2}\right)^m
\end{aligned} \tag{6.24}
$$

for each $m \in \mathbb{N}$. In particular, the periodic entropy of $\sigma|X_A$ is equal to

$$p(\sigma|X_A) = \log \frac{3 + \sqrt{5}}{2} > 0.$$

To find the number of m-periodic points of T, we first note that $T^m(x, y) = (x, y)$ if and only if

$$(B^m - \text{Id})(x, y) = (0, 0) \mod 1, \tag{6.25}$$

where $B = \left(\begin{smallmatrix} 2 & 1 \\ 1 & 1 \end{smallmatrix}\right)$. Since $\det(B^m - \text{Id}) \neq 0$, the number of solutions (x, y) of (6.25) is equal to the number of representatives of the equivalence class of $(0, 0) \in \mathbb{T}^2$ in $(B^m - \text{Id})[0, 1)^2$, which is equal to $|\det(B^m - \text{Id})|$. Therefore,

$$\text{card}\{x \in \mathbb{T}^2 : T^m(x) = x\} = |\det(B^m - \text{Id})|$$

$$= \left| \det \begin{pmatrix} \tau^m - 1 & 0 \\ 0 & \tau^{-m} - 1 \end{pmatrix} \right| \qquad (6.26)$$

$$= |(\tau^m - 1)(\tau^{-m} - 1)|$$

$$= \tau^m + \tau^{-m} - 2 = \text{tr}(B^m) - 2.$$

It follows from (6.26) that

$$p(T) = \lim_{m \to \infty} \frac{1}{m} \log(\text{tr}(B^m) - 2) = \log \frac{3 + \sqrt{5}}{2} > 0.$$

Note that $p(T) = p(\sigma | X_A)$. We emphasize that this is not a coincidence. Indeed, we shall see in Sect. 7.6 how to establish this identity for a much larger class of dynamics without computing either $p(T)$ or $p(\sigma | X_A)$.

Incidentally, by (6.24), we have

$$\text{card}\{x \in \mathbb{T}^2 : T^m(x) = x\} = \text{card}\{\omega \in X_A : \sigma^m(\omega) = \omega\} - 2.$$

We note that the discrepancy is due to the fixed points of $\sigma | X_A$ (see (6.23)). More precisely, one can easily verify that

$$\text{card}\{x \in \mathbb{T}^2 : T^m(x) = x, \ T(x) \neq x\} = \text{card}\{\omega \in X_A : \sigma^m(\omega) = \omega, \ \sigma(\omega) \neq \omega\}$$

for any $m \in \mathbb{N}$.

6.4.2 Induction of a Markov Measure

We show in this section that any Markov partition of a hyperbolic automorphism of $\mathbb{T}^2$ gives rise to a Markov measure (see Definition 3.10).

Let $T : \mathbb{T}^2 \to \mathbb{T}^2$ be a hyperbolic automorphism of $\mathbb{T}^2$. We consider a Markov partition $\{R_i : i = 1, \ldots, k\}$ of $\mathbb{T}^2$ with respect to T and the associated topological Markov chain $\sigma | X_A : X_A \to X_A$, where $A = (a_{ij})$ is the $k \times k$ transition matrix with the entries in (6.19). We also define a coding map $h : X_A \to \mathbb{T}^2$ by

$$h(\omega) = \bigcap_{n \in \mathbb{Z}} T^{-n} R_{i_n(\omega)}.$$

It is easy to verify that the statement in Proposition 6.8 also holds in this setting.

Now we define a measure μ in X_k by

$$\mu(h^{-1}C) = \lambda(C) \tag{6.27}$$

for each measurable set $C \subset \mathbb{T}^2$, where λ is the Lebesgue measure in $\mathbb{T}^2$.

Proposition 6.9. *The measure μ is a σ-invariant Markov measure in X_k.*

Proof. To show that μ is σ-invariant, we note that since $T \circ h = h \circ \sigma$ we have

$$\begin{aligned}
\sigma^{-1}(h^{-1}C) &= (h \circ \sigma)^{-1}C \\
&= (T \circ h)^{-1}C \\
&= h^{-1}(T^{-1}C)
\end{aligned}$$

for any measurable set $C \subset \mathbb{T}^2$, and thus,

$$\begin{aligned}
\mu(\sigma^{-1}(h^{-1}C)) &= \mu(h^{-1}(T^{-1}C)) \\
&= \lambda(T^{-1}C) \\
&= \lambda(C) = \mu(h^{-1}C),
\end{aligned}$$

using the T-invariance of the measure λ.

Now we show that μ is a Markov measure. We first observe that

$$\begin{aligned}
\mu(C_{i_{-n} \cdots i_n}) &= \lambda\left(\bigcap_{\ell=-n}^{n} T^{-\ell} R_{i_\ell} \right) \\
&= \lambda\left(\bigcap_{\ell=0}^{2n} T^{-\ell} R_{i_{n+\ell}} \right),
\end{aligned}$$

using the T-invariance of the measure μ. Set

$$p_i = \lambda(R_i) \quad \text{and} \quad p_{ij} = \frac{\lambda(R_i \cap T^{-1}R_j)}{\lambda(R_i)} \tag{6.28}$$

for $i, j = 1, \ldots, k$. Clearly, $p_i > 0$ for each i, and $\sum_{i=1}^{k} p_i = 1$. Moreover,

$$\begin{aligned}
\sum_{i=1}^{k} p_i p_{ij} &= \sum_{i=1}^{k} \lambda(R_i \cap T^{-1}R_j) \\
&= \lambda(T^{-1}R_j) \\
&= \lambda(R_j) = p_j.
\end{aligned}$$

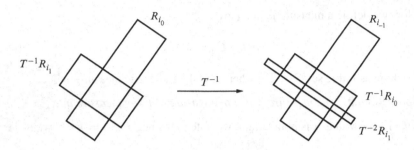

Fig. 6.9 Intersection of rectangles along the unstable direction

Therefore, (P, p) is a stochastic pair. It remains to show that the associated Markov measure is indeed μ. We note that

$$\mu(C_{i_{-1}i_0i_1}) = \lambda(R_{i_{-1}} \cap T^{-1}R_{i_0} \cap T^{-2}R_{i_1})$$

$$= \lambda(R_{i_{-1}}) \frac{\lambda(R_{i_{-1}} \cap T^{-1}R_{i_0})}{\lambda(R_{i_{-1}})} \cdot \frac{\lambda(R_{i_{-1}} \cap T^{-1}R_{i_0} \cap T^{-2}R_{i_1})}{\lambda(R_{i_{-1}} \cap T^{-1}R_{i_0})}$$

$$= p_{i_{-1}} p_{i_{-1}i_0} \frac{\lambda(R_{i_{-1}} \cap T^{-1}R_{i_0} \cap T^{-2}R_{i_1})}{\lambda(R_{i_{-1}} \cap T^{-1}R_{i_0})}.$$

We also have

$$p_{i_0i_1} = \frac{\lambda(R_{i_0} \cap T^{-1}R_{i_1})}{\lambda(R_{i_0})} = \frac{\lambda(T^{-1}R_{i_0} \cap T^{-2}R_{i_1})}{\lambda(T^{-1}R_{i_0})}.$$

By the Markov property of the Markov partition, if

$$\text{int } R_{i_{-1}} \cap \text{int } T^{-1}R_{i_0} \cap \text{int } T^{-2}R_{i_1} \neq \varnothing,$$

then the rectangle $R_{i_{-1}}$ intersects $T^{-1}R_{i_0}$ and $T^{-2}R_{i_1}$ along the whole unstable direction (see Fig. 6.9). Therefore,

$$p_{i_0i_1} = \frac{\lambda(R_{i_{-1}} \cap T^{-1}R_{i_0} \cap T^{-2}R_{i_1})}{\lambda(R_{i_{-1}} \cap T^{-1}R_{i_0})},$$

and

$$\mu(C_{i_{-1}i_0i_1}) = p_{i_{-1}} p_{i_{-1}i_0} p_{i_0i_1}.$$

Similarly, one can show that

$$\mu(C_{i_{-n}\cdots i_n}) = p_{i_{-n}} p_{i_{-n}i_{-n+1}} \cdots p_{i_{n-1}i_n}.$$

This completes the proof of the proposition. $\square$

6.5 The Case of Repellers

In the former sections, we have illustrated with the Smale horseshoe and with the hyperbolic automorphisms of $\mathbb{T}^2$ how certain symbolic representations of a hyperbolic set can be used to obtain information that otherwise would be difficult to obtain. Also as a preparation for the full generalization of this approach to an arbitrary hyperbolic set, we consider in this section noninvertible transformations and the notion of repeller, which can be considered a somewhat simpler version of hyperbolic sets in which only expansion is present. In particular, we detail the construction of Markov partitions for repellers. The corresponding construction for hyperbolic sets in Sect. 7.6 is an elaboration of this approach.

Let $f \colon U \to M$ be a C^1 map, where U is an open subset of a smooth manifold M.

Definition 6.6. Given a compact f-invariant set $J \subset U$, we say that f is *expanding* on J and that J is a *repeller* for f if there exist $c > 0$ and $\tau > 1$ such that

$$\|d_x f^n v\| \ge c\tau^n \|v\| \tag{6.29}$$

for every $x \in J$, $v \in T_x M$, and $n \in \mathbb{N}$.

Example 6.2. The expanding map of the circle $E_q(z) = z^q$, for some integer $q > 1$, satisfies $\|d_z E_q\| = q > 1$. Therefore, E_q is expanding on S^1, and S^1 is a repeller for E_q.

Example 6.3. Let T be a toral endomorphism of $\mathbb{T}^n$ induced by a matrix A with all eigenvalues with absolute value greater than 1. For example, A can be a diagonal matrix with the entries in the diagonal with values in $\mathbb{N} \setminus \{1\}$. Then T is expanding on $\mathbb{T}^n$, and $\mathbb{T}^n$ is a repeller for T.

We also consider a version of Markov partition for repellers.

Definition 6.7. Let J be a repeller for a C^1 map f. A collection of closed sets $R_1, \ldots, R_k \subset J$ is called a *Markov partition* of J (with respect to f) if:

1. $J = \bigcup_{i=1}^k R_i$, and $R_i = \overline{\operatorname{int} R_i}$ for each i.
2. $\operatorname{int} R_i \cap \operatorname{int} R_j = \varnothing$ whenever $i \ne j$.
3. If $\operatorname{int} f(R_i) \cap \operatorname{int} R_j \ne \varnothing$, then $f(R_i) \supset R_j$.

We note that the interiors are taken with respect to the induced topology on J.

We describe several examples of Markov partitions.

Example 6.4. For the expanding map of the circle E_q, a Markov partition of S^1 is composed of the sets $h([2k\pi/q, 2(k+1)\pi/q])$ for $k = 0, \ldots, q-1$, where h is the transformation in (2.11).

Fig. 6.10 Expanding
quadratic map

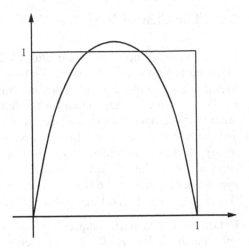

Example 6.5. Let $f: [0, 1] \to \mathbb{R}$ be the quadratic map $f(x) = ax(1-x)$ for $a > 4$ (see Fig. 6.10). We consider the f-invariant closed set

$$J = \bigcap_{k=0}^{\infty} f^{-k}[0, 1].$$

It is straightforward to verify that $f(x) = 1$ for $x = (1 \pm \sqrt{1 - 4/a})/2$, and hence,

$$|f'(x)| = a|1 - 2x| \geq a\sqrt{1 - 4/a}$$

for every $x \in J$. Therefore, if a is sufficiently large, then J is a repeller for some extension of f. A Markov partition of J is composed of the sets

$$\left[0, (1 - \sqrt{1 - 4/a})/2\right] \cap J \quad \text{and} \quad \left[(1 + \sqrt{1 - 4/a})/2, 1\right] \cap J.$$

The following result establishes the existence of Markov partitions for any repeller:

Theorem 6.1. *Any repeller has Markov partitions of arbitrarily small diameter.*

Proof. Since f is a local diffeomorphism (by the inverse function theorem) and J is compact, there exists $\delta > 0$ such that the preimage of any connected open set U with diameter diam $U < \delta$ is composed of a finite number of connected components, each of them diffeomorphic to U.

Now we construct a finite open cover $\mathcal{U}$ of J (with respect to the induced topology on J) with the property that if $f(U) \cap V \neq \varnothing$ for some $U, V \in \mathcal{U}$, then $f(U) \supset V$. Let $\mathcal{U}_0$ be a finite open cover of J by balls of radius r such that

$$2r \left(1 + \frac{1}{c(\tau - 1)} \right) < \delta. \tag{6.30}$$

Let also $\mathcal{V}_1$ be the open cover of J composed of the connected components of $f^{-1}U$ for $U \in \mathcal{U}_0$. We consider the finite open cover $\mathcal{U}_1$ of J composed of the sets

$$U^{(1)} = \bigcup_{V \in \mathcal{V}_1 : V \cap U \neq \varnothing} V$$

for $U \in \mathcal{U}_0$. Now we proceed inductively to construct a sequence of finite open covers $\mathcal{U}_m$ of J. Namely, for each $m \in \mathbb{N}$, let $\mathcal{V}_{m+1}$ be the family of connected components of $f^{-1}V$ for $V \in \mathcal{V}_m$ such that $V \cap \partial U \neq \varnothing$ for some $U \in \mathcal{U}_{m-1}$. We consider the finite open cover $\mathcal{U}_{m+1}$ of J composed of the sets

$$U^{(m+1)} = U^{(m)} \cup \bigcup_{V \in \mathcal{V}_{m+1} : V \cap U^{(m)} \neq \varnothing} V$$

for $U^{(m)} \in \mathcal{U}_m$. We have

$$\operatorname{diam} U^{(m+1)} \leq \operatorname{diam} U^{(m)} + 2rc^{-1}\tau^{-(m+1)},$$

and hence, by (6.30),

$$\operatorname{diam} U^{(m+1)} \leq 2r + 2rc^{-1} \sum_{i=1}^{m+1} \tau^{-i} \leq 2r \left(1 + \frac{1}{c(\tau - 1)} \right) < \delta$$

for each $m \in \mathbb{N}$ and $U \in \mathcal{U}_0$. Now we consider the finite open cover of J given by

$$\mathcal{U} = \{ U^{(\infty)} : U \in \mathcal{U}_0 \},$$

where

$$U^{(\infty)} = \bigcup_{m=1}^{\infty} U^{(m)}. \tag{6.31}$$

Lemma 6.1. *If $f(U) \cap V \neq \varnothing$ for some $U, V \in \mathcal{U}$, then $f(U) \supset V$.*

Proof of the lemma. It follows from (6.31) that if $U^{(\infty)} \cap f^{-1}V^{(\infty)} \neq \varnothing$, then there exists $p \in \mathbb{N}$ such that

$$U^{(m)} \cap f^{-1}V^{(m)} \neq \varnothing \quad \text{for every} \quad m \geq p.$$

Therefore, by construction, the set $U^{(m+1)}$ contains a connected component of $f^{-1}V^{(m)}$ for each $m \geq p$. This implies that $U^{(\infty)} = \bigcup_{m \geq p} U^{(m+1)}$ contains a connected component of

$$\bigcup_{m \geq p} f^{-1} V^{(m)} = f^{-1} \bigcup_{m \geq p} V^{(m)} = f^{-1} V^{(\infty)},$$

which shows that $f(U^{(\infty)}) \supset V^{(\infty)}$.   □

Now we use the cover $\mathcal{U}$ to construct a Markov partition of the set J. Write $\mathcal{U} = \{U_1, \ldots, U_p\}$ and let $\mathcal{R}$ be the family of closed sets

$$R = \overline{V_1 \cap \cdots \cap V_p}$$

that have nonempty interior, where $V_i = U_i$ or $V_i = \text{int}(J \setminus U_i)$ for each $i = 1, \ldots, p$. We show that $\mathcal{R}$ is a Markov partition. Write $\mathcal{R} = \{R_1, \ldots, R_k\}$.

Lemma 6.2. *We have $J = \bigcup_{i=1}^{k} R_i$ and $R_i = \overline{\text{int } R_i}$ for $i = 1, \ldots, k$.*

Proof of the lemma. If

$$V := J \setminus \bigcup_{i=1}^{k} R_i \neq \varnothing,$$

then there exists a nonempty open set $U \subset V$ such that $U \cap \partial U_i = \varnothing$ for $i = 1, \ldots, p$. This is due to the fact that the sets $\partial U_i = \partial(J \setminus U_i)$ have empty interior, and thus, they cannot cover J. Since

$$U \cap \partial U_i = U \cap \partial(J \setminus U_i) = \varnothing$$

for each i, we can consider the set $R = \overline{V_1 \cap \cdots \cap V_p}$ such that $U \subset V_i$ for $i = 1, \ldots, p$. Then

$$R \supset V_1 \cap \cdots \cap V_p \supset U,$$

and hence, int $R \supset U$. Since $U \neq \varnothing$, we have $R \in \mathcal{R}$, and thus,

$$U \subset \bigcup_{i=1}^{k} R_i = J \setminus V.$$

This contradiction shows that $V = \varnothing$ and $J = \bigcup_{i=1}^{k} R_i$.

For the second property, we first note that clearly $\overline{\text{int } R_i} \subset R_i$ for each i. On the other hand, if $R_i = \overline{V}$ with $V = \bigcap_{i=1}^{p} V_i$, then $R_i \supset V$, and hence, int $R_i \supset V$ (since V is open). Therefore, $\overline{\text{int } R_i} \supset \overline{V} = R_i$. This completes the proof of the lemma.   □

Now take $R_i, R_j \in \mathcal{R}$ with $i \neq j$. We write $R_i = \overline{W_i}$ and $R_j = \overline{W_j}$, where

$$W_i = V_1 \cap \cdots \cap V_p \quad \text{and} \quad W_j = V_1' \cap \cdots \cap V_p'.$$

Since $i \neq j$, there exists ℓ such that $V_\ell \neq V_\ell'$. In particular, $W_i \cap W_j = \varnothing$. Therefore,

$$W_i \cap R_j = W_i \cap \overline{W_j} = \varnothing,$$

and hence, $W_i \cap \operatorname{int} R_j = \varnothing$. This implies that

$$R_i \cap \operatorname{int} R_j = \overline{W_i} \cap \operatorname{int} R_j = \varnothing,$$

and hence, $\operatorname{int} R_i \cap \operatorname{int} R_j = \varnothing$.

To establish the last property in Definition 6.7, let $R = \overline{V_1 \cap \cdots \cap V_p} \in \mathcal{R}$. Then

$$f(R) = \overline{f(V_1) \cap \cdots \cap f(V_p)}$$

and

$$f(V_1) \cap \cdots \cap f(V_p) = \bigcap_{V_i = U_i} f(U_i) \cap \bigcap_{V_j = \operatorname{int}(J \setminus U_j)} \operatorname{int}(J \setminus f(U_j)). \qquad (6.32)$$

By Lemma 6.1, the images $f(U_i)$ and $f(U_j)$ are unions of elements of $\mathcal{U}$. It thus follows from (6.32) that $f(R)$ is a union of elements of $\mathcal{R}$. In particular, if

$$\operatorname{int} f(R) \cap \operatorname{int} S \ne \varnothing$$

for some $S \in \mathcal{R}$, then $f(R) \cap \operatorname{int} S \ne \varnothing$, and hence, $\operatorname{int} S \subset f(R)$ and $S \subset f(R)$. This completes the proof of the theorem. $\qquad \square$

In a similar manner to that in the former sections, to each Markov partition of a repeller, we can associate a coding map. Namely, let A be the $k \times k$ *transition matrix* associated to a Markov partition $\{R_1, \ldots, R_k\}$ of a repeller J, with entries

$$a_{ij} = \begin{cases} 1 & \text{if } \operatorname{int} f(R_i) \cap \operatorname{int} R_j \ne \varnothing, \\ 0 & \text{if } \operatorname{int} f(R_i) \cap \operatorname{int} R_j = \varnothing. \end{cases} \qquad (6.33)$$

We also consider the (one-sided) topological Markov chain $\sigma | X_A^+ : X_A^+ \to X_A^+$ (see Sect. 3.5.1). We define a *coding map* $h: X_A^+ \to J$ of the repeller by

$$h(\omega) = \bigcap_{n \in \mathbb{N}} f^{-n+1} R_{i_n(\omega)}. \qquad (6.34)$$

6.6 Exercises

Exercise 6.1. Give a characterization of hyperbolic periodic points analogous to that in Proposition 6.1.

Exercise 6.2. Show that a toral automorphism of $\mathbb{T}^n$ induced by a matrix B is hyperbolic if and only if B is hyperbolic.

Exercise 6.3. Show that the Smale horseshoe has empty interior and that it has no isolated points.

Exercise 6.4. Let Λ be a hyperbolic set for a diffeomorphism f and take $m \in \mathbb{N}$ such that $c\tau^m < 1$. For $v, w \in E^s(x)$, let

$$\langle v, w \rangle''_x = \sum_{k=0}^{m-1} \langle d_x f^k v, d_x f^k w \rangle,$$

and for $v, w \in E^u(x)$, let

$$\langle v, w \rangle''_x = \sum_{k=0}^{m-1} \langle d_x f^{-k} v, d_x f^{-k} w \rangle.$$

Use these expressions and (6.11) to construct a inner product $\langle \cdot, \cdot \rangle''_x$ with the properties in Proposition 6.3.

Exercise 6.5. Let Λ be a hyperbolic set for a diffeomorphism f. Show that the homeomorphism $f|\Lambda: \Lambda \to \Lambda$ is two-sided expansive (see Definition 5.5).

Exercise 6.6. Show that if Λ is a hyperbolic set for a diffeomorphism f such that f is topologically mixing on Λ and A is the transition matrix of some Markov partition of Λ, then there exists $m \in \mathbb{N}$ such that $A^m > 0$.

Exercise 6.7. Show that the transformation h in (6.22) is invertible in a dense G_δ set of full Lebesgue measure. Hint: consider the set $\bigcup_{n \in \mathbb{Z}} T^n(\bigcup_{i=1}^5 \partial R_i)$.

Exercise 6.8. Construct a Markov partition and a coding map for:

1. The hyperbolic automorphism of $\mathbb{T}^2$ induced by the matrix $\left(\begin{smallmatrix} 5 & 2 \\ 2 & 1 \end{smallmatrix}\right)$.
2. The hyperbolic automorphism of $\mathbb{T}^2$ induced by the matrix $\left(\begin{smallmatrix} 5 & 4 \\ 1 & 1 \end{smallmatrix}\right)$.

Exercise 6.9. Find explicitly the numbers p_i and p_{ij} in (6.28) for the Markov measure in (6.27) obtained from the Markov partition constructed in Sect. 6.4.1.

Exercise 6.10. Prove or disprove the following statements:

1. In the Smale horseshoe, the periodic points of period odd are dense.
2. In the Smale horseshoe, the periodic points of period prime are dense.
3. In the Smale horseshoe, the periodic points of period at least 100 are dense.

Exercise 6.11. Show that the transformation h in (6.34) is continuous and onto and that it satisfies $h \circ \sigma = f \circ h$ in X_A^+.

Exercise 6.12. Let

$$U = \mathbb{T} \times \left\{ (x, y) \in \mathbb{R}^2 : x^2 + y^2 \leq 1 \right\}.$$

Given $\lambda \in (0, 1/2)$, we define a transformation $f: U \to U$ by

$$f(\theta, x, y) = \left(2\theta, \lambda x + \frac{1}{2}\cos(2\pi\theta), \lambda y + \frac{1}{2}\sin(2\pi\theta)\right).$$

Show that:

1. f is injective.
2. The *solenoid* $\Lambda = \bigcap_{n \in \mathbb{N}} f^n(U)$ is a closed f-invariant set.
3. Λ is a hyperbolic set for f.

Exercise 6.13. Let $T: \mathbb{T}^n \to \mathbb{T}^n$ be a hyperbolic automorphism induced by a matrix B. Show that

$$h(T) = h_\lambda(T) = p(T) = \sum_{\lambda \in \mathrm{Sp}(B): |\lambda| > 1} \log|\lambda|,$$

where λ is the Lebesgue measure in $\mathbb{T}^n$ and where p is the periodic entropy (see Definition 2.6). Hint: to compute the topological entropy, use Exercise 4.17.

Exercise 6.14. Prove or disprove the following statement: if T is a hyperbolic automorphism of $\mathbb{T}^n$, then the topological entropy $h(T)$ can take any value in $\mathbb{R}^+$.

Exercise 6.15. Let Λ_i be a hyperbolic set for a diffeomorphism $f_i: M_i \to M_i$ for $i = 1, 2$. Show that the product $\Lambda_1 \times \Lambda_2$ is a hyperbolic set for the diffeomorphism $f: M_1 \times M_2 \to M_1 \times M_2$ defined by

$$f(x, y) = (f_1(x), f_2(y)).$$

Exercise 6.16. Let J be a repeller for a C^1 map f and let $\varphi: J \to \mathbb{R}$ be a continuous function. Show that for any Markov partition $R_1, \ldots, R_k$ of J, we have

$$P_{f|J}(\varphi) = \lim_{n \to \infty} \frac{1}{n} \log \sum_{i_1 \cdots i_n} \exp \sup_{R_{i_1 \cdots i_n}} \sum_{j=0}^{n-1} \varphi \circ f^k,$$

where

$$R_{i_1 \cdots i_n} = \bigcap_{j=0}^{n-1} f^{-j} R_{i_{j+1}} \tag{6.35}$$

and where the sum is taken over all $i_1, \ldots, i_n \in \{1, \ldots, k\}$ such that $R_{i_1 \cdots i_n} \neq \varnothing$.

Notes

The books [42, 44, 65, 81, 89] are excellent sources for hyperbolic dynamics. We refer to [10, 11] for the theory of nonuniform hyperbolicity, also called Pesin theory.

The study of hyperbolicity goes back to Hadamard [37], in connection with the geodesic flow on a surface with negative curvature. Perron [71] studied the existence of bounded solutions under bounded perturbations of linear equations, thus also preparing the ground for the stable manifold theorem. The notion of hyperbolic set was introduced by Smale in his seminal paper [99], where he also laid the foundations of the theory. Anosov [3] introduced the class of systems that now bears his name and in particular made several important developments in [4]. The Smale horseshoe was introduced in [98]. The construction of Markov partitions for a hyperbolic automorphism of $\mathbb{T}^2$ in Sect. 6.4.1 is due to Adler and Weiss [2].

Chapter 7
Invariant Manifolds and Markov Partitions

We continue in this chapter our study of hyperbolic dynamics, starting with the construction of stable and unstable manifolds for any point of a hyperbolic set. The optimal regularity of the invariant manifolds is obtained using invariant cone families. In particular, we use the stable and unstable manifolds to define a product structure on any locally maximal hyperbolic set. In addition, we construct Markov partitions with a substantial elaboration of the corresponding construction for repellers. The shadowing property is also presented and is used as a tool in the construction of the Markov partitions. Moreover, we describe Hopf's argument to establish the ergodicity of Lebesgue measure for a hyperbolic toral automorphism.

7.1 Introduction

We already mentioned earlier that hyperbolicity gives rise to a very rich structure and in particular to the existence of stable and unstable manifolds at each point of a hyperbolic set. This is a substantial generalization of the Hadamard–Perron theorem for a hyperbolic fixed point and has several nontrivial consequences.

As a first application, we describe Hopf's argument to establish the ergodicity of Lebesgue measure for a hyperbolic toral automorphism. This approach should be compared to the earlier one in terms of Fourier analysis. For arbitrary trans-formations, in general, one is not able to apply this other method to establish the ergodicity of an invariant measure. On the other hand, Hopf's argument can be applied with success in many other situations, although the details fall out of the scope of this book. In particular, the method depends crucially on the so-called absolute continuity property.

In another direction, the shadowing property of a locally maximal hyperbolic set tell us that there is a real orbit close to any sequence of points such that the image of each of them is sufficiently close to the next. A consequence is that if two points x and $f^m(x)$ of a given orbit in a hyperbolic set are sufficiently close, then there exists an m-periodic point close to them. The shadowing property, together

L. Barreira, *Ergodic Theory, Hyperbolic Dynamics and Dimension Theory*, Universitext,
DOI 10.1007/978-3-642-28090-0_7, © Springer-Verlag Berlin Heidelberg 2012

with the product structure defined by the stable and unstable manifolds, can be used to construct Markov partitions for any locally maximal hyperbolic set, with an elaboration of the corresponding construction for repellers.

7.2 Stable and Unstable Manifolds

We construct in this section local stable and unstable manifolds at any point of a hyperbolic set. We first consider the case of linear diffeomorphisms. Namely, let $A: \mathbb{R}^n \to \mathbb{R}^n$ be the invertible linear transformation

$$A(x, y) = (\tau x, \rho y), \tag{7.1}$$

where $(x, y) \in \mathbb{R}^p \times \mathbb{R}^q = \mathbb{R}^n$ and $\tau, \rho^{-1} \in (0, 1)$. We note that the horizontal and vertical axes $E^s = \mathbb{R}^p \times \{0\}$ and $E^u = \{0\} \times \mathbb{R}^q$ satisfy

$$E^s = \{(x, y) \in \mathbb{R}^n : \|A^m(x, y)\| \to 0 \text{ when } m \to +\infty\},$$

$$E^u = \{(x, y) \in \mathbb{R}^n : \|A^m(x, y)\| \to 0 \text{ when } m \to -\infty\}.$$

Clearly,

$$AE^s = E^s \quad \text{and} \quad AE^u = E^u,$$

that is, the linear subspaces E^s and E^u are A-invariant sets (see Sect. 2.4 for the definition).

Now we consider arbitrary diffeomorphisms and their hyperbolic sets. We start with the particular case of a hyperbolic fixed point. We recall that the stable and unstable subspaces at a point x are denoted, respectively, by $E^s(x)$ and $E^u(x)$.

Theorem 7.1 (Hadamard–Perron). *If x is a hyperbolic fixed point of a C^1 diffeomorphism f, then there exist C^1 manifolds $V^s(x)$ and $V^u(x)$ containing x such that*

$$T_x V^s(x) = E^s(x), \quad T_x V^u(x) = E^u(x)$$

and

$$f(V^s(x)) \subset V^s(x), \quad f^{-1}(V^u(x)) \subset V^u(x).$$

The sets $V^s(x)$ and $V^u(x)$ are called, respectively, *(local) stable manifold* and *(local) unstable manifold* at x. The local behavior is indicated in Fig. 7.1.

Theorem 7.1 is a particular case of Theorem 7.2, which establishes the existence of stable and unstable manifolds for arbitrary hyperbolic sets. Let Λ be a hyperbolic set for a C^1 diffeomorphism f. Given $\varepsilon > 0$, for each $x \in \Lambda$, we consider the sets

$$V^s(x) = V^s_\varepsilon(x) = \{y \in B(x, \varepsilon) : d(f^n(x), f^n(y)) < \varepsilon \text{ for every } n \geq 0\} \tag{7.2}$$

and

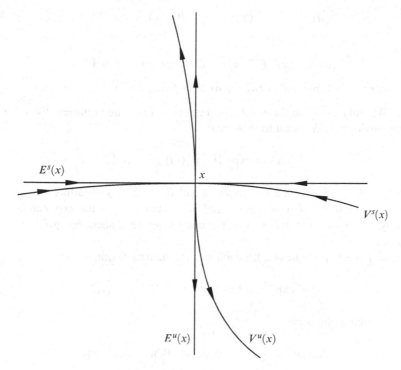

Fig. 7.1 Stable and unstable manifolds of a hyperbolic fixed point

$$V^u(x) = V_\varepsilon^u(x) = \{y \in B(x,\varepsilon) : d(f^n(x), f^n(y)) < \varepsilon \text{ for every } n \le 0\}, \quad (7.3)$$

where d is the distance in M and $B(x,\varepsilon)$ is the ball of radius ε centered at x.

Theorem 7.2. *If Λ is a hyperbolic set for a C^1 diffeomorphism f, then for any sufficiently small $\varepsilon > 0$, the following properties hold:*

1. *For each $x \in \Lambda$, the sets $V^s(x)$ and $V^u(x)$ are C^1 manifolds containing x such that*

$$T_x V^s(x) = E^s(x), \quad T_x V^u(x) = E^u(x) \quad (7.4)$$

   and

$$f(V^s(x)) \subset V^s(f(x)), \quad f^{-1}(V^u(x)) \subset V^u(f^{-1}(x)). \quad (7.5)$$

2. *There exists $\gamma = \gamma(\varepsilon) > 0$ such that*

$$V^s(x) \supset B^s(x,\gamma) \quad and \quad V^u(x) \supset B^u(x,\gamma) \quad (7.6)$$

   for every $x \in \Lambda$, where $B^s(x,\gamma)$ and $B^u(x,\gamma)$ are balls with respect to the induced distances on $V^s(x)$ and $V^u(x)$.
3. *For each $\lambda > \tau$, there exists $C > 0$ such that*

$$d(f^n(x), f^n(y)) \le C\lambda^n d(x, y), \quad y \in V^s(x) \tag{7.7}$$

and

$$d(f^{-n}(x), f^{-n}(y)) \le C\lambda^n d(x, y), \quad y \in V^u(x) \tag{7.8}$$

for every $x \in \Lambda$ and $n \in \mathbb{N}$ (with τ as in Definition 6.1).

Proof. We only establish the result for the sets $V^s(x)$. The argument for $V^u(x)$ is entirely analogous. We want to show that

$$V^s(x) = \exp_x \{(v, \psi_x^s(v)) : v \in B_\varepsilon^s\},$$

where $\psi_x^s \colon B_\varepsilon^s \to E^u(x)$ is a C^1 map in a ball $B_\varepsilon^s \subset E^s(x)$ of radius ε centered at the origin such that $\psi_x^s(0) = 0$ and $d_0\psi_x^s = 0$. Here $\exp_x$ is the exponential map, given by $\exp_x v = \gamma(1)$, where γ is the unique geodesic such that $\gamma(0) = x$ and $\gamma'(0) = v$.

Since Λ is a hyperbolic set, for each $x \in \Lambda$, the transformation

$$f_x = \exp_{f(x)}^{-1} \circ f \circ \exp_x \colon B_\varepsilon^s \times B_\varepsilon^u \to T_{f(x)}M$$

can be written in the form

$$f_x(v, w) = (\bar{A}_x v + \bar{g}_x(v, w), \bar{B}_x w + \bar{h}_x(v, w)),$$

where $(v, w) \in B_\varepsilon^s \times B_\varepsilon^u$, for some invertible linear transformations

$$\bar{A}_x \colon E^s(x) \to E^s(f(x)), \quad \bar{B}_x \colon E^u(x) \to E^u(f(x)),$$

and some maps

$$\bar{g}_x \colon B_\varepsilon^s \times B_\varepsilon^u \to E^s(x), \quad \bar{h}_x \colon B_\varepsilon^s \times B_\varepsilon^u \to E^u(x)$$

such that

$$\bar{g}_x(0, 0) = 0, \quad \bar{h}_x(0, 0) = 0, \quad d_{(0,0)}\bar{g}_x = 0, \quad \text{and} \quad d_{(0,0)}\bar{h}_x = 0. \tag{7.9}$$

By eventually considering an inner product as in Proposition 6.3, we can assume that there exists $\rho \in (\tau, 1)$ such that

$$\|\bar{A}_x\| < \rho \quad \text{and} \quad \|\bar{B}_x^{-1}\| < \rho \quad \text{for every} \quad x \in \Lambda.$$

Given $x \in \Lambda$ and $n \in \mathbb{Z}$, we consider the transformation $F_n = f_{f^n(x)}$, which we write in the form

$$F_n(v, w) = (A_n v + g_n(v, w), B_n w + h_n(v, w)),$$

where

$$A_n = \bar{A}_{f^n(x)}, \quad B_n = \bar{B}_{f^n(x)}, \quad g_n = \bar{g}_{f^n(x)}, \quad \text{and} \quad h_n = \bar{h}_{f^n(x)}.$$

Setting

$$E_n^s = E^s(f^n(x)) \quad \text{and} \quad E_n^u = E^u(f^n(x)),$$

the linear transformations

$$A_n \colon E_n^s \to E_{n+1}^s \quad \text{and} \quad B_n \colon E_n^u \to E_{n+1}^u$$

satisfy

$$\|A_n\| < \rho \quad \text{and} \quad \|B_n^{-1}\| < \rho$$

for every $n \in \mathbb{Z}$. Moreover, since f is a C^1 diffeomorphism and Λ is compact, the family $(dg_n, dh_n)_{n\in\mathbb{Z}}$ is equicontinuous in $B_{2\varepsilon}^s \times B_{2\varepsilon}^u$ (by eventually making ε smaller), and we have

$$\|d_{(v,w)}g_n\| \le K \quad \text{and} \quad \|d_{(v,w)}h_n\| \le K$$

for every $n \in \mathbb{Z}$ and $(v, w) \in B_{2\varepsilon}^s \times B_{2\varepsilon}^u$, where $K = K(\varepsilon) \to 0$ when $\varepsilon \to 0$, in view of (7.9).

Now we consider an appropriate space of Lipschitz functions. Namely, given $\kappa \in (0, 1]$, let X be the space of two-sided sequences $\varphi = (\varphi_n)_{n\in\mathbb{Z}}$ such that each function $\varphi_n \colon B_\varepsilon^s \to E_n^u$ satisfies $\varphi_n(0) = 0$ and

$$\|\varphi_n(v) - \varphi_n(w)\| \le \kappa \|v - w\|, \quad v, w \in B_\varepsilon^s.$$

One can easily verify that X is a complete metric space with the distance

$$d(\varphi, \psi) = \sup \{\|\varphi_n(v) - \psi_n(v)\| : v \in B_\varepsilon^s, \ n \in \mathbb{Z}\}.$$

Our aim is to find a sequence $\varphi \in X$ such that

$$V^s(f^n(x)) = \exp_x \{(v, \varphi_n(v)) : v \in B_\varepsilon^s\}$$

for each $n \in \mathbb{Z}$. The sequence will be obtained as the unique fixed point of a contraction operator. We first proceed formally in order to motivate the introduction of the operator. Since

$$F_n(v, \varphi_n(v)) = \big(A_n v + g_n(v, \varphi_n(v)), B_n \varphi_n(v) + h_n(v, \varphi_n(v))\big),$$

in view of (7.5), we must have

$$\varphi_{n+1}\big(A_n v + g_n(v, \varphi_n(v))\big) = B_n \varphi_n(v) + h_n(v, \varphi_n(v)),$$

that is,

$$\varphi_n(v) = B_n^{-1}\varphi_{n+1}\big(A_n v + g_n(v, \varphi_n(v))\big) - B_n^{-1} h_n(v, \varphi_n(v))$$

for each $n \in \mathbb{Z}$ and $v \in B_\varepsilon^s$. Reciprocally, if the functions φ_n satisfy this identity, then (7.5) holds. This leads us to define an operator T by

$$T(\varphi)_n(v) = B_n^{-1}\varphi_{n+1}\big(A_n v + g_n(v, \varphi_n(v))\big) - B_n^{-1} h_n(v, \varphi_n(v))$$

for each $\varphi \in X$, $n \in \mathbb{Z}$, and $v \in B_\varepsilon^s$. We want to show that T has a unique fixed point in X.

We first show that T is well defined. Take $n \in \mathbb{Z}$ and $v \in B_\varepsilon^s$. Since

$$\varphi_n(v) = \varphi_n(v) - \varphi_n(0),$$

we have $\|\varphi_n(v)\| \le \kappa\|v\| \le \varepsilon$, and hence,

$$\|A_n v + g_n(v, \varphi_n(v))\| \le \rho\varepsilon + K\|(v, \varphi_n(v))\|$$
$$< \rho\varepsilon + (1 + \kappa)K\varepsilon < \varepsilon,$$

provided that ε is sufficiently small. This allows one to compute the first term in $T(\varphi)_n(v)$. Furthermore, $T(\varphi)_n(0) = 0$, and

$$\|T(\varphi)_n(v) - T(\varphi)_n(w)\|$$
$$\le \rho\big\|\varphi_{n+1}\big(A_n v + g_n(v, \varphi_n(v))\big) - \varphi_{n+1}\big(A_n w + g_n(w, \varphi_n(w))\big)\big\|$$
$$+ \rho\|h_n(v, \varphi_n(v)) - h_n(w, \varphi_n(w))\|$$
$$\le \rho\kappa\|A_n(v - w) + g_n(v, \varphi_n(v)) - g_n(w, \varphi_n(w))\|$$
$$+ \rho K\|(v, \varphi_n(v)) - (w, \varphi_n(w))\|$$
$$\le \big(\rho^2\kappa + (1 + \kappa)^2\rho K\big)\|v - w\|$$

for each $n \in \mathbb{Z}$ and $v, w \in B_\varepsilon^s$. Provided that ε is sufficiently small, we can make K so small that

$$\|T(\varphi)_n(v) - T(\varphi)_n(w)\| \le \kappa\|v - w\|$$

for each $n \in \mathbb{Z}$ and $v, w \in B_\varepsilon^s$. This shows that $T(X) \subset X$.

Now we show that T is a contraction. Given $\varphi, \psi \in X$, $n \in \mathbb{Z}$, and $v \in B_\varepsilon^s$, we have

$$\|T(\varphi)_n(v) - T(\psi)_n(v)\|$$
$$\le \rho\big\|\varphi_{n+1}\big(A_n v + g_n(v, \varphi_n(v))\big) - \psi_{n+1}\big(A_n v + g_n(v, \psi_n(v))\big)\big\|$$
$$+ \rho\|h_n(v, \varphi_n(v)) - h_n(v, \psi_n(v))\|$$
$$\le \rho\big\|\varphi_{n+1}\big(A_n v + g_n(v, \varphi_n(v))\big) - \varphi_{n+1}\big(A_n v + g_n(v, \psi_n(v))\big)\big\|$$

$$+ \rho \| \varphi_{n+1} \big(A_n v + g_n(v, \psi_n(v)) \big) - \psi_{n+1} \big(A_n v + g_n(v, \psi_n(v)) \big) \|$$
$$+ \rho K \| \varphi_n(v) - \psi_n(v) \|$$
$$\leq \rho \kappa \| g_n(v, \varphi_n(v)) - g_n(v, \psi_n(v)) \| + \rho d(\varphi, \psi) + \rho K d(\varphi, \psi)$$
$$\leq \rho \big(1 + (1 + \kappa) K \big) d(\varphi, \psi).$$

Therefore, provided that ε is so small that $\rho' = \rho \big(1 + (1 + \kappa) K \big) < 1$, we obtain

$$d(T(\varphi), T(\psi)) \leq \rho' d(\varphi, \psi),$$

and T is a contraction in X. Hence, T has a unique fixed point $\varphi \in X$.

It remains to show that each function φ_n is of class C^1. Given $v, w \in B_\varepsilon^s$, set

$$\Delta_{v,w} \varphi_n = \frac{(w, \varphi_n(w)) - (v, \varphi_n(v))}{\| (w, \varphi_n(w)) - (v, \varphi_n(v)) \|}$$

and let $t_v \varphi_n$ be the set of vectors $w \in T_{f^n(x)} M$ such that $\Delta_{v,v_m} \varphi_n \to w$ when $m \to \infty$ for some sequence $(v_m)_{m \in \mathbb{N}}$ converging to v. We also consider the set

$$\tau_{(v, \varphi_n(v))} V_n^s = \big\{ \lambda w : w \in t_v \varphi_n \text{ and } \lambda \in \mathbb{R} \big\},$$

where

$$V_n^s = \big\{ (v, \varphi_n(v)) : v \in B_\varepsilon^s \big\}.$$

One can easily verify that φ_n is differentiable if and only if $\tau_{(v, \psi_n(v))} V_n^s$ is a subspace of dimension $\dim E_n^s$. Now we observe that $\Delta_{v,v_m} \varphi_n \to w$ when $m \to \infty$ if and only if

$$\lim_{m \to \infty} \frac{F_n(v_m, \varphi_n(v_m)) - F_n(v, \varphi_n(v))}{\| F_n(v_m, \varphi_n(v_m)) - F_n(v, \varphi_n(v)) \|} = \frac{d_{(v, \varphi_n(v))} F_n w}{\| d_{(v, \varphi_n(v))} F_n w \|}.$$

This implies that

$$\big(d_{(v, \varphi_n(v))} F_n \big) \tau_{(v, \varphi_n(v))} V_n^s = \tau_{F_n(v, \varphi_n(v))} F_n(V_n^s) \tag{7.10}$$

for every $n \in \mathbb{Z}$.

To proceed with the proof of the theorem, we need some auxiliary results. Given $\gamma \in (0, 1)$, we consider the cones

$$C_n^s = \big\{ (v, w) \in E_n^s \times E_n^u : \| w \| < \gamma \| v \| \big\} \cup \{ (0, 0) \},$$

and

$$C_n^u = \big\{ (v, w) \in E_n^s \times E_n^u : \| v \| < \gamma \| w \| \big\} \cup \{ (0, 0) \}.$$

Lemma 7.1. *For any sufficiently small $\varepsilon > 0$, we have*

$$d_{(v,w)}(F_n^{-1})C_{n+1}^s \subset C_n^s \quad \text{and} \quad d_{(v,w)}F_n C_n^u \subset C_{n+1}^u \tag{7.11}$$

for every $n \in \mathbb{Z}$ *and* $(v, w) \in B_\varepsilon^s \times B_\varepsilon^u$.

Proof of the lemma. Given $(p, q) \in E_n^s \times E_n^u$, let

$$(p', q') = d_{(v,w)}F_n(p, q) = \left(A_n p + d_{(v,w)}g_n(p, q), B_n q + d_{(v,w)}h_n(p, q)\right).$$

We obtain

$$\|p'\| \leq \rho\|p\| + K\|(p, q)\|, \tag{7.12}$$

and

$$\|q'\| \geq \rho^{-1}\|q\| - K\|(p, q)\|. \tag{7.13}$$

For $(p, q) \in C_n^u \setminus \{(0, 0)\}$, we have $\|p\| < \gamma\|q\|$, and thus,

$$\|p'\| \leq \rho\gamma\|q\| + K(1 + \gamma)\|q\|,$$

and

$$\gamma\|q'\| \geq \rho^{-1}\gamma\|q\| - K\gamma(1 + \gamma)\|q\|.$$

Thus, provided that ε is sufficiently small, we obtain $\|p'\| < \gamma\|q'\|$ for $q' \neq 0$. This establishes the second inclusion in (7.11). A similar argument yields the first inclusion. $\qquad\square$

For each $n \in \mathbb{Z}$ and $(v, w) \in B_\varepsilon^s \times B_\varepsilon^u$, we consider the intersections

$$E_n^s(v, w) = \bigcap_{j=0}^{\infty} d_{(v,w)}(F_n^{-1} \circ \cdots \circ F_{n+j-1}^{-1})\overline{C_{n+j}^s}$$

and

$$E_n^u(v, w) = \bigcap_{j=0}^{\infty} d_{(v,w)}(F_{n-1} \circ \cdots \circ F_{n-j})\overline{C_{n-j}^u}.$$

It follows from Lemma 7.1 that

$$E_n^s(v, w) \subset \overline{C_n^s} \quad \text{and} \quad E_n^u(v, w) \subset \overline{C_n^u}.$$

Lemma 7.2. *Provided that* ε *and* γ *are sufficiently small, the sets* $E_n^s(v, w)$ *and* $E_n^u(v, w)$ *are subspaces of dimensions respectively* $\dim E_n^s$ *and* $\dim E_n^u$, *vary continuously with* (v, w), *and satisfy*

$$E_n^s(v, w) \oplus E_n^u(v, w) = T_{f^n(x)}M, \tag{7.14}$$

and

$$d_{(v,w)}F_n E_n^s(v, w) = E_{n+1}^s(F_n(v, w)),$$
$$d_{(v,w)}F_n E_n^u(v, w) = E_{n+1}^u(F_n(v, w)).$$

Proof of the lemma. We observe that the set

$$H_j = d_{(v,w)}(F_{n-1} \circ \cdots \circ F_{n-j})\overline{C_{n-j}^u} \qquad (7.15)$$

contains a subspace, say F_j^u, of dimension $k = \dim E_n^u$. Let $v_{j1}, \ldots, v_{jk}$ be an orthonormal basis of F_j^u for each j. Since the closed unit ball in E_n^u is compact, there exists a sequence $(k_j)_{j \in \mathbb{N}} \subset \mathbb{N}$ such that

$$v_{k_j i} \to v_i \quad \text{when} \quad j \to \infty, \quad \text{for} \quad i = 1, \ldots, k,$$

where $v_1, \ldots, v_k$ is an orthonormal set in $E_n^u(v, w)$ (we note that the sequence of sets H_j in (7.15) is nondecreasing in j). This shows that $E_n^u(v, w)$ contains a subspace G_n^u of dimension k. A similar argument shows that $E_n^s(v, w)$ contains a subspace G_n^s of dimension $\dim M - k$. Moreover, since $C_n^s \cap C_n^u = \{(0, 0)\}$, we have

$$G_n^s \oplus G_n^u = T_{f^n(x)}M. \qquad (7.16)$$

Now we show that

$$G_n^s = E_n^s(v, w) \quad \text{and} \quad G_n^u = E_n^u(v, w).$$

For $(p, q) \in C_n^u \setminus \{(0, 0)\}$, we have $\|p\| < \gamma\|q\|$, and hence, by (7.12) and (7.13),

$$
\begin{aligned}
\|(p', q')\| &\geq \|q'\| - \|p'\| \\
&\geq \rho^{-1}\|q\| - K\|(p, q)\| - \rho\|p\| - K\|(p, q)\| \\
&\geq (\rho^{-1} - \gamma\rho)\|q\| - 2K\|(p, q)\| \\
&\geq \left(\frac{\rho^{-1} - \gamma\rho}{1 + \gamma} - 2K\right)\|(p, q)\|.
\end{aligned}
\qquad (7.17)
$$

Similarly, for $(p, q) \in C_n^s \setminus \{(0, 0)\}$, we have

$$
\begin{aligned}
\|(p', q')\| &\leq \|p'\| + \|q'\| \\
&\leq \rho\|p\| + K\|(p, q)\| + \rho^{-1}\|q\| + K\|(p, q)\| \\
&\leq (\rho + \gamma\rho^{-1})\|p\| + 2K\|(p, q)\| \\
&\leq \left(\frac{\rho + \gamma\rho^{-1}}{1 - \gamma} + 2K\right)\|(p, q)\|.
\end{aligned}
\qquad (7.18)
$$

If there exists $p \in E_n^u(v, w) \setminus G_n^u$, then it can be written in the form $p = p_s + p_u$ where $p_s \in G_n^s \setminus \{0\}$ and $p_u \in G_n^u$. By (7.17) and (7.18), we obtain

$$\|p_s\| \leq \left(\frac{\rho + \gamma\rho^{-1}}{1 - \gamma} + 2K\right)^m \left\|d_{(v,w)}(F_{n-m}^{-1} \circ \cdots \circ F_{n-1}^{-1})p_s\right\|$$

$$= \left(\frac{\rho + \gamma\rho^{-1}}{1 - \gamma} + 2K\right)^m \left\|d_{(v,w)}(F_{n-m}^{-1} \circ \cdots \circ F_{n-1}^{-1})(p - p_u)\right\|$$

$$\leq \left(\frac{(\rho + \gamma\rho^{-1})/(1 - \gamma) + 2K}{(\rho^{-1} - \gamma\rho)/(1 + \gamma) - 2K}\right)^m \|p - p_u\| \to 0$$

when $m \to \infty$, provided that ε and γ are sufficiently small. This contradiction shows that $p_s = 0$, and hence, $G_n^u = E_n^u(v, w)$. A similar argument shows that $G_n^s = E_n^s(v, w)$. By (7.16), this yields property (7.14).

It remains to establish the continuity of the spaces in (v, w). We first note that by (7.17) and (7.18),

$$\left\|d_{(v,w)}(F_{m+n-1} \circ \cdots \circ F_n)(p, q)\right\| \leq \left(\frac{\rho + \gamma\rho^{-1}}{1 - \gamma} + 2K\right)^m \|(p, q)\|$$

for every $(p, q) \in E_n^s(v, w)$ and $m \geq n$, and

$$\left\|d_{(v,w)}(F_{n-m}^{-1} \circ \cdots \circ F_{n-1}^{-1})(p, q)\right\| \leq \left(\frac{\rho^{-1} - \gamma\rho}{1 + \gamma} - 2K\right)^{-m} \|(p, q)\|$$

for every $(p, q) \in E_n^u(v, w)$ and $m \leq n$. Therefore, we can repeat the argument in the proof of Proposition 6.2 to show that the subspaces $E_n^s(v, w)$ and $E_n^u(v, w)$ vary continuously with (v, w).   □

To complete the proof of the theorem, we note that taking $\kappa = \gamma$ yields the inclusion $\tau_{(v,\varphi_n(v))} V_n^s \subset \overline{C_n^s}$, and thus, it follows from (7.10) and Lemma 7.2 that

$$\tau_{(v,\varphi_n(v))} V_n^s \subset E_n^s(v, \varphi_n(v))$$

for every $n \in \mathbb{Z}$ and $v \in B_\varepsilon^s$. On the other hand, since the set $\tau_{(v,\varphi_n(v))} V_n^s$ projects onto $E_n^s(v, \varphi_n(v))$ (because V_n^s is a graph over B_ε^s), it is a subspace with the dimension of $E_n^s(v, \varphi_n(v))$, and hence,

$$\tau_{(v,\varphi_n(v))} V_n^s = E_n^s(v, \varphi_n(v)).$$

This shows that φ_n is differentiable. Since $v \mapsto E_n^s(v, \varphi_n(v))$ is continuous, the set V_n^s is a C^1 manifold. Setting $v = 0$, we obtain $T_0 V_n^s = E_n^s$. This establishes the first two statements in Theorem 7.2. For the last statement, we observe that

$$\|F_n(v, \varphi_n(v))\| \leq \left\|\left(A_n v + g_n(v, \varphi_n(v)), \varphi_{n+1}(A_n v + g_n(v, \varphi_n(v)))\right)\right\|$$

$$\leq (1 + \gamma)\|A_n v + g_n(v, \varphi_n(v))\|$$

$$\leq (1 + \gamma)(\rho\|v\| + K\|(v, \varphi_n(v))\|)$$

$$\leq \lambda\|(v, \varphi_n(v))\|,$$

where

$$\lambda = \rho \frac{1+\gamma}{1-\gamma} + (1+\gamma)K.$$

Since λ can be made arbitrarily close to τ by taking ε and γ sufficiently small, this completes the proof of the theorem. $\qquad\square$

Properties (7.7) and (7.8) motivate the following definition:

Definition 7.1. The sets $V^s(x)$ and $V^u(x)$ are called, respectively, *(local) stable manifold* and *(local) unstable manifold* at the point x.

7.3 Ergodicity and Hopf's Argument

Here and in the following sections, we describe several applications of the existence of stable and unstable manifolds. We first describe *Hopf's argument* to establish the ergodicity of the Lebesgue measure for a hyperbolic toral automorphism.

Theorem 7.3. *The Lebesgue measure is ergodic with respect to any hyperbolic toral automorphism.*

Proof. Let λ be the Lebesgue measure in the torus $\mathbb{T}^n$. Given a continuous function $\varphi \colon \mathbb{T}^n \to \mathbb{R}$, by Birkhoff's ergodic theorem (Theorem 2.2), the functions

$$\varphi^+(x) = \lim_{m \to \infty} \frac{1}{m} \sum_{k=0}^{m-1} \varphi(T^k(x))$$

and

$$\varphi^-(x) = \lim_{m \to \infty} \frac{1}{m} \sum_{k=0}^{m-1} \varphi(T^{-k}(x))$$

are well-defined Lebesgue-almost everywhere.

Lemma 7.3. *We have $\varphi^+ = \varphi^-$ Lebesgue-almost everywhere.*

Proof of the lemma. We consider the set

$$Y = \left\{ x \in \mathbb{T}^n : \varphi^+(x) > \varphi^-(x) \right\}.$$

Since φ^+ and φ^- are invariant almost everywhere, Y is also invariant almost everywhere. Again by Theorem 2.2, we have

$$(\varphi \chi_Y)^+ = \varphi^+ \chi_Y \quad \text{and} \quad (\varphi \chi_Y)^- = \varphi^- \chi_Y$$

Lebesgue-almost everywhere. Furthermore, by (2.35),

$$\int_Y \varphi^+ \, d\lambda = \int_{\mathbb{T}^n} \varphi^+ \chi_Y \, d\lambda = \int_{\mathbb{T}^n} (\varphi \chi_Y)^+ \, d\lambda$$

$$= \int_{\mathbb{T}^n} \varphi \chi_Y \, d\lambda = \int_{\mathbb{T}^n} (\varphi \chi_Y)^- \, d\lambda$$

$$= \int_{\mathbb{T}^n} \varphi^- \chi_Y \, d\lambda = \int_Y \varphi^- \, d\lambda,$$

that is,

$$\int_Y (\varphi^+ - \varphi^-) \, d\lambda = 0.$$

Since $\varphi^+ - \varphi^- > 0$ in Y, we conclude that $\lambda(Y) = 0$. One can show in a similar manner that

$$\lambda\big(\{x \in \mathbb{T}^n : \varphi^+(x) < \varphi^-(x)\}\big) = 0.$$

Therefore, $\varphi^+ = \varphi^-$ Lebesgue-almost everywhere. $\square$

Now let $X \subset \mathbb{T}^n$ be the set of full Lebesgue measure defined by

$$X = \{x \in \mathbb{T}^n : \varphi^+(x) \text{ and } \varphi^-(x) \text{ are well defined and } \varphi^+(x) = \varphi^-(x)\}.$$

Lemma 7.4. *The following properties hold:*

1. *If $\varphi^+(x)$ is well defined, then $\varphi^+(y)$ is well defined for every $y \in V^s(x)$ and $\varphi^+(y) = \varphi^+(x)$.*
2. *If $\varphi^-(x)$ is well defined, then $\varphi^-(y)$ is well defined for every $y \in V^u(x)$ and $\varphi^-(y) = \varphi^-(x)$.*

Proof of the lemma. Since $\mathbb{T}^n$ is compact, given $\varepsilon > 0$, there exists $\delta > 0$ such that

$$|\varphi(x) - \varphi(y)| < \varepsilon \quad \text{whenever} \quad d(x, y) < \delta,$$

where d is the distance in $\mathbb{T}^n$. On the other hand, if $y \in V^s(x)$, then

$$d(T^k(y), T^k(x)) \to 0 \quad \text{when} \quad k \to \infty$$

by (7.7). Therefore, taking $\ell \in \mathbb{N}$ such that $d(T^k(y), T^k(x)) < \delta$ for $k \geq \ell$, we obtain

$$0 \leq \limsup_{m \to \infty} \frac{1}{m} \sum_{k=0}^{m-1} |\varphi(T^k(y)) - \varphi(T^k(x))|$$

$$= \limsup_{m \to \infty} \frac{1}{m} \sum_{k=\ell}^{m-1} |\varphi(T^k(y)) - \varphi(T^k(x))| \leq \varepsilon.$$

Fig. 7.2 Relation between
two stable manifolds

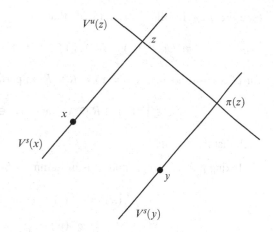

Since ε is arbitrary, we conclude that $\varphi^+(y)$ is well defined if and only if $\varphi^+(x)$ is well defined, in which case $\varphi^+(y) = \varphi^+(x)$. The desired result follows from covering $\mathbb{T}^n$ with balls of diameter δ. The proof of the second statement is analogous. $\qquad\qquad\square$

The next step is to relate two stable manifolds $V^s(x)$ and $V^s(y)$, as in Fig. 7.2.

Lemma 7.5. *For $(\lambda \times \lambda)$-almost every $(x, y) \in X \times X$, there exist points $z \in V^s(x) \cap X$ and $w \in V^s(y) \cap X$ such that $w \in V^u(z)$.*

Proof of the lemma. We consider a set $R = R^s \times R^u \subset \mathbb{T}^n$, where R^s and R^u are closed rectangles, respectively, along the stable and unstable directions. Let also m^s and m^u be the Lebesgue measures in these directions. We note that

$$\lambda(R) = \int_{R^u} m^s(V^s(x) \cap R) \, dm^u(x), \qquad (7.19)$$

and since $\lambda(X) = 1$, it follows from Lemma 7.4 that

$$\lambda(R) = \lambda(R \cap X) = \int_{R^u} m^s(V^s(x) \cap R \cap X) \, dm^u(x). \qquad (7.20)$$

We conclude from (7.19) and (7.20) that

$$m^s(R^s) = m^s(V^s(x) \cap R) = m^s(V^s(x) \cap R \cap X) \qquad (7.21)$$

for m^u-almost every $x \in R^u$.

Now for each $x, y \in R$, we define a function $\pi \colon V^s(x) \cap R \to V^s(y) \cap R$ by

$$\pi(z) = V^u(z) \cap V^s(y)$$

(see Fig. 7.2). It follows from (7.21) that

$$m^s\big(\pi(V^s(x) \cap R \cap X) \cap (V^s(y) \cap R \cap X)\big) = m^s(R^s)$$

for $(\lambda \times \lambda)$-almost every $(x, y) \in R \times R$. In particular, there exist

$$z \in V^s(x) \cap R \cap X \quad \text{and} \quad w \in V^s(y) \cap R \cap X$$

such that $\pi(z) = w$.   □

Taking points x, y, z, and w as in Lemma 7.5, it follows from Lemma 7.4 that

$$\begin{aligned}
\varphi^+(x) = \varphi^+(z) &= \varphi^-(z) \\
&= \varphi^-(w) = \varphi^+(w) = \varphi^+(y),
\end{aligned} \tag{7.22}$$

and hence, $\varphi^+(x) = \varphi^+(y)$ for $(\lambda \times \lambda)$-almost every $(x, y) \in \mathbb{T}^n \times \mathbb{T}^n$. In other words, φ^+ is constant λ-almost everywhere.

Now let $\psi \colon \mathbb{T}^n \to \mathbb{R}$ be a T-invariant function in $L^1(\mathbb{T}^n, \lambda)$. Given $\ell > 0$, there exists a continuous function $\varphi_\ell \colon \mathbb{T}^n \to \mathbb{R}$ such that

$$\int_{\mathbb{T}^n} |\psi - \varphi_\ell|\, d\lambda < 1/\ell.$$

Since ψ is T-invariant, we have

$$\psi(x) - \psi_\ell^+(x) = \lim_{m \to \infty} \frac{1}{m} \sum_{k=0}^{m-1} \big[\psi(T^k(x)) - \varphi_\ell(T^k(x))\big]$$

for μ-almost every $x \in X$, and by the dominated convergence theorem (Theorem A.3),

$$\begin{aligned}
\int_{\mathbb{T}^n} |\psi - \varphi_\ell^+|\, d\lambda &= \lim_{m \to \infty} \frac{1}{m} \int_{\mathbb{T}^n} \left| \sum_{k=0}^{m-1} (\psi - \varphi_\ell) \circ T^k \right| d\lambda \\
&\le \limsup_{m \to \infty} \frac{1}{m} \sum_{k=0}^{m-1} \int_{\mathbb{T}^n} |(\psi - \varphi_\ell) \circ T^k|\, d\lambda \\
&= \int_{\mathbb{T}^n} |\psi - \varphi_\ell|\, d\lambda < 1/\ell.
\end{aligned}$$

By (7.22), the functions φ_ℓ^+ are constant μ-almost everywhere, and hence, letting $\ell \to \infty$, we conclude that $\varphi_\ell^+ \to \psi$ μ-almost everywhere when $\ell \to \infty$ and that ψ is constant μ-almost everywhere. It follows from Proposition 2.11 that the Lebesgue measure is ergodic.   □

The alternative proof of Theorem 7.3 given by Example 2.7 (in fact in a more general setting) uses Fourier analysis. However, for arbitrary transformations, in general, one is not able to apply with success this other method to establish the ergodicity of an invariant measure. On the other hand, Hopf's argument can be applied in many other situations (see in particular Exercise 7.11), although the technical details fall out of the scope of this book. Certainly, the simpler proof of Theorem 7.3 in the special case of hyperbolic toral automorphisms substantially hides some complications.

7.4 Product Structure

Given a hyperbolic set, the families of stable and unstable manifolds $V^s(x)$ and $V^u(x)$ induce a product structure, in the following sense:

Definition 7.2. We say that a hyperbolic set Λ has a *product structure* if there exist $\varepsilon > 0$ and $\delta > 0$ such that (see Fig. 7.3)

$$\operatorname{card}(V_\varepsilon^s(x) \cap V_\varepsilon^u(y)) = 1$$

whenever $x, y \in \Lambda$ and $d(x, y) \le \delta$.

When the hyperbolic set Λ has a product structure, we write

$$[x, y] - V_\varepsilon^s(x) \cap V_\varepsilon^u(y)$$

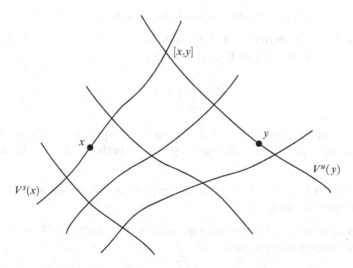

Fig. 7.3 Product structure induced by the stable and unstable manifolds

for each $x, y \in \Lambda$ with $d(x, y) \leq \delta$. We thus obtain a transformation

$$[\cdot, \cdot] : \{(x, y) \in \Lambda \times \Lambda : d(x, y) \leq \delta\} \to M. \tag{7.23}$$

A consequence of Theorem 7.2 is the following:

Proposition 7.1. *Any hyperbolic set has a product structure, and the transformation in (7.23) is continuous.*

Proof. By Proposition 6.2, the stable and unstable subspaces $E^s(x)$ and $E^u(x)$ vary continuously with $x \in \Lambda$ and are uniformly transverse. By their continuity, the same happens between any two spaces $E^s(x)$ and $E^u(y)$ with $x, y \in \Lambda$ sufficiently close. It follows from (7.4) that for ε sufficiently small, the stable and unstable manifolds $V^s(x)$ and $V^u(y)$ are transverse, and thus, the set $[x, y]$ consists of at most one point for each sufficiently close $x, y \in \Lambda$.

Furthermore, by (7.6), there exists $\gamma > 0$ such that

$$V^s(x) \supset B^s(x, \gamma) \quad \text{and} \quad V^u(y) \supset B^u(y, \gamma)$$

for $x, y \in \Lambda$. Therefore, there exists $\delta > 0$ such that if $x, y \in, \Lambda$ and $d(x, y) \leq \delta$, then $[x, y] \neq \varnothing$. This shows that the hyperbolic set has a product structure.

The continuity of the product structure follows from the continuity of the map

$$(x, y) \mapsto E^s(x) \times E^u(y)$$

and from the fact that the sets $V^s(x)$ and $V^u(y)$ are smooth manifolds tangent, respectively, to $E^s(x)$ and $E^u(y)$.   □

Now we consider a particular class of hyperbolic sets.

Definition 7.3. A hyperbolic set Λ for a diffeomorphism f is said to be *locally maximal* if there is an open set $U \supset \Lambda$ such that

$$\Lambda = \bigcap_{n \in \mathbb{Z}} f^n(U).$$

In other words, a hyperbolic set Λ for a diffeomorphism f is locally maximal if all orbits of f in a sufficiently small open neighborhood of Λ are in fact in the hyperbolic set.

Example 7.1. It is easy to verify that any hyperbolic set formed by a single fixed point is locally maximal.

For a locally maximal hyperbolic set, one can always replace M by Λ in (7.23), that is, we obtain a transformation

$$[\cdot, \cdot] : \{(x, y) \in \Lambda \times \Lambda : d(x, y) \leq \delta\} \to \Lambda. \tag{7.24}$$

Proposition 7.2. *For a locally maximal hyperbolic set* Λ, *if two points* $x, y \in \Lambda$ *are sufficiently close, then* $[x, y] \in \Lambda$.

Proof. By Proposition 7.1, the hyperbolic set Λ has a product structure. Take $x, y \in \Lambda$ with $d(x, y) \leq \delta$ such that $[x, y]$ consists of a single point in M. We must show that $[x, y] \in \Lambda$. By the continuity of the transformation $(x, y) \mapsto [x, y]$, provided that δ is sufficiently small, the image of $[\cdot, \cdot]$ is contained in some open set V with $\Lambda \subset V \subset U$, where U is any open set as in Definition 7.3. Moreover, by Property 3 in Theorem 7.2, provided that δ is sufficiently small, the open set V is so small that

$$f^n([x, y]) \in U \quad \text{for every} \quad n \in \mathbb{Z}$$

since $[x, y] \in V^s(x), V^u(y)$. This shows that $[x, y] \in \Lambda$. $\qquad\qquad\square$

7.5 The Shadowing Property

We describe in this section the so-called *shadowing property* of a locally maximal hyperbolic set. Roughly speaking, it tell us that there is a real orbit close to any sequence of points such that the image of each is sufficiently close to the next.

Let $f : M \to M$ be a C^1 diffeomorphism of a manifold M. Let also $a \in \mathbb{Z} \cup \{-\infty\}$ and $b \in \mathbb{Z} \cup \{+\infty\}$.

Definition 7.4. Given $\alpha > 0$, a sequence $(x_n)_{a \leq n \leq b} \subset M$ is called an α-*orbit* of f if

$$d(f(x_n), x_{n+1}) < \alpha \quad \text{for every} \quad n \in [a, b).$$

Definition 7.5. Given $\beta > 0$, a point $x \in M$ is said to β-*shadow* $(x_n)_{a \leq n \leq b} \subset M$ if

$$d(f^n(x), x_n) \leq \beta \quad \text{for every} \quad n \in [a, b].$$

We also say that the sequence $(x_n)_{a \leq n \leq b}$ is β-*shadowed* by the point x.

Now we establish the shadowing property.

Theorem 7.4. *Let* Λ *be a locally maximal hyperbolic set for a diffeomorphism* f. *For each* $\beta > 0$, *there exists* $\alpha > 0$ *such that each* α-*orbit* $(x_n)_{a \leq n \leq b} \subset \Lambda$ *of* f *is* β-*shadowed by some point* $x \in \Lambda$.

Proof. Let $\varepsilon > 0$ and $\delta \in (0, \varepsilon)$ be as in Definition 7.2 so that $[x, y] \in \Lambda$ whenever $x, y \in \Lambda$ (see Proposition 7.2). Let also C and λ be as in Theorem 7.2 and take $m \in \mathbb{N}$ such that $C\lambda^m \varepsilon < \delta/2$. Moreover, take $\alpha > 0$ with the property that if $(y_n)_{0 \leq n \leq m} \subset \Lambda$ is an α-orbit, then

$$d(f^n(y_0), y_n) < \delta/2 \quad \text{for every} \quad n \in [0, m].$$

Now we consider an α-orbit $(x_n)_{0 \le n \le pm}$ for some $p \in \mathbb{N}$. By the choice of m and α, we can define recursively the points $x_0' = x_0$, and

$$x_{(n+1)m}' = [x_{(n+1)m}, f^m(x_{nm}')], \quad n \in [0, p).$$

Indeed, since $x_{nm}' \in V^s(x_{nm})$, we have

$$d\big(f^m(x_{nm}'), f^m(x_{nm})\big) \le C\lambda^m \varepsilon < \delta/2,$$

and by the choice of α, we have

$$d\big(f^m(x_{nm}), x_{(n+1)m}\big) < \delta/2.$$

Therefore,

$$d\big(f^m(x_{nm}'), x_{(n+1)m}\big) < \delta,$$

and $x_{(n+1)m}'$ is well defined. By the choice of δ, we also have $x_{nm}' \in \Lambda$ for $n \in [0, p]$.

Now let $x = f^{-pm}(x_{pm}') \in \Lambda$. For each $n \in [0, pm]$, taking $q \in \mathbb{N}$ such that $n \in [qm, (q+1)m)$, we obtain

$$d\big(f^n(x), f^{n-qm}(x_{qm}')\big) \le \sum_{k=q+1}^{p} d\big(f^{n-km}(x_{km}'), f^{n-(k-1)m}(x_{(k-1)m}')\big)$$

$$\le \sum_{k=q+1}^{p} C\lambda^{km-n}\varepsilon \le \frac{C\lambda\varepsilon}{1-\lambda}$$

since $x_{km}' \in V^u(f^m(x_{(k-1)m}'))$ for each k. On the other hand, since $x_{qm}' \in V^s(x_{qm})$, and hence,

$$f^{n-qm}(x_{qm}') \in V^s(f^{n-qm}(x_{qm})),$$

we have

$$d\big(f^{n-qm}(x_{qm}'), f^{n-qm}(x_{qm})\big) < \varepsilon,$$

and by the choice of α, we have

$$d\big(f^{n-qm}(x_{qm}), x_n\big) < \delta/2 < \varepsilon/2.$$

Therefore,

$$d(f^n(x), x_n) \le d\big(f^n(x), f^{n-qm}(x_{qm}')\big) + d\big(f^{n-qm}(x_{qm}'), f^{n-qm}(x_{qm})\big)$$

$$+ d\big(f^{n-qm}(x_{qm}), x_n\big)$$

$$< \frac{C\lambda\varepsilon}{1-\tau} + \varepsilon + \frac{\varepsilon}{2} < \beta,$$

provided that ε is sufficiently small. This shows that the sequence $(x_n)_{0 \le n \le pm}$ is β-shadowed by the point x.

To establish the result for an arbitrary finite sequence, we proceed as follows. If $(x_n)_{0 \le n \le r} \subset \Lambda$ is an α-orbit, then whenever $pm \ge r$, the sequence $(x_n)_{0 \le n \le pm}$ with $x_n = f^{n-r}(x_r)$ for $n \in (r, pm]$ is also an α-orbit. Moreover, the point x constructed above for the last sequence also shadows the sequence $(x_n)_{0 \le n \le r}$. In addition, if $(x_n)_{a \le n \le b} \subset \Lambda$ is an α-orbit, then $(x_{n+a})_{0 \le n \le b-a}$ is an α-orbit, and if this sequence is β-shadowed by a point x, then the sequence $(x_n)_{a \le n \le b}$ is β-shadowed by the point $f^{-a}(x)$.

Finally, if $(x_n)_{a \le n \le b}$ is an α-orbit with $a \ge -\infty$ and $b \le +\infty$, then taking points y_ℓ that β-shadow the sequence $(x_n)_{n \in [a,b] \cap [-\ell, \ell]}$, for each $\ell \in \mathbb{N}$, it is easy to verify that $(x_n)_{a \le n \le b}$ is β-shadowed by any accumulation point of the sequence $(y_\ell)_{\ell \in \mathbb{N}}$. $\square$

The following is a consequence of Theorem 7.4:

Theorem 7.5. *Let Λ be a locally maximal hyperbolic set for a diffeomorphism f. For any sufficiently small $\alpha > 0$, if $x \in \Lambda$ and $d(f^m(x), x) < \alpha$, then there exists $y \in \Lambda$ such that $f^m(y) = y$ and*

$$d(f^n(x), f^n(y)) < \varepsilon \quad \text{for every} \quad n \in [0, m].$$

Proof. For some $\beta < \varepsilon/2$, take $\alpha \in (0, \varepsilon/2)$ as in Theorem 7.4. For each $n \in \mathbb{N}$, we set $x_n = f^k(x)$ if $n = k \mod m$ with $k \in [0, m)$. Then $(x_n)_{n \in \mathbb{N}}$ is an α-orbit of f, and by Theorem 7.4, it is β-shadowed by some point $y \in \Lambda$. We note that since the sequence x_n is m periodic, it is also β-shadowed by the point $f^m(y)$. Therefore,

$$d\big(f^n(y), f^n(f^m(y))\big) \le d(f^n(y), x_n) + d\big(x_n, f^n(f^m(y))\big) \le 2\beta < \varepsilon$$

for every $n \in \mathbb{Z}$. It follows from (7.2) and (7.3) that

$$f^m(y) \in V^s(y) \cap V^u(y),$$

and hence, $f^m(y) = y$. Moreover,

$$d(f^n(x), f^n(y)) \le d(f^n(x), x_n) + d(x_n, f^n(y))$$
$$\le d(f^m(x), x) + d(x_n, f^n(y)) < \alpha + \beta < \varepsilon$$

for every $n \in [0, m]$. This completes the proof of the theorem. $\square$

7.6 Construction of Markov Partitions

In this section, we use the product structure determined by the stable and unstable manifolds to construct Markov partitions for any locally maximal hyperbolic set.

Fig. 7.4 Intersection of a
stable manifold with a
rectangle

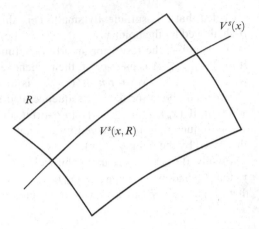

Let Λ be a locally maximal hyperbolic set for a C^1 diffeomorphism f and take $\delta > 0$ (and $\varepsilon > 0$) such that the transformation $[\cdot, \cdot]$ in (7.24) is well defined.

Definition 7.6. A closed set $R \subset \Lambda$ is called a *rectangle* if:

1. diam $R < \delta$ and $R = \overline{\text{int } R}$ (with the interior taken with respect to the induced topology on Λ).
2. $[x, y] \in R$ whenever $x, y \in R$.

This notion of rectangle includes the rectangles used in the construction of a coding map for a hyperbolic automorphism of $\mathbb{T}^2$ (see Sect. 6.4.1).

Now we introduce the notion of Markov partition. Given a rectangle R and a point $x \in R$, we consider the sets (see Fig. 7.4)

$$V^s(x, R) = V^s(x) \cap R \quad \text{and} \quad V^u(x, R) = V^u(x) \cap R.$$

Definition 7.7. A collection of rectangles $R_1, \ldots, R_k \subset \Lambda$ is called a *Markov partition* of Λ (with respect to f) if:

1. int $R_i \cap$ int $R_j = \varnothing$ whenever $i \neq j$.
2. If $x \in$ int R_i and $f(x) \in$ int R_j, then

$$f\left(V^u(x, R_i)\right) \supset V^u(f(x), R_j) \quad \text{and} \quad f^{-1}\left(V^s(f(x), R_j)\right) \supset V^s(x, R_i).$$

As for the hyperbolic automorphisms of $\mathbb{T}^2$, the second property is called the *Markov property* of the partition.

Example 7.2. The Markov partitions in Definition 6.5, for a hyperbolic automorphism of $\mathbb{T}^2$, are Markov partitions in the sense of Definition 7.7.

The following result establishes the existence of Markov partitions for any locally maximal hyperbolic set:

Theorem 7.6. *Any locally maximal hyperbolic set has Markov partitions of arbitrarily small diameter.*

Proof. Let Λ be a locally maximal hyperbolic set for a diffeomorphism f. Take $\alpha, \beta > 0$ as in Theorem 7.4. Take also $r \in (0, \alpha/2)$ such that

$$d(f(x), f(y)) < \alpha/2 \quad \text{whenever} \quad d(x, y) < r.$$

We consider a set $\Sigma = \{x_1, \ldots, x_N\}$ such that each point in Λ is at a distance at most r of some element of Σ, and we let

$$\Sigma' = \{(y_n)_{n \in \mathbb{Z}} \in \Sigma^{\mathbb{Z}} : d(f(y_n), y_{n+1}) < \alpha \text{ for } n \in \mathbb{Z}\}.$$

For each sequence $y = (y_n)_{n \in \mathbb{Z}} \in \Sigma'$, there exists a unique point $p_y \in \Lambda$ that β-shadows y. The existence follows from Theorem 7.4. For the uniqueness, we observe that if $x \in \Lambda$ is another point that β-shadows y, then

$$d\big(f^n(x'), f^n(p_y)\big) \le d\big(f^n(x'), y_n\big) + d\big(y_n, f^n(p_y)\big) \le 2\beta \tag{7.25}$$

for every $n \in \mathbb{Z}$. Provided that $2\beta < \varepsilon$, it follows from (7.2) and (7.3) that $x' \in V^s(p_y) \cap V^u(p_y)$, and thus, $x' = p_y$. Moreover, since $r < \alpha/2$, for each $x \in \Lambda$, there exists $y \in \Sigma'$ such that $x = p_y$.

Lemma 7.6. *The transformation $h \colon \Sigma' \to \Lambda$ defined by $h(y) = p_y$ is continuous.*

Proof of the lemma. Otherwise, there would exist $\delta > 0$ and for each $m \in \mathbb{N}$ points $z_m, z'_m \in \Sigma'$ such that $z_{mi} = z'_{mi}$ for every $i \in [-m, m]$, with

$$d(p_{z_m}, p_{z'_m}) \ge \delta. \tag{7.26}$$

On the other hand,

$$d\big(f^n(p_{z_m}), f^n(p_{z'_m})\big) \le 2\beta \quad \text{for} \quad n \in [-m, m]. \tag{7.27}$$

Eventually taking subsequences, we may assume that $p_{z_m} \to x$ and $p_{z'_m} \to x'$ when $m \to \infty$. It thus follows from (7.27) that

$$d\big(f^n(x), f^n(x')\big) \le 2\beta \quad \text{for} \quad n \in \mathbb{Z}.$$

As in (7.25), provided that $2\beta < \varepsilon$, this implies that $x = x'$, while (7.26) yields that $d(x, x') \ge \delta$. This contradiction shows that the transformation h is continuous. □

For each $i = 1, \ldots, N$, we consider the set

$$T_i = \{p_y : y \in \Sigma' \text{ and } y_0 = x_i\} = h\big(\{y \in \Sigma' : y_0 = x_i\}\big).$$

Since the cylinder set $\{y \in \Sigma' : y_0 = x_i\}$ is closed, it follows from Lemma 7.6 that T_i is closed for $i = 1, \ldots, N$. Moreover, for each $y, y' \in \Sigma'$ with $y_0 = y'_0$, we define a point $y \cdot y' \in \Sigma'$ by

$$(y \cdot y')_n = \begin{cases} y_n & \text{if } n \geq 0, \\ y'_n & \text{if } n \leq 0. \end{cases}$$

We note that

$$d\left(f^n(p_{y \cdot y'}), f^n(p_y)\right) \leq 2\beta \quad \text{for} \quad n \geq 0$$

and

$$d\left(f^n(p_{y \cdot y'}), f^n(p_{y'})\right) \leq 2\beta \quad \text{for} \quad n \leq 0.$$

Provided that $2\beta < \varepsilon$, this implies that

$$p_{y \cdot y'} \in V^s(p_y) \cap V^u(p_{y'}), \quad \text{that is,} \quad p_{y \cdot y'} = [p_y, p_{y'}].$$

Let $x = p_y$, $x' = p_{y'} \in T_i$, where $y_0 = y'_0 = x_i$. Then

$$[x, x'] = [p_y, p_{y'}] = p_{y \cdot y'} \in T_i. \tag{7.28}$$

To show that the sets T_i satisfy a certain Markov property, we assume that $y_1 = x_j$ and that $x' \in V^s(x, T_i) := V^s(x) \cap T_i$. Then

$$x' = [x, x'] = p_{y \cdot y'},$$

and since $\sigma(y \cdot y')_0 = x_i$, we obtain

$$f(x') = p_{\sigma(y \cdot y')} \in T_j,$$

where σ is the shift map. Therefore, $f(x') \in V^s(f(x), T_j)$, and hence,

$$f\left(V^s(x, T_i)\right) \subset V^s(f(x), T_j)). \tag{7.29}$$

One can show in a similar manner that

$$f\left(V^u(x, T_i)\right) \supset V^u(f(x), T_j). \tag{7.30}$$

Now for each $i = 1, \ldots, N$, we set

$$\partial^s T_i = \left\{x \in T_i : x \notin \text{int } V^u(x, T_i)\right\}$$

and

$$\partial^u T_i = \left\{x \in T_i : x \notin \text{int } V^s(x, T_i)\right\}.$$

Lemma 7.7. *For $i = 1, \ldots, N$, we have*

$$\partial T_i = \partial^s T_i \cup \partial^u T_i. \tag{7.31}$$

Proof of the lemma. If $x \in \operatorname{int} T_j$, then

$$V^u(x, T_i) = T_i \cap (V^u(x) \cap \Lambda)$$

is an open neighborhood of x in $V^u \cap \Lambda$, and hence, $x \in \operatorname{int} V^u(x, T_i)$. We show in a similar manner that $x \in \operatorname{int} V^s(x, T_i)$.

Now we assume that $x \in \operatorname{int} V^u(x, T_i)$ and $x \in \operatorname{int} V^s(x, T_i)$. Given $y \in \Lambda$ sufficiently close to x, the points

$$[x, y] \in V^s \cap \Lambda \quad \text{and} \quad [y, x] \in V^u(x) \cap \Lambda$$

are well defined and vary continuously with y. This implies that $[x, y], [y, x] \in T_i$ for any $y \in \Lambda$ sufficiently close to x. It follows from (7.28) that

$$y = \big[[y, x], [x, y]\big] \in T_i,$$

and hence, $x \in \operatorname{int} T_i$. This establishes (7.31). $\qquad\qquad\square$

The sets T_i shall be used to construct the Markov partition. For this, given $x \in \Lambda$, we consider the families

$$T(x) = \{T_i : x \in T_i\} \quad \text{and} \quad S(x) = \{T_j : T_j \cap T_i \neq \varnothing \text{ for some } T_i \in T(x)\}.$$

We can proceed in a similar manner to that in the proof of Lemma 7.7 to show that the set

$$A = \{x \in \Lambda : V^s(x) \cap \partial^s T = \varnothing \text{ and } V^u(x) \cap \partial^u T = \varnothing \text{ for every } T \in S(x)\}$$

is open and dense in Λ. Given $i, j = 1, \ldots, N$ with $T_i \cap T_j \neq \varnothing$, let

$$T_{ij}^1 = \{x \in T_j : V^u(x, T_i) \cap T_j \neq \varnothing, V^s(x, T_i) \cap T_j \neq \varnothing\},$$

$$T_{ij}^2 = \{x \in T_j : V^u(x, T_i) \cap T_j \neq \varnothing, V^s(x, T_i) \cap T_j = \varnothing\},$$

$$T_{ij}^3 = \{x \in T_j : V^u(x, T_i) \cap T_j = \varnothing, V^s(x, T_i) \cap T_j \neq \varnothing\},$$

$$T_{ij}^4 = \{x \in T_j : V^u(x, T_i) \cap T_j = \varnothing, V^s(x, T_i) \cap T_j = \varnothing\}.$$

We note that all these sets are open in Λ. Moreover, if $x, y \in T_i$, then

$$V^s\big([x, y], T_i\big) = V^s(x, T_i) \quad \text{and} \quad V^u\big([x, y], T_i\big) = V^u(y, T_i).$$

Therefore, each set $B = \operatorname{int} T_{ij}^k$ has the property that $[x, y] \in B$ whenever $x, y \in B$. Furthermore, each point $x \in T_i \cap A$ is in $\operatorname{int} T_{ij}^k$ for some j and k.

For each $x \in A$, we consider the open set

$$R(x) = \bigcap \{\operatorname{int} T_{ij}^k : T_i \cap T_j \neq \varnothing \text{ and } x \in T_{ij}^k\}.$$

One can easily verify that

$$[y, z] \in R(x) \quad \text{whenever} \quad y, z \in R(x).$$

Now let $y \in R(x) \cap A$. Since $R(x) = \bigcap_{x \in T_i} T_i$ and $R(x) \cap T_i = \varnothing$ for $T_i \notin T(x)$, we have $T(y) = T(x)$. Moreover, given $T_i \in T(x) = T(y)$ with $T_i \cap T_j \neq \varnothing$, the points x and y are in the same set T_{ij}^k since $T_{ij}^k \supset R(x)$. This shows that $R(y) = R(x)$. Furthermore, if $R(x) \cap R(y) \neq \varnothing$, for some $x, y \in A$, then there exists $z \in R(x) \cap R(y) \cap A$, and we obtain $R(x) = R(y) = R(z)$.

Now we consider the family

$$\mathcal{R} = \{\overline{R(x)} : x \in A\}.$$

This will be our Markov partition. For each $x, y \in A$, we have

$$R(x) = R(y) \quad \text{or} \quad R(x) \cap R(y) = \varnothing.$$

This implies that

$$(\overline{R(x)} \setminus R(x)) \cap A = \varnothing.$$

Since A is dense in Λ, the set $\overline{R(x)} \setminus R(x)$ has empty interior, and hence, $R(x) = \operatorname{int} \overline{R(x)}$. This shows that each closure $\overline{R(x)}$ is a rectangle (see Definition 7.6).

We show that $\mathcal{R}$ is a Markov partition. Given $R(x) \neq R(y)$ with $x, y \in A$, we have

$$\operatorname{int} \overline{R(x)} \cap \operatorname{int} \overline{R(y)} = R(x) \cap R(y) = \varnothing.$$

It remains to verify the Markov property. We start with an auxiliary statement.

Lemma 7.8. *For each $x, y \in A \cap f^{-1}(A)$ with $R(x) = R(y)$ and $y \in V^s(x)$, we have $R(f(x)) = R(f(y))$.*

Proof of the lemma. We first show that $T(f(x)) = T(f(y))$. Otherwise, if $f(x) \in T_i$ and $f(y) \notin T_i$, let $f(x) = p_{\sigma(y)}$, with $y_0 = x_\ell$ and $y_1 = x_i$. Then $x = p_y \in T_\ell$, and it follows from (7.29) that

$$f(y) \in f(V^s(x, T_\ell)) \subset V^s(f(x), T_i),$$

which contradicts the assumption that $f(y) \notin T_i$.

Now we show that $R(f(x)) = R(f(y))$. Let $f(x), f(y) \in T_i$ and assume that $T_i \cap T_j \neq \varnothing$. Since $f(y) \in V^s(f(x))$, we have

$$V^s(f(y), T_i) = V^s(f(x), T_i).$$

In order to proceed by contradiction, we assume that

$$V^u(f(y), T_i) \cap T_j = \varnothing \quad \text{and} \quad V^u(f(x), T_i) \cap T_j \neq \varnothing, \qquad (7.32)$$

and we take a point $f(z)$ in the second intersection. Let again $f(x) = p_y$ with $y_0 = x_\ell$ and $y_1 = x_i$. It follows from (7.30) that

$$f(z) \in V^u(f(x), T_i) \subset f(V^u(x, T_\ell)),$$

and hence, $z \in V^u(x, T_\ell)$. Let also $f(z) = p_{\sigma(y')}$ with $y'_0 = x_m$ and $y'_1 = x_j$. Then $z \in T_m$ and

$$f(V^s(z, T_m)) \subset V^s(f(z), T_j).$$

We thus have $T_\ell \in T(x) = T(y)$ and $z \in T_m \cap T_\ell \neq \varnothing$. Since $z \in V^u(x, T_\ell) \cap T_m$, there exists $w \in V^u(y, T_\ell) \cap T_m$ (note that x and y are in the same set $T^k_{\ell m}$). We thus obtain

$$[z, y] = [z, w] \in V^s(z, T_m) \cap V^u(y, T_\ell),$$

and

$$f([z, y]) = [f(z), f(y)] \in V^s(f(z), T_j) \cap V^u(f(y), T_i),$$

which contradicts (7.32). This shows that $R(f(x)) = R(f(y))$.   □

Now we consider the sets

$$C = \bigcup \left\{ V^s_\delta(z) : z \in \bigcup_{i=1}^{N} \partial^s T_i \right\}$$

and

$$D = \bigcup \left\{ V^u_\delta(z) : z \in \bigcup_{i=1}^{N} \partial^u T_i \right\},$$

for some $\delta > 0$ sufficiently small. Repeating former arguments, one can show that C and D are closed and have empty interior. Then the set $\Lambda \setminus (C \cup D) \subset \Lambda$ is open and dense. If

$$x \notin (C \cup D) \cap f^{-1}(C \cup D),$$

then $x \in \Lambda \cap f^{-1}(\Lambda)$, and the set

$$\{y \in V^s(x, R(x)) : y \in \Lambda \cap f^{-1}(\Lambda)\}$$

is open and dense in $V^s(x, \overline{R(x)})$ (with respect to the induced topology on the set $V^s(x) \cap \Lambda$). By Lemma 7.8, for each such y, we have $R(f(y)) = R(f(x))$, and thus, by continuity,

$$f(V^s(x, \overline{R(x)})) \subset \overline{R(f(x))}.$$

Since

$$f(V^s(x, \overline{R(x)})) \subset V^s(f(x)),$$

we thus obtain

$$f(V^s(x, \overline{R(x)})) \subset V^s(f(x), \overline{R(f(x))}). \tag{7.33}$$

If int $R_i \cap f^{-1}(\text{int } R_j) \neq \varnothing$ for some $R_i, R_j \in \mathcal{R}$, then there exist x and y as above such that $R_i = \overline{R(x)}$ and $R_j = \overline{R(f(x))}$. For each $y \in R_i \cap f^{-1}(R_j)$, we have

$$V^s(y, R_i) = \{[y, z] : V^s(x, R_i)\},$$

and hence, by (7.33),

$$\begin{aligned}
f(V^s(y, R_i)) &= \{[f(y), f(z)] : z \in V^s(x, R_i)\} \\
&\subset \{[f(y), w] : w \in V^s(f(x), R_j)\} \\
&\subset V^s(f(y), R_j).
\end{aligned}$$

The second inclusion in the Markov property can be obtained in a similar manner.
$\square$

In a similar manner to that for a hyperbolic automorphism of $\mathbb{T}^2$, to each Markov partition of a locally maximal hyperbolic set, we can associate a coding map. Namely, given a Markov partition $R_1, \ldots, R_k$ of a locally maximal hyperbolic set Λ for a diffeomorphism f, we define a $k \times k$ matrix $A = (a_{ij})$ by

$$a_{ij} = \begin{cases} 1 & \text{if int } f(R_i) \cap \text{int } R_j \neq \varnothing, \\ 0 & \text{if int } f(R_i) \cap \text{int } R_j = \varnothing. \end{cases}$$

Using the Markov property and the expansion and contraction along the stable and unstable manifolds, one can define a *coding map* $h: X_A \to \Lambda$ by

$$h(\omega) = \bigcap_{n \in \mathbb{Z}} f^{-n}(R_{i_n(\omega)}).$$

Theorem 7.7. *For any Markov partition of a locally maximal hyperbolic set Λ and its associated coding map h, the following properties hold:*

1. h is continuous and onto.
2. $h \circ \sigma = f \circ h$ in X_A.
3. h is injective in the set

$$\Lambda \setminus \bigcup_{n \in \mathbb{Z}} \bigcup_{i=1}^{k} f^n(\partial R_i) = \Lambda \setminus \bigcup_{n \in \mathbb{Z}} \bigcup_{i=1}^{k} f^n(\partial^s R_i \cup \partial^u R_i),$$

where

$$\partial^s R_i = \{x \in R_i : x \notin \text{int } V^u(x, R_i)\}$$

and

$$\partial^u R_i = \{x \in R_i : x \notin \text{int } V^s(x, R_i)\}.$$

4. card $h^{-1}x \le k^2$ *for every* $x \in \Lambda$.

Proof. Properties 1 and 2 can be obtained with similar arguments to those in the proof of Proposition 6.5. Property 3 is a simple consequence of the definitions (see also the proof of Lemma 7.7).

It remains to establish Property 4. We assume that card $h^{-1}x > k^2$ for some $x \in \Lambda$, and we proceed by contradiction. Take $n \in \mathbb{N}$ such that

$$(i_{-n} \cdots i_n) \ne (i'_{-n} \cdots i'_n) \tag{7.34}$$

for some sequences

$$\omega = (\cdots i_{-1}i_0i_1 \cdots), \omega' = (\cdots i'_{-1}i'_0i'_1 \cdots) \in \text{card } h^{-1}x.$$

Since the points (i_{-n}, i_n) take at most k^2 values, we can also assume that $i_{-n} = i'_{-n}$ and $i_n = i'_n$. Since $h(\omega) = h(\omega')$, we have

$$R_{i_k} \cap R_{i'_k} \ne \emptyset \quad \text{for} \quad k = -n, \ldots, n. \tag{7.35}$$

Now we take points $x, y \in \Lambda$ such that

$$f^k(x) \in \text{int } R_{i_k} \quad \text{and} \quad f^k(y) \in \text{int } R_{i'_k}$$

for $k = -n, \ldots, n$, which exist since

$$\bigcap_{k=-n}^{n} f^{-k}(\text{int } R_{i_k}) \quad \text{and} \quad \bigcap_{k=-n}^{n} f^{-k}(\text{int } R_{i'_k})$$

are nonempty open sets (with respect to the induced topology on Λ). Provided that δ is sufficiently small, it follows from (7.35) that $[f^k(x), f^k(y)]$ is well defined, and thus,

$$f^k([x, y]) = [f^k(x), f^k(y)] \quad \text{for} \quad k = -n, \ldots, n.$$

On the other hand, since $i_{-n} = i'_{-n}$ and $i_n = i'_n$, we have

$$[f^{-n}(x), f^{-n}(y)] \in \text{int } R_{i_{-n}} = \text{int } R_{i'_{-n}}$$

and

$$[f^n(x), f^n(y)] \in \text{int } R_{i_n} = \text{int } R_{i'_n}.$$

Therefore, since

$$[f^{-n}(x), f^{-n}(y)] \in V^s(f^{-n}(x), R_{i_{-n}}) \quad \text{and} \quad [f^n(x), f^n(y)] \in V^u(f^n(y), R_{i'_n}),$$

it follows from the Markov property that

$$[f^k(x), f^k(y)] = f^{k+n}([f^{-n}(x), f^{-n}(y)]) \in \text{int } V^s(f^k(x), R_{i_k}) \subset \text{int } R_{i_k}$$

and

$$[f^k(x), f^k(y)] = f^{k-n}([f^n(x), f^n(y)]) \in \text{int } V^u(f^k(x), R_{i_k'}) \subset \text{int } R_{i_k'}$$

for $k = -n, \ldots, n$. This shows that $R_{i_k} = R_{i_k'}$ for each k. On the other hand, by (7.34), there is k such that int $R_{i_k} \cap$ int $R_{i_k'} = \varnothing$. This contradiction shows that card $h^{-1}x \le k^2$. $\qquad\qquad\square$

Example 7.3. Let $T : \mathbb{T}^n \to \mathbb{T}^n$ be a hyperbolic automorphism induced by a matrix B. By Property 4 in Theorem 7.7, if A is the transition matrix associated to a Markov partition, then

$$\text{card}\left\{x \in \mathbb{T}^n : T^m(x) = x\right\} \le \text{card}\left\{\omega \in X_A : \sigma^m(\omega) = \omega\right\}$$

$$\le k^2 \,\text{card}\left\{x \in \mathbb{T}^n : T^m(x) = x\right\}$$

for each $m \in \mathbb{N}$. Therefore, by (3.35),

$$p(T) = p(\sigma|X_A) = \log \rho(A).$$

It follows from Exercise 6.13 that

$$h(T) = h_\lambda(T) = p(T) = \log \rho(A) = \sum_{\lambda \in \text{Sp}(B):|\lambda|>1} \log|\lambda|,$$

where λ is the Lebesgue measure in $\mathbb{T}^n$.

7.7   Exercises

Exercise 7.1. Show that any local stable manifold and any local unstable manifold contain at most one periodic point.

Exercise 7.2. Let T be a hyperbolic automorphism of $\mathbb{T}^2$ induced by a matrix A.

1. Show that the eigenvalues of A are real if and only if $|\text{tr } A| \ge 2$. In particular, if all entries of A are positive, then both eigenvalues are real.
2. Show that the stable and unstable manifolds are dense in $\mathbb{T}^2$.

Exercise 7.3. Find all invariant curves of the linear transformation A in (7.1) with $p = q = 1$. Hint: show that the graph

Fig. 7.5 Invariant curves of a
linear transformation

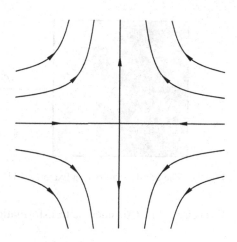

$$C = \{(x, \varphi(x)) : x \in \mathbb{R}\}$$

of a function $\varphi \colon \mathbb{R} \to \mathbb{R}$ is A-invariant if and only if

$$\varphi(\tau x) = \rho \varphi(x) \quad \text{for every} \quad x \in \mathbb{R}$$

and consider the functions $\varphi(x) = cx^{\log \rho / \log \tau}$ for $c \in \mathbb{R}$ (see Fig. 7.5).

Exercise 7.4. Let Λ be a hyperbolic set for a diffeomorphism $f \colon M \to M$. Show that there exists $\varepsilon > 0$ such that if $x \in \Lambda$ and $y \in M$ are distinct, then

$$d(f^n(x), f^n(y)) > \varepsilon \quad \text{for some} \quad n \in \mathbb{Z}.$$

Exercise 7.5. Show that the Smale horseshoe is a locally maximal hyperbolic set.

Exercise 7.6. Construct a Markov partition for the Smale horseshoe.

Exercise 7.7. Let Λ be a hyperbolic set with product structure $[\cdot, \cdot]$ such that if two points $x, y \in \Lambda$ are sufficiently close, then $[x, y] \in \Lambda$. Show that Λ is locally maximal.

Exercise 7.8. Show that the solenoid Λ in Exercise 6.12 is locally maximal.

Exercise 7.9. Construct a Markov partition for the solenoid.

Exercise 7.10. Let Λ be a locally maximal hyperbolic set for a diffeomorphism f. Show that for each number $\beta > 0$, there exists $\alpha > 0$ such that if $x \in \Lambda$ satisfies $d(f^n(x), x) < \alpha$ for some $n \in \mathbb{N}$, then there exists $y \in \Lambda$ with $f^n(y) = y$ and

$$d\big(f^k(y), f^k(x)\big) \le \beta \quad \text{for every} \quad k \in [0, n].$$

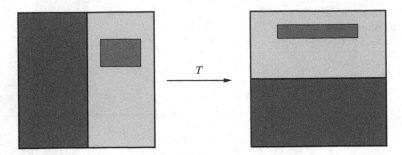

Fig. 7.6 The transformation T in Example 7.11

Exercise 7.11. Consider the transformation $T: [0, 1)^2 \to [0, 1)^2$ defined by

$$T(x, y) = \begin{cases} (2x, y/2) & \text{if } 0 \le x < 1/2, \\ (2x - 1, y/2 + 1/2) & \text{if } 1/2 \le x < 1 \end{cases}$$

(see Fig. 7.6).

1. Show that T preserves the Lebesgue measure λ in $[0, 1)^2$.
2. Use Hopf's argument to show that λ is ergodic.
3. Say if the pair (T, λ) is equivalent to a shift map with a one-sided or a two-sided Bernoulli measure.
4. Compute the entropies $h_\lambda(T)$ and $h(T)$.

Exercise 7.12. Let $f: \mathbb{T}^m \to \mathbb{T}^m$ be a C^1 diffeomorphism such that $\mathbb{T}^m$ is a hyperbolic set for f. Let also μ be an f-invariant probability measure in $\mathbb{T}^m$ absolutely continuous with respect to the Lebesgue measure, with continuous and positive Radon–Nikodym derivative ρ.

1. Show that

$$\lim_{n \to \infty} \frac{\rho(f^n(y))}{\rho(f^n(x))} = 1$$

for every $x \in \mathbb{T}^m$ and $y \in V^s(x)$. Hint: note that

$$\rho(f^n(y)) \le \rho(f^n(x)) + |\rho(f^n(y)) - \rho(f^n(x))|.$$

2. Show that if ρ is Hölder continuous, then there exists $\tau > 0$ such that

$$\limsup_{n \to \infty} \frac{1}{n} \log \left| \frac{\rho(f^n(y))}{\rho(f^n(x))} - 1 \right| < -\tau$$

for every $x \in \mathbb{T}^m$ and $y \in V^s(x)$.

3. Show that

$$\frac{\rho(y)}{\rho(x)} = \prod_{k=0}^{\infty} \frac{\det d_{f^k(y)} f}{\det d_{f^k(x)} f} = \lim_{n \to \infty} \frac{\det d_y f^n}{\det d_x f^n}$$

for Lebesgue-almost every $(x, y) \in \mathbb{T}^m \times \mathbb{T}^m$ such that $y \in V^s(x)$. Hint: use Exercise 2.21.

Notes

We refer to Anosov [4] for a historical account of Theorem 7.1. Theorem 7.2 is due to Hirsch and Pugh [40]. For further developments of Hopf's argument, we refer to Anosov and Sinai [5] for Anosov systems (see also [44]) and to Pesin [72] for nonuniformly hyperbolic systems (see also [10, 11]). Our proof of the shadowing property (Theorem 7.4) is based on [20]. Theorem 7.5 is due to Anosov [4] in the case of Anosov diffeomorphisms and to Bowen [19] in the general case. The construction of Markov partitions is due to Sinai [94, 95] in the case of Anosov diffeomorphisms and to Bowen [18] in the general case (using the shadowing property). Our proof of the existence of Markov partitions for a locally maximal hyperbolic set (Theorem 7.6) is based on [20]. The argument for the proof of the last statement in Theorem 7.7 is taken from [89].

Part IV
Dimension Theory

Chapter 8
Basic Notions and Examples

This chapter is a self-contained introduction to the basic notions and results of dimension theory. The presentation is oriented primarily towards the applications to dimension theory of hyperbolic dynamics in the following chapter. We first introduce the notions of Hausdorff dimension and of lower and upper box dimensions, both for sets and measures. We then consider the notions of lower and upper pointwise dimensions, and show how they can be used to estimate and sometimes compute the dimension of a measure. All the notions and results are illustrated with various examples, including model examples of repellers and hyperbolic sets.

8.1 Introduction

The notion of Hausdorff dimension, introduced by Hausdorff in 1918, attributes a possibly noninteger dimension to a set that may not be an Euclidean space or even a manifold. In particular, it plays an important role in geometric measure theory and in the theory of dynamical systems. Certainly, being a single number, it fails to give information about any nonuniform mass distribution on the set. On the other hand, it is still able to distinguish many sets with a complicated structure.

A practical drawback of the Hausdorff dimension, which is defined in terms of Hausdorff measures, is that it is often difficult to compute. This partly caused the introduction of several other characteristics of dimension type that sometimes are easier to compute. This is the case, for example, of the lower and upper box dimensions of a set. On the other hand, these various notions of dimension need not coincide, and a considerable part of the theory is dedicated to describe criteria either for their coincidence or for their noncoincidence. In this chapter, mostly we shall be concentrated in obtaining general relations between the Hausdorff dimension and the lower and upper box dimensions, in view of the applications to dimension theory of dynamical systems. In particular, we shall avoid discussing specific classes of sets that have no dynamical nature.

L. Barreira, *Ergodic Theory, Hyperbolic Dynamics and Dimension Theory*, Universitext, DOI 10.1007/978-3-642-28090-0_8, © Springer-Verlag Berlin Heidelberg 2012

8.2 Dimension of Sets

We introduce in this section the Hausdorff dimension and the lower and upper box dimensions of a given set. We also give several examples that illustrate the computation of the dimensions. Sometimes this is a daunting task, which may also depend on the particular class of sets under consideration.

8.2.1 Basic Notions

Let X be a subset of a smooth manifold, and let d be the distance in X. We define the *diameter* of a collection $\mathcal{U}$ of subsets of X by

$$\operatorname{diam} \mathcal{U} = \sup \{\operatorname{diam} U : U \in \mathcal{U}\},$$

where

$$\operatorname{diam} U = \sup \{d(x, y) : x, y \in U\}$$

is the *diameter* of the set U. Given $Z \subset X$ and $\alpha \in \mathbb{R}$, we define the α-*dimensional Hausdorff measure* of Z by

$$m(Z, \alpha) = \lim_{\varepsilon \to 0} \inf_{\mathcal{U}} \sum_{U \in \mathcal{U}} (\operatorname{diam} U)^{\alpha}, \qquad (8.1)$$

where the infimum is taken over all finite or countable covers $\mathcal{U}$ of the set Z with $\operatorname{diam} \mathcal{U} \le \varepsilon$. One can easily verify that the function $\alpha \mapsto m(Z, \alpha)$ jumps from $+\infty$ to 0 at a single point, and thus, the notion of Hausdorff dimension can be introduced as follows:

Definition 8.1. The *Hausdorff dimension* of a set $Z \subset X$ is defined by

$$\dim_H Z = \inf \{\alpha \in \mathbb{R} : m(Z, \alpha) = 0\}$$
$$= \sup \{\alpha \in \mathbb{R} : m(Z, \alpha) = +\infty\}.$$

The *lower* and *upper box dimensions* of $Z \subset X$ are defined, respectively, by

$$\underline{\dim}_B Z = \liminf_{\varepsilon \to 0} \frac{\log N(Z, \varepsilon)}{-\log \varepsilon} \quad \text{and} \quad \overline{\dim}_B Z = \limsup_{\varepsilon \to 0} \frac{\log N(Z, \varepsilon)}{-\log \varepsilon},$$

where $N(Z, \varepsilon)$ denotes the least number of balls of radius ε that are needed to cover the set Z.

Proposition 8.1. *We have*

$$\dim_H Z \le \underline{\dim}_B Z \le \overline{\dim}_B Z. \qquad (8.2)$$

Proof. The second inequality is clear. For the first, we note that by considering only the covers $\mathcal{V}$ of the set Z by balls of radius ε we obtain

$$m(Z,\alpha) \leq \liminf_{\varepsilon \to 0} \sum_{V \in \mathcal{V}} (\text{diam } U)^\alpha$$

$$= \liminf_{\varepsilon \to 0} \sum_{V \in \mathcal{V}} (2\varepsilon)^\alpha$$

$$= \lim_{\varepsilon \to 0} \left[(2\varepsilon)^\alpha N(Z,\varepsilon) \right].$$

Now take $\alpha < \dim_H Z$. Then $m(Z,\alpha) = +\infty$, and in particular,

$$(2\varepsilon)^\alpha N(Z,\varepsilon) < 1$$

for any sufficiently small $\varepsilon > 0$. Taking logarithms, we obtain

$$\underline{\dim}_B Z = \liminf_{\varepsilon \to 0} \frac{\log N(Z,\varepsilon)}{-\log \varepsilon}$$

$$\geq \liminf_{\varepsilon \to 0} \frac{\alpha \log(2\varepsilon)}{\log \varepsilon} = \alpha.$$

Letting $\alpha \to \dim_H Z$ thus yields $\underline{\dim}_B Z \geq \dim_H Z$. $\qquad\square$

8.2.2 Examples

The following are examples of the computation of the Hausdorff dimension and the lower and upper box dimensions. Sometimes this may be an overwhelming task, particularly since it is difficult to use the same method for different classes of sets.

Example 8.1. We consider the set $J \subset [0,1]$ composed of the points with a base-3 representation without the digit 1. One can easily verify that J is a closed set with empty interior and without isolated points. We shall compute the box dimensions of J. Take $n \in \mathbb{N}$. Given $\varepsilon \in (3^{-(n+1)}, 3^{-n}]$, we have

$$2^{n+1} > N(J,\varepsilon) \geq 2^n.$$

Therefore,

$$\frac{\log(2^n)}{-\log(3^{-(n+1)})} < \frac{\log N(J,\varepsilon)}{-\log \varepsilon} < \frac{\log(2^{n+1})}{-\log(3^{-n})},$$

that is,

$$\frac{n}{n+1} \cdot \frac{\log 2}{\log 3} < \frac{\log N(J,\varepsilon)}{-\log \varepsilon} < \frac{n+1}{n} \cdot \frac{\log 2}{\log 3},$$

for each $n \in \mathbb{N}$. Letting $\varepsilon \to 0$ and thus $n \to \infty$, we obtain

$$\underline{\dim}_B J = \overline{\dim}_B J = \frac{\log 2}{\log 3}. \tag{8.3}$$

We show in Example 8.3 that $\dim_H J = \log 2 / \log 3$.

Example 8.2. Let $\Lambda \subset \mathbb{R}^2$ be the Smale horseshoe (see Sect. 6.3.1 for the definition). Proceeding in a similar manner to that in Example 8.1, for each $n \in \mathbb{N}$ and $\varepsilon \in (3^{-(n+1)}, 3^{-n}]$, we have

$$4^{n+1} > N(\Lambda, \varepsilon) \geq 4^n.$$

Therefore,

$$\frac{n}{n+1} \cdot \frac{\log 4}{\log 3} < \frac{\log N(\Lambda, \varepsilon)}{-\log \varepsilon} < \frac{n+1}{n} \cdot \frac{\log 4}{\log 3}$$

for each $n \in \mathbb{N}$, and hence,

$$\underline{\dim}_B \Lambda = \overline{\dim}_B \Lambda = \frac{\log 4}{\log 3}.$$

We show in Example 8.8 that $\dim_H \Lambda = \log 4 / \log 3$.

Now we consider a more elaborate example. We construct a subset of $[0, 1]$ as follows. Given numbers $a_1, a_2 \in [0, 1)$ and $\lambda_1, \lambda_2 \in (0, 1)$, we consider the functions

$$f_i(x) = \lambda_i x + a_i \quad \text{for} \quad i = 1, 2. \tag{8.4}$$

We always assume that

$$f_i([0, 1]) \subset [0, 1] \quad \text{for} \quad i = 1, 2$$

and

$$f_1([0, 1]) \cap f_2([0, 1]) = \varnothing. \tag{8.5}$$

Up to reordering the functions f_1 and f_2, this is equivalent to assume that

$$\lambda_1 + a_1 < a_2 \quad \text{and} \quad \lambda_2 + a_2 \leq 1.$$

We also consider the compact set

$$J = \bigcap_{n=1}^{\infty} \bigcup_{i_1 \cdots i_n} \Delta_{i_1 \cdots i_n}, \tag{8.6}$$

where the union is taken over all $i_1, \ldots, i_n \in \{1, 2\}$ and where

$$\Delta_{i_1 \cdots i_n} = (f_{i_1} \circ \cdots \circ f_{i_n})([0, 1]).$$

We notice that $\Delta_{i_1 \cdots i_n}$ is a closed interval of length $\lambda_{i_1} \cdots \lambda_{i_n}$ and that

$$\Delta_{i_1 \cdots i_n} \cap \Delta_{j_1 \cdots j_n} = \varnothing \quad \text{whenever} \quad (i_1 \cdots i_n) \neq (j_1 \cdots j_n). \tag{8.7}$$

One can also show that

$$J = \bigcup_{\omega \in \Sigma_2^+} h(\omega),$$

where

$$h(\omega) = \lim_{n \to \infty} (f_{i_1} \circ \cdots \circ f_{i_n})(p) \tag{8.8}$$

for any $p \in [0, 1]$ (i.e., the limit is independent of p).

Proposition 8.2. *We have*

$$\dim_H J = \underline{\dim}_B J = \overline{\dim}_B J = s, \tag{8.9}$$

where $s \in (0, 1)$ is the unique root of the equation

$$\lambda_1^s + \lambda_2^s = 1. \tag{8.10}$$

Proof. To verify that s is well defined, it is sufficient to observe that the function

$$\varphi(s) = \lambda_1^s + \lambda_2^s$$

is strictly decreasing and that $\varphi(0) = 2$ and $\varphi(1) = \lambda_1 + \lambda_2 < 1$, in view of (8.5).

In order to show that s is a lower bound for $\dim_H J$, we define a probability measure μ in J by requiring that

$$\mu(\Delta_{i_1 \cdots i_n}) = (\lambda_{i_1} \cdots \lambda_{i_n})^s \tag{8.11}$$

for every $n \in \mathbb{N}$ and $i_1, \ldots, i_n \in \{1, 2\}$. Since

$$\bigcup_{i_n=1}^{2} \Delta_{i_1 \cdots i_n} = \Delta_{i_1 \cdots i_{n-1}}$$

and

$$\mu\left(\bigcup_{i_n=1}^{2} \Delta_{i_1 \cdots i_n} \right) = \sum_{i_n=1}^{2} \mu(\Delta_{i_1 \cdots i_n})$$

$$= \sum_{i_n=1}^{2} (\lambda_{i_1} \cdots \lambda_{i_n})^s$$

$$= (\lambda_{i_1} \cdots \lambda_{i_{n-1}})^s \sum_{i_n=1}^{2} \lambda_{i_n}^s$$

$$= \mu(\Delta_{i_1 \cdots i_{n-1}}),$$

the measure μ is well defined. Moreover, by (8.7) and (8.10), we have

$$\mu(J) = \sum_{i_1 \cdots i_n} \mu(\Delta_{i_1 \cdots i_n})$$

$$= \sum_{i_1 \cdots i_n} (\lambda_{i_1} \cdots \lambda_{i_n})^s$$

$$= (\lambda_1^s + \lambda_2^s)^n = 1.$$

We also construct a special cover of the set J. Given a sequence $\omega = (i_1 i_2 \cdots) \in \Sigma_2^+$ and $r \in (0, 1)$, let $n = n(\omega, r)$ be the unique integer such that

$$\lambda_{i_1} \cdots \lambda_{i_n} < r \le \lambda_{i_1} \cdots \lambda_{i_{n-1}}. \tag{8.12}$$

We denote by $\Delta(\omega, r)$ the set $\Delta_{i_1 \cdots i_n}$ with $n = n(\omega, r)$. Let also $\tilde{\Delta}(\omega, r)$ be the largest interval containing $h(\omega)$ (see (8.8)) such that:

1. $\tilde{\Delta}(\omega, r) = \Delta(\omega', r)$ for some $\omega' \in \Sigma_2^+$ with $h(\omega') \in \tilde{\Delta}(\omega, r)$.
2. $\Delta(\omega', r) \subset \tilde{\Delta}(\omega, r)$ whenever $h(\omega') \in \tilde{\Delta}(\omega, r)$.

By construction, for a fixed r, the intervals $\tilde{\Delta}(\omega, r)$ are pairwise disjoint. Moreover, they form a cover of the set J. On the other hand, it follows from (8.12) that

$$r/c \le \operatorname{diam} \tilde{\Delta}(\omega, r) < r, \tag{8.13}$$

where $c = 1/\min\{\lambda_1, \lambda_2\}$. This implies that for any interval I of length r, there is at most a number c of sets $\tilde{\Delta}(\omega, r)$ that intersect I, say $D_1, \dots, D_k$ with $k \le c$. By (8.11) and (8.12), we have

$$\mu(D_i) < r^s \quad \text{for} \quad i = 1, \dots, k.$$

Therefore,

$$\mu(I) \le \sum_{i=1}^{k} \mu(D_i) < \sum_{i=1}^{k} r^s \le cr^s.$$

This implies that

$$\mu(U) \le c(\operatorname{diam} U)^s \tag{8.14}$$

for any set U intersecting J. Hence, if $\mathcal{U}$ is a countable cover of J, then

$$1 = \mu(J) \leq \sum_{U \in \mathcal{U}} \mu(U) \leq c \sum_{U \in \mathcal{U}} (\operatorname{diam} U)^s.$$

By (8.1), we conclude that $m(J, s) \geq 1/c$, and hence, $\dim_H J \geq s$.

Now we show that s is an upper bound for $\overline{\dim}_B J$. Given $r \in (0, 1)$, we consider again the cover of J formed by the pairwise disjoint sets $\tilde{\Delta}(\omega, r)$. We note that there are finitely many of these sets, say $\tilde{\Delta}_1, \ldots, \tilde{\Delta}_{N(r)}$. This follows from the fact that J is compact together with (8.7). Moreover, by (8.10), we obtain

$$\sum_{i_{m+1} \cdots i_n} \left(\operatorname{diam} \Delta_{i_1 \cdots i_n} \right)^s = \left(\operatorname{diam} \Delta_{i_1 \cdots i_m} \right)^s \sum_{i_{m+1} \cdots i_n} \left(\lambda_{i_{m+1}} \cdots \lambda_{i_n} \right)^s$$

$$= \left(\operatorname{diam} \Delta_{i_1 \cdots i_m} \right)^s \left(\lambda_1^s + \lambda_2^s \right)^{n-m} = \left(\operatorname{diam} \Delta_{i_1 \cdots i_n} \right)^s,$$

and hence,

$$\sum_{i=1}^{N(r)} \left(\operatorname{diam} \tilde{\Delta}_i \right)^s = 1. \tag{8.15}$$

By (8.13), we have

$$r/c \leq \operatorname{diam} \tilde{\Delta}_i < r \quad \text{for each} \quad i = 1, \ldots, N(r),$$

and since the sets $\tilde{\Delta}_i$ are pairwise disjoint, it follows from (8.15) that

$$N(J, r) \leq N(r) \leq (c/r)^s.$$

Therefore,

$$\overline{\dim}_B J \leq \limsup_{r \to 0} \frac{\log N(J, r)}{-\log r}$$

$$\leq \limsup_{r \to 0} \frac{s \log(c/r)}{-\log r} = s.$$

The desired result follows now readily from (8.2). □

We refer to Example 9.3 for a more general construction that includes the set J in (8.6) as a particular case.

Example 8.3. Let J be the set in Example 8.1. We note that it is given by (8.6) with $k = 2$, provided that

$$f_1(x) = \frac{1}{3}x \quad \text{and} \quad f_2(x) = \frac{1}{3}x + \frac{2}{3}.$$

The constants λ_i in (8.4) are thus $\lambda_1 = \lambda_2 = 1/3$. Therefore, it follows from Proposition 8.2 that

$$\dim_H J = \underline{\dim}_B J = \overline{\dim}_B J = s,$$

where s is the unique root of (8.10). Hence, we obtain $2(1/3)^s = 1$, which yields $s = \log 2 / \log 3$. This value was already obtained for the lower and upper box dimensions in (8.3).

8.3 Dimension of Measures

We consider in this section the Hausdorff dimension and the lower and upper box dimensions of a given measure. We also show how the pointwise dimension can be used to estimate and sometimes even compute the dimensions of a measure.

8.3.1 Basic Notions and Examples

We first introduce the notions of Hausdorff dimension and of lower and upper box dimensions of a measure. Let μ be a finite measure in X.

Definition 8.2. The *Hausdorff dimension* and the *lower* and *upper box dimensions* of μ are defined, respectively, by

$$\dim_H \mu = \inf \left\{ \dim_H Z : \mu(X \setminus Z) = 0 \right\},$$

$$\underline{\dim}_B \mu = \lim_{\delta \to 0} \inf \left\{ \underline{\dim}_B Z : \mu(Z) \geq \mu(X) - \delta \right\},$$

$$\overline{\dim}_B \mu = \lim_{\delta \to 0} \inf \left\{ \overline{\dim}_B Z : \mu(Z) \geq \mu(X) - \delta \right\}.$$

One can easily verify that

$$\dim_H \mu = \lim_{\delta \to 0} \inf \left\{ \dim_H Z : \mu(Z) \geq \mu(X) - \delta \right\}. \tag{8.16}$$

Indeed, let c be the right-hand side of (8.16). Clearly,

$$\dim_H \mu \geq \inf \left\{ \dim_H Z : \mu(Z) \geq \mu(X) - \delta \right\}$$

for every δ, and hence, $\dim_H \mu \geq c$. On the other hand, there exists a sequence of sets Z_n such that $\mu(Z_n) \to \mu(X)$ and $\dim_H Z_n \to c$ when $n \to \infty$. Therefore,

$$\dim_H \mu \leq \dim_H \bigcup_{n \in \mathbb{N}} Z_n = \sup_{n \in \mathbb{N}} \dim_H Z_n = c.$$

It follows from (8.2) and (8.16) that

$$\dim_H \mu \leq \underline{\dim}_B \mu \leq \overline{\dim}_B \mu. \qquad (8.17)$$

The following quantities allow us to formulate a criterion for the coincidence of the three dimensions in (8.17):

Definition 8.3. The *lower* and *upper pointwise dimensions* of the measure μ at the point $x \in X$ are defined by

$$\underline{d}_\mu(x) = \liminf_{r \to 0} \frac{\log \mu(B(x,r))}{\log r} \quad \text{and} \quad \overline{d}_\mu(x) = \limsup_{r \to 0} \frac{\log \mu(B(x,r))}{\log r},$$

where $B(x,r)$ is the ball of radius r centered at x.

Example 8.4. If μ is the Lebesgue measure in $\mathbb{R}^n$, then $\mu(B(x,r)) = c_n r^n$ for every $x \in \mathbb{R}^n$ and $r > 0$, where c_n is a constant depending only on n. Then

$$\underline{d}_\mu(x) = \overline{d}_\mu(x) = n \quad \text{for every} \quad x \in \mathbb{R}^n.$$

Example 8.5. Let $J \subset [0,1]$ be the set in Example 8.1. We construct a probability measure μ in J as follows. For each interval

$$I_{i_1 \cdots i_n} = \left[0.i_1 \cdots i_n, 0.i_1 \cdots i_n 22 \cdots \right] \cap J,$$

with $i_1, \ldots, i_n \in \{0, 2\}$, we set $\mu(I_{i_1 \cdots i_n}) = 2^{-n}$. Since

$$I_{i_1 \cdots i_n} = I_{i_1 \cdots i_n 0} \cup I_{i_1 \cdots i_n 2} \quad \text{and} \quad I_{i_1 \cdots i_n 0} \cap I_{i_1 \cdots i_n 2} = \varnothing,$$

the measure μ is well defined.

Now take $x \in J$, $n \in \mathbb{N}$ and $r \in (3^{-(n+1)}, 3^{-n}]$. Then $B(x,r) \subset B(x, 3^{-n})$, and since the ball $B(x, 3^{-n})$ intersects at most one set $I_{i_1 \cdots i_n}$, we have

$$\mu(B(x,r)) \leq 2^{-n}. \qquad (8.18)$$

One the other hand, we have $B(x,r) \supset B(x, 3^{-(n+1)})$, and one can easily verify that $B(x, 3^{-(n+1)})$ contains some set $I_{j_1 \cdots j_{n+2}}$. Therefore,

$$\mu(B(x,r)) \geq 2^{-(n+2)}. \qquad (8.19)$$

It follows from (8.18) and (8.19) that

$$\frac{n}{n+1} \cdot \frac{\log 2}{\log 3} < \frac{\log \mu(B(x,r))}{\log r} < \frac{n+2}{n} \cdot \frac{\log 2}{\log 3}.$$

Finally, letting $r \to 0$ and thus $n \to \infty$, we obtain

$$\underline{d}_\mu(x) = \overline{d}_\mu(x) = \frac{\log 2}{\log 3} \quad \text{for every} \quad x \in J. \tag{8.20}$$

8.3.2 Dimension Estimates via Pointwise Dimension

The following result relates the Hausdorff dimension with the lower pointwise dimension:

Theorem 8.1. *The following properties hold:*

1. *If $\underline{d}_\mu(x) \geq \alpha$ for μ-almost every $x \in X$, then $\dim_H \mu \geq \alpha$.*
2. *If $\underline{d}_\mu(x) \leq \alpha$ for every $x \in Z \subset X$, then $\dim_H Z \leq \alpha$.*
3. *We have*
$$\dim_H \mu = \operatorname{ess\,sup}\{\underline{d}_\mu(x) : x \in X\}.$$

Proof. Set
$$Y = \{x \in X : \underline{d}_\mu(x) \geq \alpha\}.$$

Given $\varepsilon > 0$, for each $x \in Y$, there exists $r(x) > 0$ such that

$$\mu(B(x, r)) \leq (2r)^{\alpha - \varepsilon} \tag{8.21}$$

for every $r \in (0, r(x))$. Given $\rho > 0$, set

$$Y_\rho = \{x \in Y : r(x) \geq \rho\}.$$

Clearly,
$$Y_{\rho_1} \subset Y_{\rho_2} \quad \text{for} \quad \rho_1 \geq \rho_2, \quad \text{and} \quad Y = \bigcup_{\rho > 0} Y_\rho.$$

Since $\mu(X \setminus Y) = 0$, there exists $\rho > 0$ such that $\mu(Y_\rho) \geq \mu(X)/2$. Now let $Z \subset Y$ be a set of full μ-measure and let $\mathcal{U}$ be a cover of $Z \cap Y_\rho$ by open balls. Without loss of generality, we assume that $U \cap Y_\rho \neq \varnothing$ for every $U \in \mathcal{U}$. Then for each $U \in \mathcal{U}$, there exists $x_U \in U \cap Y_\rho$, and we consider the new cover

$$\mathcal{V} = \{B(x_U, \operatorname{diam} U) : U \in \mathcal{U}\}$$

of the set $Z \cap Y_\rho$. It follows from (8.21) that

$$\sum_{U \in \mathcal{U}} (\operatorname{diam} U)^{\alpha - \varepsilon} = 2^{\varepsilon - \alpha} \sum_{V \in \mathcal{V}} (\operatorname{diam} V)^{\alpha - \varepsilon}$$

$$\geq 2^{\varepsilon - \alpha} \sum_{V \in \mathcal{U}} \mu(V)$$

$$\geq 2^{\varepsilon - \alpha} \mu(Y_\rho) \geq 2^{\varepsilon - \alpha} \mu(X)/2.$$

Since $\mathcal{U}$ is arbitrary, we obtain

$$m(Z \cap Y_\rho, \alpha - \varepsilon) \geq 2^{\varepsilon-\alpha}\mu(X)/2.$$

This implies that $\dim_H(Z \cap Y_\rho) \geq \alpha - \varepsilon$, and by the arbitrariness of ε, we conclude that

$$\dim_H Z \geq \dim_H(Z \cap Y_\rho) \geq \alpha.$$

Now we establish the second property. For each $x \in Z$ and $\varepsilon > 0$, there exists a sequence $r_n = r_n(x, \varepsilon) \searrow 0$ when $n \to \infty$ such that

$$\mu(B(x, r_n)) \geq (2r_n)^{\alpha+\varepsilon} \tag{8.22}$$

for every $n \in \mathbb{N}$. Consider a cover

$$\mathcal{U} \subset \{B(x, r_n(x, \varepsilon)) : x \in Z \text{ and } n \in \mathbb{N}\}$$

of the set Z. We note that the diameter of $\mathcal{U}$ can be made arbitrarily small. To proceed with the proof, we need the following result:

Lemma 8.1 (Besicovitch's covering lemma). *Given a set $Z \subset \mathbb{R}^m$ and a bounded function $r \colon Z \to \mathbb{R}^+$, the cover $\{B(x, r(x)) : x \in Z\}$ of Z contains a countable subcover $\mathcal{V}$ of finite multiplicity, that is, there exists a constant $K > 0$ such that*

$$\operatorname{card}\{V \in \mathcal{V} : x \in V\} \leq K \quad \text{for every} \quad x \in Z.$$

By Lemma 8.1 and Whitney's embedding theorem, there exists a subcover $\mathcal{V} \subset \mathcal{U}$ of Z and a constant $K > 0$ such that

$$\operatorname{card}\{V \in \mathcal{V} : x \in V\} \leq K$$

for every $x \in Z$. Therefore, by (8.22),

$$\sum_{V \in \mathcal{V}} (\operatorname{diam} V)^{\alpha+\varepsilon} \leq \sum_{V \in \mathcal{V}} \mu(B(x, r)) \leq K\mu(X),$$

and since the diameter of the cover $\mathcal{V}$ can be made arbitrarily small, we conclude that

$$m(Z, \alpha + \varepsilon) \leq K\mu(X).$$

This implies that $\dim_H Z \leq \alpha + \varepsilon$, and since ε is arbitrary, we obtain $\dim_H Z \leq \alpha$.

Finally, for the third property, let

$$\alpha = \operatorname{ess sup}\{\underline{d}_\mu(x) : x \in X\} \quad \text{and} \quad Z = \{x \in X : \underline{d}_\mu(x) \leq \alpha\}.$$

We have $\mu(Z) = \mu(X)$, and by the second property,

$$\dim_H \mu \leq \dim_H Z \leq \alpha.$$

Now given $\varepsilon > 0$, let

$$Z_\varepsilon = \{x \in X : \underline{d}_\mu(x) \geq \alpha - \varepsilon\}.$$

We have $\mu(Z_\varepsilon) > 0$, and it follows from the first property that

$$\dim_H \mu \geq \dim_H (\mu|Z_\varepsilon) \geq \alpha - \varepsilon.$$

Since ε is arbitrary, we obtain $\dim_H \mu \geq \alpha$. $\square$

We give several applications of Theorem 8.1.

Example 8.6. Let J be the set in (8.6). For the measure μ in J defined by (8.11), we have the estimate in (8.14) for any set U intersecting J. In particular,

$$\mu(B(x,r)) \leq c(2r)^s,$$

and hence,

$$\underline{d}_\mu(x) \geq \liminf_{r \to 0} \frac{\log[c(2r)^s]}{\log r} = s$$

for every $x \in J$. Therefore, it follows from Theorem 8.1 that

$$\dim_H J \geq \dim_H \mu \geq s.$$

This shows that the argument used in the proof of Proposition 8.2 to obtain the lower bound for the Hausdorff dimension can be reformulated in terms of the lower pointwise dimension.

Example 8.7. Let J and μ be the set and the measure in Example 8.5. By (8.20), we have

$$\underline{d}_\mu(x) = \log 2/\log 3 \quad \text{for every} \quad x \in J.$$

Thus, it follows from Theorem 8.1 that

$$\dim_H J = \dim_H \mu = \frac{\log 2}{\log 3}. \tag{8.23}$$

This yields another proof that the set J in Example 8.1 has Hausdorff dimension $\log 2/\log 3$.

Example 8.8. Let $\Lambda \subset \mathbb{R}^2$ be the Smale horseshoe. We have already shown in Example 8.2 that the lower and upper box dimensions of Λ are equal to $\log 4/\log 3$. Now we compute $\dim_H \Lambda$. In a similar manner to that in Example 8.5, we first construct a measure μ in Λ. Set

$$\mu(Q_n(\omega)) = 4^{-n} \quad \text{for each} \quad \omega \in X_2, \ n \in \mathbb{N},$$

with $Q_n(\omega)$ as in (6.15). Proceeding also in a similar manner to that in Example 8.5, one can show that

$$\underline{d}_\mu(x) = \overline{d}_\mu(x) = \frac{\log 4}{\log 3} \quad \text{for every} \quad x \in \Lambda.$$

Thus, it follows from Theorems 8.1 and 8.2 together with Example 8.2 that

$$\dim_H \Lambda = \underline{\dim}_B \Lambda = \overline{\dim}_B \Lambda = \frac{\log 4}{\log 3}$$

and

$$\dim_H \mu = \underline{\dim}_B \mu = \overline{\dim}_B \mu = \frac{\log 4}{\log 3}.$$

We also give a criterion for the coincidence of the Hausdorff dimension and the lower and upper box dimensions of a measure.

Theorem 8.2 (Young). *If μ is a finite measure in X and there exists $d \geq 0$ such that*

$$\lim_{r \to 0} \frac{\log \mu(B(x, r))}{\log r} = d \tag{8.24}$$

for μ-almost every $x \in X$, then

$$\dim_H \mu = \underline{\dim}_B \mu = \overline{\dim}_B \mu = d.$$

Proof. By Theorem 8.1, we have $\dim_H \mu \geq d$. Now we show that $\overline{\dim}_B \mu \leq d$ and thus the result follows from (8.17). Set

$$Z = \{x \in X : \overline{d}_\mu(x) \leq d\}.$$

Given $\varepsilon > 0$, for each $x \in Z$, there exists $r(x) > 0$ such that if $r \in (0, r(x))$, then

$$\mu(B(x, r)) \geq (2r)^{d+\varepsilon}.$$

Given $\rho > 0$, we consider the set

$$Y_\rho = \{x \in Z : r(x) \geq \rho\}.$$

Clearly,

$$Y_{\rho_1} \subset Y_{\rho_2} \quad \text{for} \quad \rho_1 \geq \rho_2, \quad \text{and} \quad Z = \bigcup_{\rho > 0} Y_\rho.$$

Therefore, since $\mu(X \setminus Z) = 0$, we have $\mu(Y_\rho) \nearrow \mu(X)$ when $\rho \to 0$. For each $r < \rho$, the balls $B(x, r)$ form a cover $\mathcal{U}$ of the set Y_ρ. By Lemma 8.1, there exists a subcover $\mathcal{V} \subset \mathcal{U}$ of Y_ρ of finite multiplicity K. Therefore,

$$\sum_{V \in \mathcal{V}} (\operatorname{diam} V)^{d+\varepsilon} \le \sum_{V \in \mathcal{V}} \mu(B(x, r)) \le K\mu(X).$$

Since

$$\sum_{V \in \mathcal{V}} (\operatorname{diam} V)^{d+\varepsilon} = (2r)^{d+\varepsilon} \operatorname{card} \mathcal{V} \ge (2r)^{d+\varepsilon} N(Y_\rho, r),$$

we obtain

$$N(Y_\rho, r) \le \frac{K\mu(X)}{(2r)^{d+\varepsilon}},$$

and hence,

$$\overline{\dim}_B Y_\rho = \limsup_{r \to 0} \frac{\log N(Y_\rho, r)}{-\log r} \le d + \varepsilon.$$

Since $\mu(Y_\rho) \nearrow \mu(X)$ when $\rho \to 0$, we conclude that

$$\overline{\dim}_B \mu \le \limsup_{\rho \to 0} \overline{\dim}_B Y_\rho \le d + \varepsilon.$$

Finally, since ε is arbitrary, we obtain $\overline{\dim}_B \mu \le d$. This completes the proof of the theorem. $\square$

Definition 8.4. The limit in (8.24), when it exists, is called the *pointwise dimension* of the measure μ at x, and we denote it by $d_\mu(x)$.

The following is an application of Theorem 8.2:

Example 8.9. Let again J and μ be as in Example 8.5. By (8.20), we have

$$d_\mu(x) = \log 2 / \log 3 \quad \text{for every} \quad x \in J.$$

Thus, it follows from Theorem 8.2 together with (8.3) and (8.23) that

$$\dim_H J = \underline{\dim}_B J = \overline{\dim}_B J = \frac{\log 2}{\log 3},$$

and

$$\dim_H \mu = \underline{\dim}_B \mu = \overline{\dim}_B \mu = \frac{\log 2}{\log 3}.$$

8.4 Exercises

Exercise 8.1. Show that

$$\dim_H Z_1 \le \dim_H Z_2 \quad \text{whenever} \quad Z_1 \subset Z_2$$

and

$$\dim_H \bigcup_{i \in \mathbb{N}} Z_i = \sup_{i \in \mathbb{N}} \dim_H Z_i.$$

Exercise 8.2. Show that:

1. If $Z_1 \subset Z_2$, then

$$\underline{\dim}_B Z_1 \le \underline{\dim}_B Z_2 \quad \text{and} \quad \overline{\dim}_B Z_1 \le \overline{\dim}_B Z_2.$$

2.

$$\underline{\dim}_B \bigcup_{i \in \mathbb{N}} Z_i \ge \sup_{i \in \mathbb{N}} \underline{\dim}_B Z_i \quad \text{and} \quad \overline{\dim}_B \bigcup_{i \in \mathbb{N}} Z_i \ge \sup_{i \in \mathbb{N}} \overline{\dim}_B Z_i.$$

Exercise 8.3. Show that for any set Z,

$$\underline{\dim}_B Z = \underline{\dim}_B \overline{Z} \quad \text{and} \quad \overline{\dim}_B Z = \overline{\dim}_B \overline{Z}.$$

Exercise 8.4. Show that

$$\overline{\dim}_B \bigcup_{i=1}^{n} Z_i = \max_{i=1,\dots,n} \overline{\dim}_B Z_i.$$

Exercise 8.5. Verify that if $\underline{\dim}_B Z = \overline{\dim}_B Z = s$, then

$$s = \lim_{n \to \infty} \frac{1}{n} \log N(Z, e^{-n}).$$

Exercise 8.6. Compute the Hausdorff dimension and the lower and upper box dimensions of the set $\mathbb{Q} \cap [0, 1]$.

Exercise 8.7. Show that the inequalities

$$\dim_H X + \dim_H Y \le \dim_H (X \times Y)$$

and

$$\overline{\dim}_B (X \times Y) \le \overline{\dim}_B X + \overline{\dim}_B Y$$

hold for any subsets X and Y of $\mathbb{R}^m$.

Exercise 8.8. Let $f: X \to Y$ be a Lipschitz map. Show that $\dim_H f(A) \le \dim_H A$ for any set $A \subset X$.

Exercise 8.9. Let $f: X \to Y$ be a diffeomorphism. Show that $\dim_H f(A) = \dim_H A$ for any set $A \subset X$.

Exercise 8.10. Show that

$$\underline{d}_\mu(x) = \liminf_{n \to \infty} \frac{\log \mu(B(x, ae^{-n}))}{-n}$$

and

$$\overline{d}_\mu(x) = \limsup_{n \to \infty} \frac{\log \mu(B(x, ae^{-n}))}{-n}$$

for each $a > 0$ and $x \in X$. Hint: for each $r > 0$, take $n = n(r) \in \mathbb{N}$ such that

$$ae^{-(n+1)} \le r < ae^{-n} < 1.$$

Exercise 8.11. Show that if $T: X \to X$ is a diffeomorphism and μ is a T-invariant probability measure in X, then

$$\underline{d}_\mu(f(x)) = \underline{d}_\mu(x) \quad \text{and} \quad \overline{d}_\mu(f(x)) = \overline{d}_\mu(x)$$

for every $x \in X$.

Exercise 8.12. Show that if there exist a finite Borel measure μ in a set Z and constants $c, d > 0$ such that $\mu(B(x, r)) \ge cr^d$ for every $x \in Z$ and all sufficiently small $r > 0$, then $\dim_H Z \le d$.

Exercise 8.13. Given numbers $\lambda_1, \lambda_2 \in (0, 1)$ with $\lambda_1 + \lambda_2 < 1$, for each $n \in \mathbb{N}$ and $i_1, \dots, i_n \in \{1, 2\}$, let $\Delta_{i_1 \cdots i_n} \subset [0, 1]$ be a closed interval of diameter

$$\operatorname{diam} \Delta_{i_1 \cdots i_n} = \lambda_{i_1} \cdots \lambda_{i_n} \tag{8.25}$$

satisfying condition (8.7). Show that (8.9) holds for the set J in (8.6), with the same s. This shows that the location of the sets $\Delta_{i_1 \cdots i_n}$ plays no role in the statement of Proposition 8.2.

Exercise 8.14. Given numbers $\lambda_1, \dots, \lambda_k \in (0, 1)$ with $\sum_{i=1}^k \lambda_i < 1$, for each $n \in \mathbb{N}$ and $i_1, \dots, i_n \in \{1, \dots, k\}$, let $\Delta_{i_1 \cdots i_n} \subset [0, 1]$ be a closed interval of diameter as in (8.25) satisfying condition (8.7). Show that (8.9) holds for the set J in (8.6), where s is the unique root of the equation $\sum_{i=1}^k \lambda_i^s = 1$.

Exercise 8.15. Let f be a C^1 diffeomorphism in some open neighborhood of the square $[0, 1]^2$ and let H_i and V_i for $i = 1, 2$ be, respectively, horizontal and vertical strips such that:

1. $f(H_i) = V_i$ for $i = 1, 2$.
2. There exist $\lambda_i, \mu_i \in (0, 1)$ and $(a_i, b_i) \in [0, 1)$ with

$$(f|H_i)(x, y) = (\lambda_i x + a_i, \mu_i y + b_i) \quad \text{for} \quad i = 1, 2.$$

Show that:

1. $\Lambda = \bigcap_{n \in \mathbb{Z}} f^n(H_1 \cup H_2)$ is a hyperbolic set for f.
2.

$$\dim_H \Lambda = \underline{\dim}_B \Lambda = \overline{\dim}_B \Lambda = \alpha + \beta,$$

where α and β are, respectively, the unique roots of the equations

$$\lambda_1^\alpha + \lambda_2^\alpha = 1 \quad \text{and} \quad \mu_1^\beta + \mu_2^\beta = 1.$$

Hint: use Exercise 8.7.

Exercise 8.16. Use Exercise 8.15 to show that the Hausdorff dimension of the Smale horseshoe is $\log 4 / \log 3$.

Exercise 8.17. Compute the Hausdorff dimension and the lower and upper box dimensions of the solenoid in Exercise 6.12.

Notes

We refer to the books [29, 33, 58, 73] for further topics of dimension theory. See, for example, [58] for Besicovitch's covering lemma (Lemma 8.1). Theorem 8.2 is due to Young [105].

Chapter 9
Dimension Theory of Hyperbolic Dynamics

We study in this chapter the dimension of hyperbolic invariant sets of conformal transformations, both invertible and noninvertible. This means that the derivative of the map along the stable and unstable directions is a multiple of an isometry at every point. More precisely, we compute the Hausdorff dimension and the lower and upper box dimensions of repellers and hyperbolic sets for a conformal dynamics. The dimension of the invariant sets is expressed, as explicitly as possible, in terms of the topological pressure. It turns out that Markov partitions are a principal element of the proofs. In particular, they allow us to reduce effectively some of the arguments and computations to the special case of symbolic dynamics.

9.1 Introduction

The dimension theory of dynamical systems progressively developed during the last 2 decades into an independent and quite active field of research. The theory concerns the study of dynamical systems from the point of view of dimension. In other words, one does not consider topics of dimension theory that are not of dynamical nature, of course, independently of their importance. Roughly speaking, the main objective is to measure the complexity from the dimensional point of view of the objects that remain invariant under the dynamics, such as invariant sets and invariant measures.

The thermodynamic formalism has a privileged relation to the dimension theory of dynamical systems. This is due to the fact that the unique solution s of the equation

$$P(s\varphi) = 0, \tag{9.1}$$

where φ is a certain function associated to a given invariant set, is often related to the Hausdorff dimension of the set. This equation was introduced by Bowen and is usually called Bowen's equation. It is also appropriate to call it the Bowen–Ruelle equation, taking into account not only the fundamental role of the thermodynamic formalism developed by Ruelle but also his study of the Hausdorff dimension of

L. Barreira, *Ergodic Theory, Hyperbolic Dynamics and Dimension Theory*, Universitext,   253
DOI 10.1007/978-3-642-28090-0_9, © Springer-Verlag Berlin Heidelberg 2012

repellers of a conformal dynamics (this means that the derivative of the map is a multiple of an isometry at every point). For a repeller J of a conformal map f, he showed that

$$\dim_H J = s,$$

where s is the solution of (9.1) for the function $\varphi \colon J \to \mathbb{R}$ defined by

$$\varphi(x) = -\log \|d_x f\|.$$

To a certain extent, the study of the dimension of hyperbolic sets is analogous. Indeed, assuming that the derivatives of the diffeomorphism along the stable and unstable directions are multiples of isometries, starting with the work of McCluskey and Manning, it was possible to develop a corresponding theory. However, there are nontrivial differences between the theory for repellers and for hyperbolic sets. For example, each conformal repeller J has a unique invariant measure of full Hausdorff dimension, that is, an invariant probability measure μ such that

$$\dim_H \mu = \inf \big\{ \dim_H Z : \mu(J \setminus Z) = 0 \big\} = \dim_H J.$$

On the other hand, unless some cohomology relations hold, there is no invariant measure of full dimension concentrated on a conformal hyperbolic set.

Virtually all known equations used to compute or estimate the dimension of an invariant set, both for an invertible and an noninvertible dynamics, are particular cases of (9.1) or of appropriate generalizations. Nevertheless, despite many significant developments, only the case of conformal dynamics is completely understood.

Some motivations for the study of dimension in the context of the theory of dynamical systems come from the study of attractors in infinite-dimensional spaces. The longtime behavior of many dynamical systems, such as those coming from partial differential equations and delay differential equations, can essentially be described in terms of a global attractor. An important question is how many degrees of freedom are necessary to specify the dynamics on the attractor. It turns out that a large class of attractors have finite Hausdorff dimension or even finite box dimension. Hence, the dynamics on the attractor is essentially finite-dimensional.

9.2 Repellers

We start our study with the somewhat simpler case of repellers. After introducing the notion of conformal map, we give several examples of repellers of conformal maps. We then compute the Hausdorff dimension and the lower and upper box dimensions of any such repeller in terms of the topological pressure. More precisely, the dimension is the root of an equation involving the topological pressure.

9.2.1 Conformal Maps and Examples

Let $f: U \to M$ be a C^1 map in an open subset of a smooth manifold M and let $J \subset U$ be a repeller for f (see Definition 6.6).

Definition 9.1. We say that f is *conformal* on J if $d_x f$ is a multiple of an isometry for every $x \in J$.

We give several examples of repellers of conformal maps.

Example 9.1. Consider:

1. k disjoint closed intervals $\Delta_1, \ldots, \Delta_k \subset \mathbb{R}$.
2. A C^1 map $f: U \to \mathbb{R}$, where U is an open neighborhood of $\Delta = \bigcup_{i=1}^k \Delta_i$.

We assume that f is expanding on U and that $f(\partial \Delta) \subset \partial \Delta$ and $\Delta_i \subset f(\Delta_j)$ whenever $\partial \Delta_i \cap \partial f(\Delta_j) \neq \varnothing$. Then the set

$$J = \bigcap_{j=1}^{\infty} f^{-j+1} \Delta$$

is a repeller for f. Moreover, since $d_x f$ is a single number for each x, the map f is conformal on J.

Example 9.2 (Hyperbolic Julia sets). Let S be the Riemann sphere and let $R: S \to S$ be a rational map of degree greater than one. Since R is holomorphic, it is conformal. We say that an n-periodic point $z \in S$ of R is *repelling* if $|(R^n)'(z)| > 1$. The *Julia set* J of R is the closure of the set of the repelling periodic points of R. If R is expanding on J, then the Julia set is a repeller for a conformal map.

Example 9.3. We construct a repeller in $\mathbb{R}$ as follows. Let $\Delta_1, \ldots, \Delta_k \subset [0, 1]$ be disjoint closed intervals of lengths, respectively, $\lambda_1, \ldots, \lambda_k > 0$. Let also $g_i: \Delta_i \to [0, 1]$ be affine maps such that $g_i(\Delta_i) = [0, 1]$ for $i = 1, \ldots, k$. We define a map

$$f: \bigcup_{i=1}^k \Delta_i \to [0, 1]$$

by $f|\Delta_i = g_i$ for $i = 1, \ldots, k$. We note that f is the restriction of an expanding map to the set $\bigcup_{i=1}^k \Delta_i$, and that

$$J = \bigcap_{n=0}^{\infty} f^{-n} \left(\bigcup_{i=1}^k \Delta_i \right) \tag{9.2}$$

is a repeller for that map. Indeed, $|f'|\Delta_i| = \lambda_i^{-1}$ for each i. We also note that when $k = 2$, we recover the set J in (8.6) by taking

$$f_i = g_i^{-1} = (f|\Delta_i)^{-1} \quad \text{for} \quad i = 1, 2.$$

9.2.2 Dimension of Repellers

For conformal maps, the following result gives a formula for the Hausdorff dimension and the lower and upper box dimensions of any repeller in terms of the topological pressure. We recall that f is said to be of class $C^{1+\alpha}$, for some $\alpha \in (0,1]$, if f is of class C^1 and that there exists $c > 0$ such that

$$\|d_x f - d_y f\| \le cd(x,y)^\alpha \tag{9.3}$$

for every $x, y \in M$.

Theorem 9.1 (Dimension of repellers). *If J is a repeller for a $C^{1+\alpha}$ transformation f, for some $\alpha \in (0,1]$, such that f is conformal on J, then*

$$\dim_H J = \underline{\dim}_B J = \overline{\dim}_B J = s, \tag{9.4}$$

where s is the unique real number such that

$$P_{f|J}(s\varphi) = 0 \tag{9.5}$$

for the function $\varphi \colon J \to \mathbb{R}$ defined by

$$\varphi(x) = -\log\|d_x f\|. \tag{9.6}$$

Proof. The argument is an elaboration of the proof of Proposition 8.2, although the required modifications are substantial.

The uniqueness of the number s follows from the strict monotonicity of the function $s \mapsto P_{f|J}(s\varphi)$ that can be established as follows. Take $m \in \mathbb{N}$ such that $c\tau^m > 1$, with c and τ as in (6.29). For each $n \in \mathbb{N}$, we write $n = qm + r$, where $q = \lfloor n/m \rfloor$ and $r \in \{0, \ldots, m-1\}$. Then

$$\sum_{k=0}^{n-1} \varphi(f^k(x)) = -\sum_{k=0}^{n-1} \log\|d_{f^k(x)} f\|$$

$$= -\log \prod_{k=0}^{qm+r-1} \|d_{f^k(x)} f\|$$

$$= -\log \|d_x f^{qm+r}\|$$

$$\le -\log\left[\|d_x f^r\|(c\tau^m)^q\right]$$

$$\le -\log \min_r \min_{x \in J} \|d_x f^r\| - \left(\frac{n}{m} - 1\right)\log(c\tau^m)$$

for every $x \in J$ and $n \in \mathbb{N}$. Therefore, given $t \leq s$ and $n \in \mathbb{N}$, we have

$$s \sum_{k=0}^{n-1} \varphi \circ f^k \leq t \sum_{k=0}^{n-1} \varphi \circ f^k + (s-t)\left(a - \frac{n}{m}\log(c\tau^m)\right),$$

where

$$a = -\log \min_r \min_{x \in J} \|d_x f^r\| + \log(c\tau^m).$$

Writing $P = P_{f|J}$, this implies that

$$P(s\varphi) \leq P(t\varphi) - (s-t)\frac{1}{m}\log(c\tau^m), \tag{9.7}$$

and thus, the function $s \mapsto P(s\varphi)$ is strictly decreasing. It also follows from (9.7) that

$$\lim_{s \to +\infty} P(s\varphi) = -\infty \quad \text{and} \quad \lim_{t \to -\infty} P(t\varphi) = +\infty.$$

Therefore, there exists a unique real number s such that $P(s\varphi) = 0$.

Now let $R_1, \ldots, R_k$ be the elements of a Markov partition of J (by Theorem 6.1, any repeller has Markov partitions). We also consider the associated topological Markov chain $\sigma|X_A^+ : X_A^+ \to X_A^+$, with the entries of the $k \times k$ transition matrix A given by (6.33), and the coding map $h: X_A^+ \to J$ defined by (6.34). For each $\omega = (i_1 i_2 \cdots) \in \Sigma_A^+$ and $r \in (0, 1)$, let $n = n(\omega, r) \in \mathbb{N}$ be the unique integer such that

$$\|d_{h(\omega)} f^n\|^{-1} < r \leq \|d_{h(\omega)} f^{n-1}\|^{-1} \tag{9.8}$$

and set

$$\Delta(\omega, r) = \bigcap_{j=0}^{n-1} f^{-j} R_{i_{j+1}}.$$

By (9.3), for each $x, y \in \Delta(\omega, r)$, it follows from (6.29) that

$$\frac{\|d_x f^n\|}{\|d_y f^n\|} = \prod_{j=0}^{n-1} \frac{\|d_{f^j(x)} f\|}{\|d_{f^j(y)} f\|}$$

$$\leq \prod_{j=0}^{n-1} \left(1 + \frac{\|d_{f^j(x)} f - d_{f^j(y)} f\|}{\|d_{f^j(y)} f\|}\right)$$

$$\leq \prod_{j=0}^{n-1} \left(1 + Cd(f^j(x), f^j(y))^\alpha\right)$$

$$\leq \prod_{j=0}^{n-1} \left(1 + Cc^\alpha d(f^j(x), f^j(y))^\alpha \tau^{\alpha(j-n)}\right),$$

where $C = 1/\inf_{z \in J} \|d_z f\|$. Since

$$d\left(f^j(x), f^j(y)\right) \le \max_{i=1,\dots,k} \operatorname{diam} R_i =: \gamma,$$

there exists a constant $C' > 0$ such that

$$\frac{\|d_x f^n\|}{\|d_y f^n\|} \le \prod_{j=0}^{n-1} \left(1 + C' \tau^{\alpha(j-n)}\right)$$

$$\le \prod_{l=1}^{\infty} \left(1 + C' \tau^{-\alpha l}\right) =: D < \infty. \tag{9.9}$$

Now let g be the local inverse of $f^n | \Delta(\omega, r)$, where $n = n(\omega, r)$. Since $(d_x f^n)^{-1} = d_{f^n(x)} g$ and f is conformal on J, we have

$$\|d_{f^n(x)} g\| = \|d_x f^n\|^{-1}. \tag{9.10}$$

By (9.9), we thus obtain

$$\begin{aligned}
\operatorname{diam} \Delta(\omega, r) &= \sup_{x, y \in \Delta(\omega, r)} d(x, y) \\
&\le \sup_{x, y \in \Delta(\omega, r)} d\left(f^n(x), f^n(y)\right) \sup_{z \in f^n(\Delta(\omega, r))} \|d_z g\| \\
&\le \gamma \sup_{w \in \Delta(\omega, r)} \left(\|d_w f^n\|^{-1}\right) \\
&\le \gamma D \|d_x f^n\|^{-1} < r
\end{aligned} \tag{9.11}$$

for every $x \in \Lambda(\omega, r)$, provided that γ is sufficiently small, that is, provided that the diameter of the Markov partition is sufficiently small.

The next step is the construction of a special cover of J. This is an elaboration of a corresponding construction in the proof of Proposition 8.2. We first note that $h(\omega) \in \Delta(\omega, r)$ and that if $h(\omega') \in \Delta(\omega, r)$ and $n(\omega', r) \ge n(\omega, r)$, then $\Delta(\omega', r) \subset \Delta(\omega, r)$. For each $\omega \in \Sigma_A^+$ and $r > 0$, let $\tilde{\Delta}(\omega, r)$ be the largest set containing $h(\omega)$ such that:

1. $\tilde{\Delta}(\omega, r) = \Delta(\omega', r)$ for some $\omega' \in \Sigma_A^+$ with $h(\omega') \in \tilde{\Delta}(\omega, r)$.
2. $\Delta(\omega', r) \subset \tilde{\Delta}(\omega, r)$ whenever $h(\omega') \in \tilde{\Delta}(\omega, r)$.

By construction, for a fixed r, the sets $\tilde{\Delta}(\omega, r)$ intersect at most along their boundaries. The cover formed by the sets $\tilde{\Delta}(\omega, r)$ is called a *Moran cover* of J.

Since each set R_i is the closure of its interior, there exists $\rho > 0$ such that R_i contains a ball of radius ρ for $i = 1, \dots, k$. Using again (9.10), this implies that each set $\tilde{\Delta}(\omega, r)$ contains a ball of radius $\rho D^{-1} \|d_x f^n\|^{-1}$ for some point $x \in \tilde{\Delta}(\omega, r)$.

Moreover, by (9.8), we conclude that there exists a constant $\kappa \in (0, 1)$ (independent of ω and r) such that each set $\tilde{\Delta}(\omega, r)$ contains a ball of radius κr. On the other hand, it also follows from (9.8) that there exists a constant $\kappa' > 1$ such that each set $\tilde{\Delta}(\omega, r)$ has diameter at most $\kappa' r$. Since the sets $\tilde{\Delta}(\omega, r)$ intersect at most along their boundaries, there exists a constant $C > 0$ (independent of r) such that each ball $B(x, r)$ intersects at most a number C of the sets $\tilde{\Delta}(\omega, r)$.

Now we establish the identities in (9.4) involving the Hausdorff dimension and the lower and upper box dimensions. Given $\delta, r, \theta > 0$, there exists a countable cover $\mathcal{U}$ of J with $\operatorname{diam} \mathcal{U} < r$ such that

$$\sum_{U \in \mathcal{U}} (\operatorname{diam} U)^{\dim_H J + \delta} < \theta.$$

For each $U \in \mathcal{U}$ we consider the family of sets

$$\mathcal{U}_U = \{\Delta(\omega, \operatorname{diam} U) : \Delta(\omega, \operatorname{diam} U) \cap U \neq \varnothing \text{ and } \omega \in \Sigma_A^+\},$$

which is a cover of U. Then

$$\mathcal{V} = \{V : V \in \mathcal{U}_U \text{ and } U \in \mathcal{U}\}$$

is a cover of J, and

$$
\begin{aligned}
\sum_{V \in \mathcal{V}} (\operatorname{diam} V)^{\dim_H J + \varepsilon} &\leq \sum_{U \in \mathcal{U}} \sum_{V \in \mathcal{U}_U} (\operatorname{diam} V)^{\dim_H J + \varepsilon} \\
&\leq \sum_{U \in \mathcal{U}} \sum_{V \in \mathcal{U}_U} (\operatorname{diam} U)^{\dim_H J + \varepsilon} \\
&\leq \sum_{U \in \mathcal{U}} C (\operatorname{diam} U)^{\dim_H J + \varepsilon} < C\theta.
\end{aligned}
\tag{9.12}
$$

Now given $\alpha \in \mathbb{R}$ and $N \in \mathbb{N}$, we set

$$M(\alpha, N) = \inf_{\mathcal{C}} \sum_{\Delta(\omega, r) \in \mathcal{C}} \exp\left(-\alpha\, n(\omega, r) + \sup_{\Delta(\omega, r)} \sum_{j=0}^{n(\omega, r)-1} \varphi \circ f^j\right),$$

where the infimum is taken over all finite covers $\mathcal{C}$ of J by sets in the family

$$\mathcal{H}_N = \{\Delta(\omega, r) : n(\omega, r) > N\}.$$

Lemma 9.1. *We have*

$$P(\varphi) = \inf\left\{\alpha : \lim_{N \to \infty} M(\alpha, N) = 0\right\}.$$

Proof of the lemma. Clearly,

$$M(\alpha, N) \leq \sum_{i_1 \cdots i_N} \exp\left(-\alpha N + \sup_{\Delta_{i_1 \cdots i_N}} \sum_{j=0}^{N-1} \varphi \circ f^j\right), \qquad (9.13)$$

where

$$\Delta_{i_1 \cdots i_N} = \bigcap_{j=0}^{N-1} f^{-j} R_{i_{j+1}}.$$

On the other hand, given $\varepsilon > 0$, there exists $C > 0$ such that

$$\sum_{i_1 \cdots i_N} \exp \sup_{\Delta_{i_1 \cdots i_N}} \sum_{j=0}^{N-1} \varphi \circ f^j \leq C e^{(P(\varphi)+\varepsilon)N}$$

for every $N \in \mathbb{N}$. Therefore, provided that $\alpha > P(\varphi)$ and ε is sufficiently small, it follows from (9.13) that

$$M(\alpha, N) \leq e^{-\alpha N} C e^{(P(\varphi)+\varepsilon)N} \to 0$$

when $N \to \infty$. Hence,

$$p := \inf\left\{\alpha : \lim_{N \to \infty} M(\alpha, N) = 0\right\} \leq P(\varphi).$$

Now let $\alpha > p$. There exist $N \in \mathbb{N}$ and a finite cover $\mathcal{C} \subset \mathcal{H}_N$ of J, say with elements $\tilde{\Delta}_1, \ldots, \tilde{\Delta}_q$, such that

$$N(\alpha, \mathcal{C}) := \sum_{\Delta \in \mathcal{C}} \exp\left(-\alpha |\Delta| + \sup_{\Delta} \sum_{j=0}^{|\Delta|-1} \varphi \circ f^j\right) < 1, \qquad (9.14)$$

using the notation $|\Delta_{i_1 \cdots i_N}| = N$. Let $I_1, \ldots, I_q$ be the finite sequences such that

$$\tilde{\Delta}_i = \Delta_{I_i} \quad \text{for} \quad i = 1, \ldots, q. \qquad (9.15)$$

For each $n \in \mathbb{N}$, we consider the cover $\mathcal{C}_n$ of J formed by the sets $\Delta_{I_{i_1} \cdots I_{i_n}}$ with $i_1, \ldots, i_n \in \{1, \ldots, q\}$, and we write $\Gamma_{i_1 \cdots i_n} = \Delta_{I_{i_1} \cdots I_{i_n}}$. Since

$$\sup_{\Gamma_{i_1 \cdots i_n}} \sum_{j=0}^{|\Gamma_{i_1 \cdots i_n}|-1} \varphi \circ f^j \leq \sum_{\ell=1}^{n} \sup_{\Gamma_{i_\ell \cdots i_n}} \sum_{j=0}^{|\Gamma_{i_\ell}|-1} \varphi \circ f^j$$

$$\leq \sum_{\ell=1}^{n} \sup_{\Gamma_{i_\ell}} \sum_{j=0}^{|\Gamma_{i_\ell}|-1} \varphi \circ f^j,$$

it follows from (9.15) that

$$N(\alpha, \mathcal{C}_n) \leq \prod_{\ell=1}^{n} \sum_{\Delta \in \mathcal{C}} \exp\left(-\alpha |\Delta| + \sup_{\Delta} \sum_{j=0}^{|\Delta|-1} \varphi \circ f^j\right) = N(\alpha, \mathcal{C})^n.$$

Therefore, by (9.14), we obtain

$$N(\alpha, \mathcal{C}_\infty) \leq \sum_{n \in \mathbb{N}} N(\alpha, \mathcal{C})^n < \infty,$$

where $\mathcal{C}_\infty = \bigcup_{n \in \mathbb{N}} \mathcal{C}_n$. Now let m be the maximal length of the finite sequences I_i. For each $n \in \mathbb{N}$ and $\omega \in \Sigma_A^+$, there exists $\Delta_I \in \mathcal{C}_\infty$ such that

$$h(\omega) \in \Delta_I \quad \text{and} \quad n \leq |\Delta_I| < n + m.$$

Let also $\mathcal{C}^*$ be the family of all sets Δ_K such that K is a finite sequence consisting of the first n elements of some finite sequence I such that $\Delta_I \in \mathcal{C}_\infty$. We have

$$N(\alpha, \mathcal{C}^*) \leq N(\alpha, \mathcal{C}_\infty) e^{m \sup |\varphi|} \max\{1, e^{\alpha m}\} < \infty.$$

Since

$$N(\alpha, \mathcal{C}^*) = \sum_{i_1 \cdots i_n} e^{-\alpha n} \exp \sup_{\Delta_{i_1 \cdots i_n}} \sum_{j=0}^{n-1} \varphi \circ f^j,$$

we obtain $P(\varphi) - \alpha \leq 0$, and thus, $p \geq P(\varphi)$.   □

The lemma allows us to show that s is a lower bound for the Hausdorff dimension of J. Indeed, it follows from (9.12) that

$$\lim_{N \to \infty} M(0, N) = 0,$$

and by Lemma 9.1, we obtain

$$P\big((\dim_H J + \delta)\varphi\big) \geq 0.$$

Since the function $t \mapsto P(t\varphi)$ is strictly decreasing, we conclude that $\dim_H J + \delta \geq s$, and it follows from the arbitrariness of δ that $\dim_H J \geq s$.

Now we show that s is an upper bound for the upper box dimension of J. It follows from (9.8) that

$$n(\omega, r) \leq \frac{\log r}{-\log \tau} + a,$$

with τ as in (6.29), and where $a = 1 - \log c / \log \tau$ (one can always assume that $c \le 1$, and hence that $a \ge 1$). Let $\Delta(\omega_j, r)$ for $j = 1, \ldots, N$ be the elements of the Moran cover. It follows from (9.11) that diam $\Delta(\omega_j, r) < r$ for each j. Therefore,

$$\sum_{m \in \mathbb{N}} \operatorname{card} \{ j : n(\omega_j, r) = m \} \ge N(F, r).$$

This implies that there exists $m = m(r)$ such that

$$\operatorname{card} \{ j : n(\omega_j, r) = m \} \ge \frac{N(J, r)}{- \log r / \log \tau + a}.$$

On the other hand, for each $\delta > 0$, there exists a sequence $r_n \searrow 0$ such that

$$N(J, r_n) > r_n^{\delta - \overline{\dim}_B J}.$$

Setting $m = m(r_n)$ and using again (9.8), we obtain

$$k \sum_{i_1 \cdots i_{m-1}} \sup_{x \in \Delta_{i_1 \cdots i_{m-1}}} \| d_x f^{m-1} \|^{-\alpha} \ge \sum_{i_1 \cdots i_m} \sup_{x \in \Delta_{i_1 \cdots i_{m-1}}} \| d_x f^{m-1} \|^{-\alpha}$$

$$\ge r_n^\alpha \frac{r_n^{\delta - \overline{\dim}_B J}}{- \log r_n / \log \tau + a}$$

$$\ge \frac{\log \tau}{2a} r_n^{2\delta + \alpha - \overline{\dim}_B J}$$

for all sufficiently large n (since $1 / - \log r \ge r^\alpha$ for all sufficiently small $r > 0$). Therefore, provided that $\alpha < \overline{\dim}_B J - 2\delta$, we obtain

$$\sum_{i_1 \cdots i_{m-1}} \exp \sup_{x \in \Delta_{i_1 \cdots i_{m-1}}} \left(\alpha \sum_{l=0}^{m-2} \varphi \circ f^l \right) \ge 1 \qquad (9.16)$$

for all sufficiently large n, where again $m = m(r_n)$. Furthermore, it also follows from (9.8) that

$$n(\omega, r) \ge \frac{\log r}{- \log \tau'}, \qquad (9.17)$$

where

$$\tau' = \sup \{ \| d_x f \| : x \in J \}.$$

We note that $\tau' > 1$. Otherwise, we would have $\| d_x f^n v \| \le \| v \|$ for every $n \in \mathbb{N}$, $x \in J$ and $v \in T_x M$, which contradicts (6.29). It follows from (9.17) that

$$\min_j n(\omega_j, r) \to \infty \quad \text{when} \quad r \to 0,$$

and thus, $m(r_n) \to \infty$ when $n \to \infty$. Since n can be taken arbitrarily large, it follows from (9.16) that $P(\alpha\varphi) \geq 0$, and thus, $\alpha \leq s$ whenever $\alpha < \overline{\dim}_B J - 2\delta$. This implies that $\overline{\dim}_B J - 2\delta \leq s$, and it follows from the arbitrariness of δ that $\overline{\dim}_B J \leq s$. □

Now we give several examples.

Example 9.4. Let J be a repeller for a $C^{1+\alpha}$ transformation and let φ be the function in (9.6). We note that

$$\sum_{k=0}^{n-1} \varphi(f^k(x)) = -\sum_{k=0}^{n-1} \log \|d_{f^k(x)} f\|$$

$$= -\log \prod_{k=0}^{n-1} \|d_{f^k(x)} f\|$$

$$= -\log \|d_x f^n\|.$$

Now let $R_1, \ldots, R_k$ be a Markov partition of J. It follows from Exercise 6.16 that

$$P_{f|J}(s\varphi) = \lim_{n\to\infty} \frac{1}{n} \log \sum_{i_1\cdots i_n} \exp \sup_{R_{i_1\cdots i_n}} \left(s \sum_{k=0}^{n-1} \varphi \circ f^k \right),$$

where the sum is taken over all finite sequences $i_1 \cdots i_n$ such that $R_{i_1\cdots i_n} \neq \varnothing$, with $R_{i_1\cdots i_n}$ as in (6.35). Therefore,

$$P_{f|J}(s\varphi) = \lim_{n\to\infty} \frac{1}{n} \log \sum_{i_1\cdots i_n} \sup_{R_{i_1\cdots i_n}} \left(\|d_x f^n\|^{-s} \right). \tag{9.18}$$

Example 9.5. Let $J \subset [0, 1]$ be the set constructed in Example 8.1. We first note that J is a repeller. Indeed, let $f : [0, 1/3] \cup [2/3, 1] \to \mathbb{R}$ be defined by

$$f(x) = \begin{cases} 3x, & x \in [0, 1/3], \\ 3x - 2, & x \in [2/3, 1]. \end{cases}$$

This is the restriction of a $C^{1+\alpha}$ transformation to the set

$$K = [0, 1/3] \cup [2/3, 1].$$

Moreover, $f'(x) = 3$ for every x, and

$$J = \bigcap_{n=0}^{\infty} f^{-n} K.$$

This shows that J is a repeller for some extension of f. The function $\varphi: J \to \mathbb{R}$ in (9.6) is given by

$$\varphi(x) = -\log|f'(x)| = -\log 3.$$

Now we observe that the sets

$$R_1 = J \cap [0, 1/3] \quad \text{and} \quad R_2 = J \cap [2/3, 1]$$

form a Markov partition of J. Thus, it follows from (9.18) that

$$P_{f|J}(s\varphi) = \lim_{n\to\infty} \frac{1}{n} \log \sum_{i_1\cdots i_n} \exp(-ns\log 3),$$

where $i_1, \ldots, i_n \in \{1, 2\}$. Therefore,

$$P_{f|J}(s\varphi) = \lim_{n\to\infty} \frac{1}{n} \log\left(2^n e^{-ns\log 3}\right) = \log(2 \cdot 3^{-s}).$$

Solving the equation $P_{f|J}(s\varphi) = 0$ yields $s = \log 2/\log 3$, and thus, it follows from Theorem 9.1 that

$$\dim_H J = \underline{\dim}_B J = \overline{\dim}_B J = \frac{\log 2}{\log 3}.$$

We note that this is also a consequence of Examples 8.1 and 8.3.

Example 9.6. We consider again the repeller J in (9.2). With the notation in Example 9.3, one can easily verify that the sets $R_i = J \cap \Delta_i$, for $i = 1, \ldots, k$, form a Markov partition of J. For the function φ in (9.6), we have

$$\varphi|R_i = -\log|f'|\Delta_i| = \log\lambda_i,$$

and it follows from (9.18) that

$$P_{f|J}(s\varphi) = \lim_{n\to\infty} \frac{1}{n} \log \sum_{i_1\cdots i_n} \exp \sup_{R_{i_1\cdots i_n}} \left(s\sum_{j=0}^{n-1} \varphi \circ f^j\right)$$

$$= \lim_{n\to\infty} \frac{1}{n} \log \sum_{i_1\cdots i_n} \exp\left(s\sum_{j=0}^{n-1} \varphi|R_{i_{j+1}\cdots i_n}\right)$$

$$= \lim_{n\to\infty} \frac{1}{n} \log \sum_{i_1\cdots i_n} \exp\left(s\sum_{j=1}^{n} \log\lambda_{i_j}\right)$$

$$= \lim_{n\to\infty} \frac{1}{n} \log \sum_{i_1\cdots i_n} \prod_{j=1}^{n} \lambda_{i_j}^s$$

$$= \lim_{n \to \infty} \frac{1}{n} \log \left(\sum_{j=1}^{k} \lambda_j^s \right)^n$$

$$= \log \sum_{j=1}^{k} \lambda_j^s.$$

Thus, in this case, the equation $P_{f|J}(s\varphi) = 0$ in (9.5) reduces to $\sum_{j=1}^{k} \lambda_j^s = 1$. When $k = 2$, this formula was already obtained in Proposition 8.2.

9.3 Hyperbolic Sets

We study in this section the dimension of the hyperbolic sets of a conformal diffeomorphism. Here the conformality means that the diffeomorphism is conformal along the stable and unstable directions. More precisely, for a locally maximal hyperbolic set, we compute the Hausdorff dimension and the lower and upper box dimensions along the stable and unstable manifolds. We then use this result to compute the dimension of the hyperbolic set.

9.3.1 Dimension Along the Invariant Manifolds

Let $f: M \to M$ be a diffeomorphism and let $\Lambda \subset M$ be a hyperbolic set for f.

Definition 9.2. We say that f is *conformal* on Λ if the linear transformations $d_x f | E^s(x)$ and $d_x f | E^u(x)$ are multiples of isometries for every $x \in \Lambda$.

We note that if M is a surface and

$$\dim E^s(x) = \dim E^u(x) = 1 \quad \text{for every} \quad x \in \Lambda,$$

then f is conformal. More explicit examples are:

1. The Smale horseshoe in Sect. 6.3.1.
2. The solenoid in Example 6.12.
3. The general horseshoes in Exercise 8.15.

The following result is a version of Theorem 9.1 along the stable and unstable manifolds of a locally maximal hyperbolic set. Let $\varphi_s: \Lambda \to \mathbb{R}$ and $\varphi_u: \Lambda \to \mathbb{R}$ be the functions defined by

$$\varphi_s(x) = \log \|d_x f | E^s(x)\| \quad \text{and} \quad \varphi_u(x) = -\log \|d_x f | E^u(x)\|.$$

By Proposition 6.2, the spaces $E^s(x)$ and $E^u(x)$ vary continuously with x, and thus, the functions φ_s and φ_u are continuous (because they are compositions of continuous functions). We also recall the notion of topologically mixing transformation in Definition 3.13.

Theorem 9.2. *Let Λ be a locally maximal hyperbolic set for a $C^{1+\alpha}$ diffeomorphism, for some $\alpha \in (0, 1]$, such that f is conformal and topologically mixing on Λ. Then*

$$\dim_H(V^s(x) \cap \Lambda) = \underline{\dim}_B(V^s(x) \cap \Lambda) = \overline{\dim}_B(V^s(x) \cap \Lambda) = t_s, \qquad (9.19)$$

and

$$\dim_H(V^u(x) \cap \Lambda) = \underline{\dim}_B(V^u(x) \cap \Lambda) = \overline{\dim}_B(V^u(x) \cap \Lambda) = t_u, \qquad (9.20)$$

where t_s and t_u are the unique real numbers such that

$$P_{f|\Lambda}(t_s\varphi_s) = P_{f|\Lambda}(t_u\varphi_u) = 0. \qquad (9.21)$$

Proof. Let $R_1, \ldots, R_k$ be the elements of a Markov partition of Λ with diameter smaller than the size of the local stable and unstable manifolds (by Theorem 7.6 the set Λ has Markov partitions of arbitrarily small diameter). Given $x, y \in \Lambda$ in the same rectangle R_i, let $H^s_{x,y} : V^u(x) \to V^u(y)$ be the *holonomy map* defined by

$$H^s_{x,y}(z) = [z, y].$$

By Proposition 7.1, the map $H^s_{x,y}$ depends continuously on x and y. Furthermore, by results of Hasselblatt in [39], since f is conformal on Λ, the product structure is a Lipschitz homeomorphism with Lipschitz inverse. This implies that the holonomy map $H^s_{x,y}$ is Lipschitz.

Now we establish identities (9.19) and (9.20) for the dimensions along the stable and unstable manifolds. We only consider $V^u(x)$ since the arguments for $V^s(x)$ are entirely analogous. For $i = 1, \ldots, k$, let V_i be segments of local unstable manifolds such that

$$V_i \cap \Lambda = V^u(x_i) \cap \Lambda \cap R_i \qquad (9.22)$$

for some point $x_i \in R_i$. We note that

$$V_i \cap \mathrm{int}\, R_j = \varnothing \quad \text{whenever} \quad j \neq i.$$

Since the holonomy maps are Lipschitz, the numbers $\dim_H(V_i \cap \Lambda)$, $\underline{\dim}_B(V_i \cap \Lambda)$, and $\overline{\dim}_B(V_i \cap \Lambda)$ are independent of x_i. Furthermore, since f is topologically mixing on Λ, we have

$$\dim_H(V_i \cap \Lambda) = \dim_H(V_j \cap \Lambda),$$

$$\underline{\dim}_B(V_i \cap \Lambda) = \underline{\dim}_B(V_j \cap \Lambda), \qquad (9.23)$$

$$\overline{\dim}_B(V_i \cap \Lambda) = \overline{\dim}_B(V_j \cap \Lambda)$$

for every i and j. Set $V = \bigcup_{i=1}^{k} V_i$. We define

$$R_{i_0 \cdots i_n} = \bigcap_{j=0}^{n} f^{-j} R_{i_j} \quad \text{and} \quad V_{i_0 \cdots i_n} = V \cap R_{i_0 \cdots i_n}$$

for every $(\cdots i_{-1} i_0 i_1 \cdots) \in \Sigma_A$ and $n \in \mathbb{N}$, where A is the transition matrix obtained from the Markov partition.

We first show that t_u is an upper bound for the upper box dimension of Λ. Note that

$$f^n(V_{i_0 \cdots i_n}) \subset V^u(f^n(x_{i_0})) \cap R_{i_n},$$

and

$$H^s_{f^n(x_{i_0}), x_{i_n}} f^n(V_{i_0 \cdots i_n} \cap \Lambda) = V_{i_n} \cap \Lambda.$$

Furthermore, each point in $V_i \cap \Lambda$ has exactly one preimage under the holonomy map $H^s_{f^n(x_{i_0}), x_{i_n}}$. Hence, if $\mathcal{U}$ is a cover of $V_{i_n} \cap \Lambda$, then

$$\bigcup_{U \in \mathcal{U}} f^{-n} \left(H^s_{x_{i_n}, f^n(x_{i_0})} U \right) \supset V_{i_0 \cdots i_n} \cap \Lambda.$$

This implies that

$$N(V_{i_0 \cdots i_n} \cap \Lambda, r) \le N(V_{i_n} \cap \Lambda, K^{-1} r \overline{\lambda}_{i_0 \cdots i_n})$$

for every $r > 0$, where $K > 1$ is a Lipschitz constant for the holonomy map and where

$$\overline{\lambda}_{i_0 \cdots i_n} = \max \left\{ \|d_x f^{-n}|E^u(x)\| : x \in R_{i_0 \cdots i_n} \right\}.$$

Therefore,

$$N(V \cap \Lambda, r) \le \sum_{i_0 \cdots i_n} N(V_{i_0 \cdots i_n} \cap \Lambda, r)$$

$$\le \sum_{i_0 \cdots i_n} N(V \cap \Lambda, K^{-1} r \overline{\lambda}_{i_0 \cdots i_n}).$$

Now take $s > \overline{\dim}_B(V \cap \Lambda)$. Then there exists $r_0 > 0$ such that

$$N(V \cap \Lambda, r) < r^{-s} \quad \text{for every} \quad r \in (0, r_0).$$

Setting

$$c_n(s) = \sum_{i_1 \cdots i_n} \overline{\lambda}_{i_0 \cdots i_n}^{\,s},$$

we thus obtain

$$N(V \cap \Lambda, r) \le r^{-s} K^s c_n(s)$$

for every $r < \lambda_n K r_0$, where $\lambda_n = \min_{i_0 \cdots i_n} \overline{\lambda}_{i_0 \cdots i_n}$. By induction, we conclude that

$$N(V \cap \Lambda, r) \le r^{-s} K^s c_n(s)^m$$

for every $r < (\lambda_n K)^m r_0$. Now we note that $\lambda_n K < 1$ for all sufficiently large n. Therefore,

$$\frac{\log N(V \cap \Lambda, r)}{-\log r} \le s + \frac{m \log c_n(s)}{-\log r}$$

$$\le s + \frac{m \log c_n(s)}{-\log[(\lambda_n K)^m r_0]},$$

and

$$\overline{\dim}_B(V \cap \Lambda) \le s + \limsup_{m \to +\infty} \frac{m \log c_n(s)}{-\log[(\lambda_n K)^m r_0]}$$

$$= s - \frac{\log c_n(s)}{\log(\lambda_n K)}.$$

Letting $s \searrow \overline{\dim}_B(V \cap \Lambda)$, we conclude that

$$c_n(\overline{\dim}_B(V \cap \Lambda)) \ge 1.$$

On the other hand, for each $s > \overline{\dim}_B(V \cap \Lambda)$, we have

$$c_n(s) = \sum_{i_0 \cdots i_n} \overline{\lambda}_{i_0 \cdots i_n}{}^s = \sum_{i_0 \cdots i_n} \exp \max_{x \in R_{i_0 \cdots i_n}} \left(s \sum_{k=0}^{n} \varphi_u(f^k(x)) \right),$$

and hence,

$$P(s\varphi_u) = \lim_{n \to \infty} \frac{1}{n} \log c_n(s) \ge 0 = P(t_u \varphi_u),$$

where $P = P_{f|\Lambda}$. Since the function $s \mapsto P(s\varphi_u)$ is strictly decreasing, it follows that $s \le t_u$ for every $s > \overline{\dim}_B(V \cap \Lambda)$, and thus,

$$\overline{\dim}_B(V \cap \Lambda) \le t_u. \tag{9.24}$$

Now we show that t_u is a lower bound for the Hausdorff dimension of Λ. Assume on the contrary that $\dim_H(V \cap \Lambda) < t_u$ and let s be a positive number such that

$$\dim_H(V \cap \Lambda) < s < t_u. \tag{9.25}$$

Then $m(V \cap \Lambda, s) = 0$, and since $V \cap \Lambda$ is compact, for each $\delta > 0$, there is a finite cover $\mathcal{U}$ of $V \cap \Lambda$ by open balls such that

$$\sum_{U \in \mathcal{U}} (\operatorname{diam} U)^s < \delta^s. \qquad (9.26)$$

For each $n \in \mathbb{N}$, let δ_n be a positive number such that

$$p_n(U) := \operatorname{card} \left\{ (i_0 \cdots i_n) : U \cap R_{i_0 \cdots i_n} \neq \varnothing \right\} < k$$

whenever $\operatorname{diam} U < \delta_n$ (the existence of δ_n follows from the properties of the Markov partition). We note that $\delta_n \to 0$ when $n \to \infty$. It follows from (9.26) with $\delta = \delta_n$ that $\operatorname{diam} \mathcal{U} < \delta_n$ and hence that

$$p_n(U) < k \quad \text{for every} \quad U \in \mathcal{U}.$$

Let $N = n + m - 1$, for some $m \in \mathbb{N}$ such that $A^m > 0$ (recall that f is topologically mixing on Λ). For each $(\cdots i_{-1} i_0 i_1 \cdots) \in \Sigma_A$ and $n \in \mathbb{N}$, we consider the cover $\mathcal{U}_{i_0 \cdots i_N}$ of V composed of the projections along the stable leaves into V of the sets $f^N(U)$ with $U \in \mathcal{U}$ such that $U \cap R_{i_0 \cdots i_n} \neq \varnothing$. We have

$$\sum_{U \in \mathcal{U}_{i_0 \cdots i_N}} (\operatorname{diam} U)^s \leq \underline{\lambda}_{i_0 \cdots i_N}^{-s} \sum_{U \in \mathcal{U}, U \cap R_{i_0 \cdots i_n} \neq \varnothing} (\operatorname{diam} U)^s,$$

where

$$\underline{\lambda}_{i_0 \cdots i_n} = \min \left\{ \| d_x f^{-n} | E^u(x) \| : x \in R_{i_0 \cdots i_n} \right\}.$$

Now let us assume that

$$\sum_{U \in \mathcal{U}_{i_0 \cdots i_N}} (\operatorname{diam} U)^s \geq \delta_n^{\ s}$$

for every $(\cdots i_{-1} i_0 i_1 \cdots) \in \Sigma_A$ and $n \in \mathbb{N}$. Then

$$k\delta_n^{\ s} > k \sum_{U \in \mathcal{U}} (\operatorname{diam} U)^s \geq \sum_{U \in \mathcal{U}} p_n(U)(\operatorname{diam} U)^s$$

$$= \sum_{i_0 \cdots i_n} \sum_{U \in \mathcal{U}, U \cap R_{i_0 \cdots i_n} \neq \varnothing} (\operatorname{diam} U)^s$$

$$\geq k^{-m+1} \sum_{i_0 \cdots i_N} \sum_{U \in \mathcal{U}, U \cap R_{i_0 \cdots i_n} \neq \varnothing} (\operatorname{diam} U)^s \qquad (9.27)$$

$$\geq k^{-m+1} \sum_{i_0 \cdots i_N} \left(\underline{\lambda}_{i_0 \cdots i_N} \sum_{U \in \mathcal{U}_{i_0 \cdots i_N}} (\operatorname{diam} U)^s \right)$$

$$\geq k^{-m+1} \delta_n^{\ s} \sum_{i_0 \cdots i_N} \underline{\lambda}_{i_0 \cdots i_N}^{\ s}.$$

On the other hand, by a property analogous to that in (9.9), there is a constant $C > 0$ (independent of $n \in \mathbb{N}$ and $(i_0 \cdots i_n)$) such that

$$C^{-1} \leq \frac{\overline{\lambda}_{i_0 \cdots i_n}}{\underline{\lambda}_{i_0 \cdots i_n}} \leq C.$$

Therefore, by (9.27),

$$
\begin{aligned}
P(s\varphi_u) &= \lim_{N \to \infty} \frac{1}{N} \sum_{i_0 \cdots i_N} \overline{\lambda}_{i_0 \cdots i_n}{}^s \\
&\leq \lim_{N \to \infty} \frac{1}{N} \sum_{i_0 \cdots i_N} \underline{\lambda}_{i_0 \cdots i_n}{}^s \leq 0.
\end{aligned}
\tag{9.28}
$$

Since the function $s \mapsto P(s\varphi_u)$ is strictly decreasing and $P(t_u\varphi_u) = 0$, we conclude that (9.28) contradicts (9.25). Hence, we must have

$$\sum_{U \in \mathcal{U}_{i_0 \cdots i_N}} (\operatorname{diam} U)^s < \delta_n{}^s \tag{9.29}$$

for some sequence $(i_0 \cdots i_N)$ and all sufficiently large n (recall that $N = n+m-1$). Now we restart the process using the cover $\mathcal{V}_1 = \mathcal{U}_{i_0 \cdots i_N}$ to find inductively finite covers $\mathcal{V}_\ell$ of $V \cap \Lambda$ for each $\ell \in \mathbb{N}$. By (9.29), we have diam $\mathcal{V}_\ell < \delta_n$, and hence, $p_n(U) < k$ for every $U \in \mathcal{V}_\ell$. This implies that card $\mathcal{V}_{\ell+1} <$ card $\mathcal{V}_\ell$, and hence, card $\mathcal{V}_\ell = 1$ for some $\ell = \ell(n)$. Writing $\mathcal{V}_{\ell(n)} = \{U_n\}$, we thus obtain

$$\operatorname{diam}(V \cap \Lambda) \leq \operatorname{diam} U_n < \delta_n \to 0 \quad \text{when} \quad n \to \infty,$$

which is impossible. This contradiction shows that

$$\dim_H(V \cap \Lambda) \geq t_u. \tag{9.30}$$

By (9.24) and (9.30), we obtain

$$\dim_H(V \cap \Lambda) = \underline{\dim}_B(V \cap \Lambda) = \overline{\dim}_B(V \cap \Lambda) = t_u.$$

By (9.22) and (9.23), this establishes the identities in (9.19).   □

We emphasize that all dimensions in identities (9.19) and (9.20) are independent of the point x.

Example 9.7. Let Λ be the general horseshoe constructed in Exercise 8.15. At each point $x \in \Lambda$, the stable and unstable manifolds of sufficiently small size (possibly depending on x) are respectively horizontal and vertical segments. Moreover, the projection of the intersection $V^s(x) \cap \Lambda$ to the horizontal axis coincides with the set J constructed in Example 9.3 by taking $k = 2$ and

$$g_i(x) = \frac{x - a_i}{\lambda_i} \quad \text{for} \quad i = 1, 2.$$

By Example 9.6, we have

$$\dim_H(V^s(x) \cap \Lambda) = \underline{\dim}_B(V^s(x) \cap \Lambda) = \overline{\dim}_B(V^s(x) \cap \Lambda) = \alpha, \quad (9.31)$$

where α is the unique root of the equation $\lambda_1^\alpha + \lambda_2^\alpha = 1$. Similarly, one can show that

$$\dim_H(V^u(x) \cap \Lambda) = \underline{\dim}_B(V^u(x) \cap \Lambda) = \overline{\dim}_B(V^u(x) \cap \Lambda) = \beta, \quad (9.32)$$

where β is the unique root of the equation $\mu_1^\beta + \mu_2^\beta = 1$.

9.3.2 Dimension of Hyperbolic Sets

We show in this section that the Hausdorff dimension and the lower and upper box dimensions of a locally maximal hyperbolic set are equal to the sum of the dimensions along the stable and unstable manifolds.

Theorem 9.3 (Dimension of hyperbolic sets). *Let Λ be a locally maximal hyperbolic set for a $C^{1+\alpha}$ diffeomorphism, for some $\alpha \in (0, 1]$, such that f is conformal and topologically mixing on Λ. Then*

$$\dim_H \Lambda = \underline{\dim}_B \Lambda = \overline{\dim}_B \Lambda = t_s + t_u, \quad (9.33)$$

with t_s and t_u as in (9.21).

Proof. Since f is conformal on Λ, the product structure is a Lipschitz homeomorphism with Lipschitz inverse, and thus,

$$\dim_H[V^s(x) \cap \Lambda, V^u(x) \cap \Lambda] = \dim_H((V^s(x) \cap \Lambda) \times (V^u(x) \cap \Lambda)),$$
$$\underline{\dim}_B[V^s(x) \cap \Lambda, V^u(x) \cap \Lambda] = \underline{\dim}_B((V^s(x) \cap \Lambda) \times (V^u(x) \cap \Lambda)), \quad (9.34)$$
$$\overline{\dim}_B[V^s(x) \cap \Lambda, V^u(x) \cap \Lambda] = \overline{\dim}_B((V^s(x) \cap \Lambda) \times (V^u(x) \cap \Lambda)).$$

By Exercise 8.7, it follows from (9.19), (9.20), and (9.34) that

$$\dim_H[V^s(x) \cap \Lambda, V^u(x) \cap \Lambda] = \underline{\dim}_B[V^s(x) \cap \Lambda, V^u(x) \cap \Lambda]$$
$$= \overline{\dim}_B[V^s(x) \cap \Lambda, V^u(x) \cap \Lambda] \quad (9.35)$$
$$= t_s + t_u.$$

On the other hand, since Λ is locally maximal, we have

$$[V^s(x) \cap \Lambda, V^u(x) \cap \Lambda] \subset \Lambda$$

for every $x \in \Lambda$, and thus, there exist points $x_1, \ldots, x_N \in \Lambda$ such that

$$\Lambda = \bigcup_{n=1}^{N} [V^s(x_n) \cap \Lambda, V^u(x_n) \cap \Lambda].$$

The identities in (9.33) follow now immediately from (9.35). $\square$

Example 9.8. Let Λ be the general horseshoe constructed in Exercise 8.15. It follows from Theorem 9.3 together with (9.31) and (9.32) that the Hausdorff dimension and the lower and upper box dimensions of Λ are equal to $\alpha + \beta$. This value was already obtained in Exercise 8.15.

9.4   Exercises

Exercise 9.1. For a repeller J of a $C^{1+\alpha}$ transformation f such that f is conformal on J, show that if

$$a \leq \|d_x f\| \leq b \quad \text{for every} \quad x \in J, \tag{9.36}$$

then

$$\frac{h(f|Z)}{b} \leq \dim_H J \leq \frac{h(f|J)}{a}. \tag{9.37}$$

Exercise 9.2. For a repeller J as in Exercise 9.1, show that the inequalities in (9.37) remain true when condition (9.36) is replaced by

$$c^{-1} a^n \leq \|d_x f^n\| \leq c b^n \quad \text{for every} \quad x \in J, n \in \mathbb{N}.$$

Exercise 9.3. Show that a repeller for a map f is also a repeller for any power f^n, with $n \in \mathbb{N}$.

Exercise 9.4. Find a nonconformal map f such that f^2 is conformal.

Exercise 9.5. Let J be a repeller for a $C^{1+\alpha}$ transformation f. Show that if some power f^n of the map f is conformal on J, then (9.4) holds, where s is now the unique root of the equation

$$P_{f^n|J}\left(-s \log \|df^n\|\right) = 0.$$

Hint: use Exercise 9.3.

Exercise 9.6. For a repeller J of a $C^{1+\alpha}$ transformation such that f is conformal on J, show that if μ is an equilibrium measure for the function $\varphi \colon J \to \mathbb{R}$ defined by $\varphi(x) = -\log \|d_x f\|$, then

$$\dim_H J = \underline{\dim}_B J = \overline{\dim}_B J = \frac{h_\mu(f)}{\int_J \log \|df\| \, d\mu}.$$

Exercise 9.7. Use Theorem 9.3 to show that the Hausdorff dimension of the Smale horseshoe is $\log 4 / \log 3$.

Exercise 9.8. Show that for each $d \in (0, 2)$, there is a horseshoe Λ such that

$$\dim_H \Lambda = \underline{\dim}_B \Lambda = \overline{\dim}_B \Lambda = d.$$

Hint: change the construction of the Smale horseshoe.

Exercise 9.9. Show that for each $d_s, d_u \in (0, 1)$, there is a horseshoe Λ such that

$$\dim_H (V^s(x) \cap \Lambda) = \underline{\dim}_B (V^s(x) \cap \Lambda) = \overline{\dim}_B (V^s(x) \cap \Lambda) = d_s$$

and

$$\dim_H (V^u(x) \cap \Lambda) = \underline{\dim}_B (V^u(x) \cap \Lambda) = \overline{\dim}_B (V^u(x) \cap \Lambda) = d_u$$

for every $x \in \Lambda$.

Exercise 9.10. For a repeller J of a $C^{1+\alpha}$ transformation f such that f is conformal on J, let $R_1, \ldots, R_k$ be the elements of a Markov partition and let

$$\Delta_{i_1 \cdots i_n} = \bigcap_{j=0}^{n-1} f^{-j} R_{i_{j+1}}$$

for each $n \in \mathbb{N}$ and $i_1, \ldots, i_n \in \{1, \ldots, k\}$. We assume that there exist a probability measure μ in J and constants $D_1, D_2 > 0$ such that

$$D_1 \leq \frac{\mu(\Delta_{i_1 \cdots i_n})}{\|d_x f^n\|^{-1}} \leq D_2$$

for every $n \in \mathbb{N}$ and $x \in \Delta_{i_1 \cdots i_n}$. Show that:

1. μ is equivalent to the Hausdorff measure $m(\cdot, s)$, where $s = \dim_H J$, with Radon–Nikodym derivative bounded and bounded away from zero.
2. $0 < m(J, s) < \infty$.
3. $\underline{d}_\mu(x) = \overline{d}_\mu(x) = s$ for every $x \in J$.
4. $\dim_H \mu = \underline{\dim}_B \mu = \overline{\dim}_B \mu = s$.

Exercise 9.11. For a repeller J of a $C^{1+\alpha}$ transformation f such that f is conformal on J, let $R_1, \ldots, R_k$ be the elements of a Markov partition with transition matrix A. We assume that there exists $q \in \mathbb{N}$ such that all entries of the matrix A^q are positive. Given a Hölder continuous function $\varphi \colon J \to \mathbb{R}$, we consider the numbers

$$a_{i_1 \cdots i_n} = \max \left\{ \exp \varphi_n(x) : x \in \Delta_{i_1 \cdots i_n} \right\},$$

where $\varphi_n = \sum_{l=0}^{n-1} \varphi \circ f^l$, and also,

$$\alpha_n = \sum_{i_1 \cdots i_n} a_{i_1 \cdots i_n}.$$

Show that:

1. $\alpha_{n+l} \leq \alpha_n \alpha_l$ for every $n, l \in \mathbb{N}$.
2. There exists $D > 0$ such that

$$|\varphi_n(x) - \varphi_n(y)| < D$$

   for every $n \in \mathbb{N}$ and $x, y \in \Delta_{i_1 \cdots i_n}$.
3. There exists $D' > 0$ such that $\alpha_{n+l} \geq D' \alpha_n \alpha_l$ for every $n, l \in \mathbb{N}$.
4.

$$P_{f|J}(\varphi) = \lim_{n \to \infty} \frac{1}{n} \log \alpha_n = \inf_{n \in \mathbb{N}} \frac{1}{n} \log \alpha_n.$$

5. For each $l \in \mathbb{N}$ and any probability measure ν_l satisfying

$$\nu_l(\Delta_{i_1 \cdots i_l}) = a_{i_1 \cdots i_l} / \alpha_l$$

   for each $(i_1 i_2 \cdots) \in \Sigma_A^+$, there exist $C_1, C_2 > 0$ such that

$$C_1 \leq \frac{\nu_l(\Delta_{i_1 \cdots i_n})}{\exp\left(-n P_{f|J}(\varphi) + \varphi_n(x)\right)} \leq C_2$$

   for every $n, l \in \mathbb{N}$ and $x \in \Delta_{i_1 \cdots i_n}$.
6. Any sublimit ν of the sequence $(\nu_l)_l$ satisfies

$$C_1 \leq \frac{\nu(\Delta_{i_1 \cdots i_n})}{\exp\left(-n P_{f|J}(\varphi) + \varphi_n(x)\right)} \leq C_2$$

   for every $n \in \mathbb{N}$ and $x \in \Delta_{i_1 \cdots i_n}$.
7. For any sublimit μ of the sequence of measures

$$\mu_n = \frac{1}{n} \sum_{j=0}^{n-1} \nu \circ f^{-j},$$

where ν is any sublimit of the sequence $(\nu_l)_l$, there exist $C_1', C_2' > 0$ such that

$$C_1' \le \frac{\mu(\Delta_{i_1 \cdots i_n})}{\exp\left(-n P_{f|J}(\varphi) + \varphi_n(x)\right)} \le C_2'$$

for every $n \in \mathbb{N}$ and $x \in \Delta_{i_1 \cdots i_n}$.

8. μ is an equilibrium measure for φ, that is, μ is an f-invariant probability measure in J and

$$P_{f|J}(\varphi) = h_\mu(f) + \int_J \varphi \, d\mu.$$

Notes

We refer to the books [8, 9, 73] for further topics of dimension theory of dynamical systems. Equation (9.5) was introduced by Bowen [21] in his study of quasicircles. For a repeller J of a conformal map f, Ruelle [85] showed that $\dim_H J = s$ (under the assumption that f is topologically mixing on J). The equality between the Hausdorff and box dimensions is due to Falconer [31]. Our proof of Theorem 9.1 is based on [8]. McCluskey and Manning [59] obtained (9.19) and (9.20) for the Hausdorff dimension. The equality between the Hausdorff and box dimensions of a hyperbolic set is due to Takens [101] for C^2 diffeomorphisms and to Palis and Viana [68] in the general case. Our proof of Theorem 9.2 is based on [7], which is inspired in arguments in [101].

Appendix A
Notions from Measure Theory

We recall in this appendix all the necessary material from measure theory.

A.1 Measure Spaces

Let $\mathcal{A}$ be a family of subsets of a set X. We say that $\mathcal{A}$ is a *σ-algebra* of X if:

1. $\varnothing, X \in \mathcal{A}$.
2. $B \in \mathcal{A}$ whenever $X \setminus B \in \mathcal{A}$.
3. $\bigcup_{k=1}^{\infty} B_k \in \mathcal{A}$ whenever $B_k \in \mathcal{A}$ for every $k \in \mathbb{N}$.

Each element of a σ-algebra $\mathcal{A}$ is called an *$\mathcal{A}$-measurable set*. Given a family $\mathcal{A}$ of subsets of X, the smallest σ-algebra containing $\mathcal{A}$, that is, the intersection of all σ-algebras containing $\mathcal{A}$, is called the *σ-algebra of X generated by $\mathcal{A}$*.

Now let $\mathcal{A}$ be a σ-algebra of X. We say that a function $\mu\colon \mathcal{A} \to [0, +\infty]$ is a *measure* in X with respect to $\mathcal{A}$ provided that:

1. $\mu(\varnothing) = 0$.
2. If $B_k \in \mathcal{A}$ for every $k \in \mathbb{N}$ and $B_k \cap B_\ell = \varnothing$ whenever $k \neq \ell$, then

$$\mu\left(\bigcup_{k=1}^{\infty} B_k\right) = \sum_{k=1}^{\infty} \mu(B_k).$$

The triple $(X, \mathcal{A}, \mu)$ is then called a *measure space*. We note that a measure is completely determined by its values in any family of sets generating the σ-algebra. We say that the measure space $(X, \mathcal{A}, \mu)$ is:

1. *Finite* if $\mu(X) < \infty$.
2. *Infinite* if $\mu(X) = \infty$.
3. *σ-finite* if there exist sets $B_k \in \mathcal{A}$ for $k \in \mathbb{N}$ such that $X = \bigcup_{k=1}^{\infty} B_k$ and $\mu(B_k) < \infty$ for every $k \in \mathbb{N}$.
4. A *probability space* and in this case that μ is a *probability measure* if $\mu(X) = 1$.

L. Barreira, *Ergodic Theory, Hyperbolic Dynamics and Dimension Theory*, Universitext,   277
DOI 10.1007/978-3-642-28090-0, © Springer-Verlag Berlin Heidelberg 2012

We say that a given property holds μ-*almost everywhere* if the set of points where it does not hold has zero μ-measure.

A.2   Outer Measures and Measurable Sets

Let $\mathcal{A}$ be the family of all subsets of a set X. A function $\mu \colon \mathcal{A} \to [0, +\infty]$ is called an *outer measure* in X provided that:

1. $\mu(\varnothing) = 0$.
2. $\mu(B) \leq \mu(C)$ whenever $B \subset C$.
3. If $B_k \in \mathcal{A}$ for every $k \in \mathbb{N}$, then

$$\mu \left(\bigcup_{k=1}^{\infty} B_k \right) \leq \sum_{k=1}^{\infty} \mu(B_k).$$

In particular, when $(X, \mathcal{A}, \mu)$ is a measure space, one can define an outer measure in X by

$$\mu^*(B) = \inf \left\{ \sum_{k=1}^{\infty} \mu(B_k) : B \subset \bigcup_{k=1}^{\infty} B_k \text{ and } B_k \in \mathcal{A} \text{ for every } k \in \mathbb{N} \right\}. \qquad \text{(A.1)}$$

We call μ^* the *outer measure associated to* $(X, \mathcal{A}, \mu)$.

Given an outer measure μ in X we say that a set $A \subset X$ is μ-*measurable* if

$$\mu(B) = \mu(B \cap A) + \mu(B \setminus A)$$

for any $B \subset X$. If μ is an outer measure in X, then the following properties hold:

1. The family $\mathcal{A}$ of μ-measurable sets is a σ-algebra, and the restriction of μ to $\mathcal{A}$ is a measure in X.
2. If μ is an outer measure associated to a measure space $(X, \mathcal{A}', \mu')$, then $\mathcal{A}' \subset \mathcal{A}$ and $\mu'(B) = \mu(B)$ for every $B \in \mathcal{A}'$.

A.3   Measures in Topological Spaces

Now let X be a topological space, and let $\mathcal{A}$ be the family of all open subsets of X. The σ-algebra of X generated by $\mathcal{A}$ is called the *Borel σ-algebra* of X, and its elements are called *Borel-measurable sets*. Given a measure μ in X its *support* is the complement of the largest open set U with $\mu(U) = 0$, and we denote it by $\operatorname{supp} \mu$.

We also consider the particular case of the set $X = \mathbb{R}^n$ with its usual topology. Let $\mathcal{A}' \subset \mathcal{A}$ be the family of open rectangles, that is, the family of sets of the form

$$(a_1, b_1) \times \cdots \times (a_n, b_n).$$

Since each open subset of $\mathbb{R}^n$ is a countable union of element of $\mathcal{A}'$, the σ-algebra of $\mathbb{R}^n$ generated by $\mathcal{A}'$ is the Borel σ-algebra of $\mathbb{R}^n$. In particular, the Borel σ-algebra of $\mathbb{R}$ is generated by the open intervals. If $\mathcal{B}$ is the Borel σ-algebra of $\mathbb{R}^n$, then there exists a unique measure $\mu: \mathcal{B} \to [0, +\infty]$ such that

$$\mu\big((a_1, b_1) \times \cdots \times (a_n, b_n)\big) = \prod_{i=1}^{n}(b_i - a_i)$$

for every $a_k < b_k$ and $k = 1, \ldots, n$. Using (A.1), one can construct an outer measure μ^* associated to μ. The σ-algebra of μ^*-measurable sets is called the *Lebesgue σ-algebra* of $\mathbb{R}^n$, and its elements are called *Lebesgue-measurable sets*. The restriction of μ^* to this σ-algebra is called the *Lebesgue measure* in $\mathbb{R}^n$. For any set $B \subset \mathbb{R}^n$ with $\mu^*(B) < \infty$, the following properties are equivalent:

1. The set B is Lebesgue-measurable.
2. Given $\varepsilon > 0$, there exist a compact set $K \subset B$ and an open set $U \supset B$ such that $\mu(U \setminus K) < \varepsilon$.
3. Given $\varepsilon > 0$, there exist rectangles $R_1, \ldots, R_m$ such that

$$\mu^*\left(B \setminus \bigcup_{i=1}^{m} R_i\right) < \varepsilon \quad \text{and} \quad \mu^*\left(\bigcup_{i=1}^{m} R_i \setminus B\right) < \varepsilon.$$

In particular, if $B \subset \mathbb{R}^n$ is a Lebesgue-measurable set with $\mu^*(B) < \infty$, then

$$\mu^*(B) = \inf\{\mu(U) : U \supset B \text{ is open}\}$$

and

$$\mu^*(B) = \sup\{\mu(K) : K \subset B \text{ is compact}\}.$$

A.4 Measurable Functions, Integration, and Convergence

Let X be a set and let $\mathcal{A}$ be a σ-algebra of X. A transformation $T: X \to X$ is said to be $\mathcal{A}$-*measurable* if $T^{-1}B \in \mathcal{A}$ whenever $B \in \mathcal{A}$. A function $\varphi: X \to \mathbb{R}$ is said to be $\mathcal{A}$-*measurable* if $\varphi^{-1}B \in \mathcal{A}$ whenever B is a Borel-measurable set.

Given $\mathcal{A}$-measurable sets $B_1, \ldots, B_n \subset X$ and numbers $a_1, \ldots, a_n \in \mathbb{R}$, we define a *simple function* by

$$\varphi = \sum_{k=1}^{n} a_k \chi_{B_k},$$

where χ_B is the *characteristic function* $\chi_B \colon X \to \{0, 1\}$ of the set $B \subset X$, given by

$$\chi_B(x) = \begin{cases} 1 & \text{if } x \in B, \\ 0 & \text{if } x \notin B. \end{cases}$$

Given a measure μ in X with respect to $\mathcal{A}$ and an $\mathcal{A}$-measurable function $\varphi \geq 0$, we define the *Lebesgue integral* of φ in X by

$$\int_X \varphi \, d\mu = \sup \left\{ \sum_{k=1}^n a_k \mu(B_k) : \sum_{k=1}^n a_k \chi_{B_k} \leq \varphi \right\}.$$

For an arbitrary $\mathcal{A}$-measurable function φ, we say that φ is μ-*integrable* if

$$\int_X \varphi^+ \, d\mu < \infty \quad \text{and} \quad \int_X \varphi^- \, d\mu < \infty,$$

where

$$\varphi^+ = \max\{\varphi, 0\} \quad \text{and} \quad \varphi^- = \max\{-\varphi, 0\}.$$

In this case, we define the *Lebesgue integral* of φ in X by

$$\int_X \varphi \, d\mu = \int_X \varphi^+ \, d\mu - \int_X \varphi^- \, d\mu.$$

One can show that φ is μ-integrable if and only if

$$\int_X |\varphi| \, d\mu < \infty.$$

We denote by $L^1(X, \mu)$ the set of μ-integrable functions $\varphi \colon X \to \mathbb{R}$. We note that

$$L^1(X, \mu) \ni \varphi \mapsto \int_X \varphi \, d\mu$$

is a linear transformation.

Theorem A.1 (Fatou's lemma).   *If $\varphi_n \colon X \to \mathbb{R}$ is a sequence of nonnegative measurable functions, then*

$$\int_X \liminf_{n \to \infty} \varphi_n \, d\mu \leq \liminf_{n \to \infty} \int_X \varphi_n \, d\mu.$$

Theorem A.2 (Monotone convergence theorem).   *If $\varphi_n \colon X \to \mathbb{R}_0^+$ is a nondecreasing sequence of measurable functions, then*

$$\int_X \lim_{n \to \infty} \varphi_n \, d\mu = \lim_{n \to \infty} \int_X \varphi_n \, d\mu.$$

Theorem A.3 (Dominated convergence theorem). *If $\varphi_n: X \to \mathbb{R}$ is a sequence of measurable functions such that the limit*

$$\varphi = \lim_{n \to \infty} \varphi_n$$

exists, and there is a μ-integrable function $\psi: X \to \mathbb{R}_0^+$ such that $|\varphi_n| \leq \psi$ for every $n \in \mathbb{N}$, then φ is μ-integrable and

$$\int_X \varphi \, d\mu = \lim_{n \to \infty} \int_X \varphi_n \, d\mu.$$

When X is a topological space, we denote by $C(X)$ the space of continuous functions $\varphi: X \to \mathbb{R}$. We recall that an operator $J: C(X) \to \mathbb{R}$ is said to be *positive* if $J(\varphi) \geq 0$ whenever $\varphi \geq 0$.

Theorem A.4 (Riesz's representation theorem). *If X is a compact metric space and $J: C(X) \to \mathbb{R}$ is a positive continuous linear operator, then there exists exactly one measure μ in the Borel σ-algebra of X such that*

$$J(\varphi) = \int_X \varphi \, d\mu$$

for every $\varphi \in C(X)$. If, in addition, $J(1) = 1$, then μ is a probability measure.

A.5 Absolutely Continuous Measures

Let μ and ν be measures in X with respect to the same σ-algebra. We say that ν is *absolutely continuous* with respect to μ, and we write $\nu \ll \mu$, if $\nu(B) = 0$ for every $B \in \mathcal{A}$ with $\mu(B) = 0$. For example, if $\varphi \in L^1(X, \mu)$, then the measure ν defined by

$$\nu(B) = \int_B \varphi \, d\mu$$

for each $B \in \mathcal{A}$ is absolutely continuous with respect to μ.

Theorem A.5 (Radon–Nikodym). *If $(X, \mathcal{A}, \mu)$ and $(X, \mathcal{A}, \nu)$ are σ-finite measure spaces and ν is absolutely continuous with respect to μ, then there exists an $\mathcal{A}$-measurable function $\varphi: X \to \mathbb{R}$, uniquely determined in a set of full μ-measure, such that*

$$\nu(B) = \int_B \varphi \, d\mu$$

for every $B \in \mathcal{A}$ with $\nu(B) < \infty$.

Any function φ as in Theorem A.5 is called a *Radon–Nikodym derivative* of ν with respect to μ, and we represent it by $d\nu/d\mu$. We note that if $(X, \mathcal{A}, \mu)$ and $(X, \mathcal{A}, \nu)$ are finite measure spaces and $\nu \ll \mu$, then $d\nu/d\mu \in L^1(X, \mu)$.

Theorem A.6. *Let $(X, \mathcal{A}, \mu)$ be a σ-finite measure space and let $\mathcal{F} \subset \mathcal{A}$ be a σ-subalgebra. For each μ-integrable $\mathcal{A}$-measurable function $\varphi: X \to \mathbb{R}$ there exists an $\mathcal{F}$-measurable function $\varphi_{\mathcal{F}}: X \to \mathbb{R}$ such that*

$$\int_B \varphi_{\mathcal{F}} \, d\mu = \int_B \varphi \, d\mu \quad \text{for every} \quad B \in \mathcal{F}.$$

The function $\varphi_{\mathcal{F}}$ is called a *conditional expectation* of φ with respect to $\mathcal{F}$. See Proposition 2.9 for a proof of Theorem A.6 in the case of finite measure spaces.

Theorem A.7 (Increasing martingale theorem [69]). *If $\varphi: X \to \mathbb{R}$ is a μ-integrable function, and $\mathcal{F}_n$ is a nondecreasing sequence of σ-algebras such that the union $\bigcup_{n=1}^{\infty} \mathcal{F}_n$ generates the σ-algebra $\mathcal{F}$, then $\varphi_{\mathcal{F}_n} \to \varphi_{\mathcal{F}}$ μ-almost everywhere and*

$$\int_X |\varphi_{\mathcal{F}_n} - \varphi_{\mathcal{F}}| \, d\mu \to 0 \quad \text{when} \quad n \to \infty.$$

A.6   Product Spaces

Now let $(X, \mathcal{A}, \mu)$ and $(Y, \mathcal{B}, \nu)$ be measure spaces. We consider the *product space* $(X \times Y, \mathcal{C}, \mu \times \nu)$, where $\mathcal{C}$ is the σ-algebra generated by the family of sets

$$\mathcal{A} \times \mathcal{B} = \{A \times B : A \in \mathcal{A} \text{ and } B \in \mathcal{B}\}$$

and where $\mu \times \nu$ is the measure in $\mathcal{C}$ such that

$$(\mu \times \nu)(A \times B) = \mu(A)\nu(B)$$

for every $A \times B \in \mathcal{A} \times \mathcal{B}$. The measure $\mu \times \nu$ is called a *product measure*.

Theorem A.8 (Fubini). *If $\varphi \in L^1(X \times Y, \mu \times \nu)$, then:*

1. *The function $x \mapsto \varphi(x, y)$ is μ-integrable for ν-almost every $y \in Y$.*
2. *The function $y \mapsto \varphi(x, y)$ is ν-integrable for μ-almost every $x \in X$.*
3. *We have*

$$\int_{X \times Y} \varphi \, d(\mu \times \nu) = \int_Y \left[\int_X \varphi(x, y) \, d\mu(x) \right] d\nu(y)$$

$$= \int_X \left[\int_Y \varphi(x, y) \, d\nu(y) \right] d\mu(x).$$

References

1. R. Adler, A. Konheim, M. McAndrew, Topological entropy. Trans. Am. Math. Soc. **114**, 309–319 (1965)
2. R. Adler, B. Weiss, Similarity of automorphisms of the torus. Mem. Am. Math. Soc. **98** (1970)
3. D. Anosov, Roughness of geodesic flows on compact Riemannian manifolds of negative curvature. Dokl. Akad. Nauk SSSR **145**, 707–709 (1962)
4. D. Anosov, Geodesic flows on closed Riemann manifolds with negative curvature. Proc. Steklov Inst. Math. **90**, 1–235 (1967)
5. D. Anosov, Ya. Sinai, Certain smooth ergodic systems. Russ. Math. Surv. **22**, 103–167 (1967)
6. M. Artin, B. Mazur, On periodic points. Ann. Math. (2) **81**, 82–99 (1965)
7. L. Barreira, A non-additive thermodynamic formalism and applications to dimension theory of hyperbolic dynamical systems. Ergod. Theor. Dyn. Syst. **16**, 871–927 (1996)
8. L. Barreira, *Dimension and Recurrence in Hyperbolic Dynamics*. Progress in Mathematics, vol. 272 (Birkhäuser, Basel, 2008)
9. L. Barreira, *Thermodynamic Formalism and Applications to Dimension Theory*. Progress in Mathematics, vol. 294 (Birkhäuser, Basel, 2011)
10. L. Barreira, Ya. Pesin, *Lyapunov Exponents and Smooth Ergodic Theory*. University Lecture Series, vol. 23 (American Mathematical Society, RI, 2002)
11. L. Barreira, Ya. Pesin, *Nonuniform Hyperbolicity*. Encyclopedia of Mathematics and Its Applications, vol. 115 (Cambridge University Press, London, 2007)
12. L. Barreira, Ya. Pesin, J. Schmeling, Dimension and product structure of hyperbolic measures. Ann. Math. (2) **149**, 755–783 (1999)
13. L. Barreira, B. Saussol, Hausdorff dimension of measures via Poincaré recurrence. Comm. Math. Phys. **219**, 443–463 (2001)
14. L. Barreira, J. Schmeling, Sets of "non-typical" points have full topological entropy and full Hausdorff dimension. Isr. J. Math. **116**, 29–70 (2000)
15. G. Birkhoff, Proof of the ergodic theorem. Proc. Acad. Sci. USA **17**, 656–660 (1931)
16. M. Boshernitzan, Quantitative recurrence results. Invent. Math. **113**, 617–631 (1993)
17. H. Bothe, The Hausdorff dimension of certain solenoids. Ergod. Theor. Dyn. Syst. **15**, 449–474 (1995)
18. R. Bowen, Markov partitions for Axiom A diffeomorphisms. Am. J. Math. **92**, 725–747 (1970)
19. R. Bowen, Topological entropy and axiom A. In *Global Analysis* (Proc. Sympos. Pure Math. XIV, Berkeley, 1968) (American Mathematical Society, RI, 1970), pp. 23–41
20. R. Bowen, *Equilibrium States and Ergodic Theory of Anosov Diffeomorphisms*. Lecture Notes in Mathematics, vol. 470 (Springer, Berlin, 1975)
21. R. Bowen, Hausdorff dimension of quasi-circles. Inst. Hautes Études Sci. Publ. Math. **50**, 259–273 (1979)

22. L. Breiman, The individual ergodic theorem of information theory. Ann. Math. Stat. **28**, 809–811 (1957)
23. M. Brin, G. Stuck, *Introduction to Dynamical Systems* (Cambridge University Press, London, 2002)
24. G. Choe, *Computational Ergodic Theory*. Algorithms and Computation in Mathematics, vol. 13 (Springer, Berlin, 2005)
25. P. Collet, J. Lebowitz, A. Porzio, The dimension spectrum of some dynamical systems. J. Stat. Phys. **47**, 609–644 (1987)
26. I. Cornfeld, S. Fomin, Ya. Sinai, *Ergodic Theory*. Grundlehren der mathematischen Wissenchaften, vol. 245 (Springer, Berlin, 1982)
27. E. Dinaburg, On the relations among various entropy characteristics of dynamical systems. Math. USSR-Izv. **5**, 337–378 (1971)
28. A. Douady, J. Oesterlé, Dimension de Hausdorff des attracteurs. C. R. Acad. Sc. Paris **290**, 1135–1138 (1980)
29. K. Falconer, *The Geometry of Fractal Sets*. Cambridge Tracts in Mathematics, vol. 85 (Cambridge University Press, London, 1986)
30. K. Falconer, The Hausdorff dimension of self-affine fractals. Math. Proc. Camb. Phil. Soc. **103**, 339–350 (1988)
31. K. Falconer, Dimensions and measures of quasi self-similar sets. Proc. Am. Math. Soc. **106**, 543–554 (1989)
32. K. Falconer, Bounded distortion and dimension for non-conformal repellers. Math. Proc. Camb. Phil. Soc. **115**, 315–334 (1994)
33. K. Falconer, *Fractal Geometry. Mathematical Foundations and Applications* (Wiley, NY, 2003)
34. T. Goodman, Relating topological entropy and measure entropy. Bull. Lond. Math. Soc. **3**, 176–180 (1971)
35. T. Goodman, Maximal measures for expansive homeomorphisms. J. Lond. Math. Soc. (2) **5**, 439–444 (1972)
36. L. Goodwyn, Topological entropy bounds measure-theoretic entropy. Proc. Am. Math. Soc. **23**, 679–688 (1969)
37. J. Hadamard, Les surfaces à courbures opposées et leur lignes géodesiques. J. Math. Pure. Appl. **4**, 27–73 (1898)
38. T. Halsey, M. Jensen, L. Kadanoff, I. Procaccia, B. Shraiman, Fractal measures and their singularities: The characterization of strange sets. Phys. Rev. A (3) **34**, 1141–1151 (1986); errata in **34**, 1601 (1986)
39. B. Hasselblatt, Regularity of the Anosov splitting and of horospheric foliations. Ergod. Theor. Dyn. Syst. **14**, 645–666 (1994)
40. M. Hirsch, C. Pugh, Stable manifolds and hyperbolic sets. In *Global Analysis* (Proc. Sympos. Pure Math. XIV, Berkeley, 1968) (American Mathematical Society, RI, 1970), pp. 133–163
41. H. Hu, Box dimensions and topological pressure for some expanding maps. Comm. Math. Phys. **191**, 397–407 (1998)
42. M. Irwin, *Smooth Dynamical Systems* (Academic Press, NY, 1980)
43. A. Katok, Lyapunov exponents, entropy and periodic orbits for diffeomorphisms. Inst. Hautes Études Sci. Publ. Math. **51**, 137–173 (1980)
44. A. Katok, B. Hasselblatt, *Introduction to the Modern Theory of Dynamical Systems*. Encyclopedia of Mathematics and Its Applications, vol. 54 (Cambridge University Press, London, 1995)
45. G. Keller, *Equilibrium States in Ergodic Theory*. London Mathematical Society Student Texts, vol. 42 (Cambridge University Press, London, 1998)
46. A. Khinchin, On the basic theorems of information theory. Uspehi Mat. Nauk (N.S.) **11**, 17–75 (1956)
47. A. Khinchin, *Mathematical Foundations of Information Theory* (Dover, NY, 1957)
48. B. Kitchens, *Symbolic Dynamics, One-Sided, Two-Sided and Countable State Markov Shifts*. Universitext (Springer, Berlin, 1998)

49. A. Kolmogorov, A new metric invariant of transient dynamical systems and automorphisms in Lebesgue spaces. Dokl. Akad. Nauk SSSR (N.S.) **119**, 861–864 (1958)
50. A. Kolmogorov, Entropy per unit time as a metric invariant of automorphisms. Dokl. Akad. Nauk SSSR **124**, 754–755 (1959)
51. U. Krengel, *Ergodic Theorems* (de Gruyter, Berlin, 1985)
52. N. Kryloff, N. Bogoliouboff, La théorie générale de la mesure dans son application à l'étude des systèmes dynamiques de la mécanique non linéaire. Ann. Math. (2) **38**, 65–113 (1937)
53. H. Lebesgue, Intégrale, longueur, aire. Ann. Mat. Pura Appl. (3) **7**, 231–359 (1902)
54. F. Ledrappier, L.-S. Young, The metric entropy of diffeomorphisms II. Relations between entropy, exponents and dimension. Ann. Math. (2) **122**, 540–574 (1985)
55. D. Lind, B. Marcus, *An Introduction to Symbolic Dynamics and Coding* (Cambridge University Press, London, 1995)
56. A. Lopes, The dimension spectrum of the maximal measure. SIAM J. Math. Anal. **20**, 1243–1254 (1989)
57. R. Mañé, *Ergodic Theory and Differentiable Dynamics*. Ergebnisse der Mathematik und ihrer Grenzgebiete, vol. 8 (Springer, Berlin, 1987)
58. P. Mattila, *Geometry of Sets and Measures in Euclidean Spaces. Fractals and Rectifiability*. Cambridge Studies in Advanced Mathematics, vol. 44 (Cambridge University Press, London, 1995)
59. H. McCluskey, A. Manning, Hausdorff dimension for horseshoes. Ergod. Theor. Dyn. Syst. **3**, 251–260 (1983)
60. B. McMillan, The basic theorems of information theory. Ann. Math. Stat. **24**, 196–219 (1953)
61. M. Misiurewicz, A short proof of the variational principle for a Z_+^N action on a compact space. In *International Conference on Dynamical Systems in Mathematical Physics*, Rennes, 1975. Astérisque, vol. 40 (Soc. Math. France, Montrouge, 1976), pp. 147–157
62. P. Moran, Additive functions of intervals and Hausdorff measure. Proc. Camb. Phil. Soc. **42**, 15–23 (1946)
63. M. Morse, A one-to-one representation of geodesics on a surface of negative curvature. Am. J. Math. **43**, 33–51 (1921)
64. M. Morse, G. Hedlund, Symbolic dynamics. Am. J. Math. **60**, 815–866 (1938)
65. Z. Nitecki, *Differentiable Dynamics* (MIT Press, MA, 1971)
66. D. Ornstein, *Ergodic Theory, Randomness, and Dynamical Systems*. Yale Mathematical Monographs, vol. 5 (Yale University Press, CT, 1974)
67. D. Ornstein, B. Weiss, Entropy and data compression schemes. IEEE Trans. Inform. Theor. **39**, 78–83 (1993)
68. J. Palis, M. Viana, On the continuity of Hausdorff dimension and limit capacity for horseshoes. In *Dynamical Systems* (Valparaiso, 1986), ed. by R. Bamón, R. Labarca, J. Palis. Lecture Notes in Mathematics, vol. 1331 (Springer, Berlin, 1988), pp. 150–160
69. W. Parry, *Topics in Ergodic Theory* (Cambridge University Press, London, 1981)
70. W. Parry, M. Pollicott, *Zeta Functions and the Periodic Orbit Structure of Hyperbolic Dynamics*. Astérisque, vol. 187–188 (Soc. Math. France, Montrouge, 1990)
71. O. Perron, Die Stabilitätsfrage bei Differentialgleichungen. Math. Z. **32**, 703–728 (1930)
72. Ya. Pesin, Characteristic exponents and smooth ergodic theory. Russ. Math. Surv. **32**, 55–114 (1977)
73. Ya. Pesin, *Dimension Theory in Dynamical Systems. Contemporary Views and Applications*. Chicago Lectures in Mathematics (Chicago University Press, IL, 1997)
74. Ya. Pesin, B. Pitskel', Topological pressure and the variational principle for noncompact sets. Funct. Anal. Appl. **18**, 307–318 (1984)
75. Ya. Pesin, H. Weiss, On the dimension of deterministic and random Cantor-like sets, symbolic dynamics, and the Eckmann–Ruelle conjecture. Comm. Math. Phys. **182**, 105–153 (1996)
76. Ya. Pesin, H. Weiss, A multifractal analysis of Gibbs measures for conformal expanding maps and Markov Moran geometric constructions. J. Stat. Phys. **86**, 233–275 (1997)
77. K. Petersen, *Ergodic Theory* (Cambridge University Press, London, 1983)

78. H. Poincaré, Sur le problème des trois corps et les équations de la dynamique. Acta Math. **13**, 1–270 (1890)
79. M. Pollicott, M. Yuri, *Dynamical Systems and Ergodic Theory*. London Mathematical Society Student Texts, vol. 40 (Cambridge University Press, London, 1998)
80. D. Rand, The singularity spectrum $f(\alpha)$ for cookie-cutters. Ergod. Theor. Dyn. Syst. **9**, 527–541 (1989)
81. C. Robinson, *Dynamical Systems. Stability, Symbolic Dynamics, and Chaos*. Studies in Advanced Mathematics (CRC Press, FL, 1995)
82. D. Rudolph, *Fundamentals of Measurable Dynamics: Ergodic Theory on Lebesgue Spaces* (Oxford University Press, London, 1990)
83. D. Ruelle, Statistical mechanics on a compact set with $\mathbb{Z}^\nu$ action satisfying expansiveness and specification. Trans. Am. Math. Soc. **185**, 237–251 (1973)
84. D. Ruelle, *Thermodynamic Formalism*. Encyclopedia of Mathematics and Its Applications, vol. 5 (Addison-Wesley, MA, 1978)
85. D. Ruelle, Repellers for real analytic maps. Ergod. Theor. Dyn. Syst. **2**, 99–107 (1982)
86. J. Schmeling, On the completeness of multifractal spectra. Ergod. Theor. Dyn. Syst. **19**, 1595–1616 (1999)
87. C. Shannon, A mathematical theory of communication. Bell Syst. Tech. J. **27**, 379–423, 623–656 (1948)
88. M. Shereshevsky, A complement to Young's theorem on measure dimension: The difference between lower and upper pointwise dimension. Nonlinearity **4**, 15–25 (1991)
89. M. Shub, *Global Stability of Dynamical Systems* (Springer, Berlin, 1986)
90. K. Simon, The Hausdorff dimension of the Smale–Williams solenoid with different contraction coefficients. Proc. Am. Math. Soc. **125**, 1221–1228 (1997)
91. K. Simon, B. Solomyak, Hausdorff dimension for horseshoes in $\mathbb{R}^3$. Ergod. Theor. Dyn. Syst. **19**, 1343–1363 (1999)
92. D. Simpelaere, Dimension spectrum of axiom A diffeomorphisms. II. Gibbs measures. J. Stat. Phys. **76**, 1359–1375 (1994)
93. Ya. Sinai, On the concept of entropy for a dynamic system. Dokl. Akad. Nauk SSSR **124**, 768–771 (1959)
94. Ya. Sinai, Construction of Markov partitions. Funct. Anal. Appl. **2**, 245–253 (1968a)
95. Ya. Sinai, Markov partitions and C-diffeomorphisms. Funct. Anal. Appl. **2**, 61–82 (1968b)
96. Ya. Sinai, *Introduction to Ergodic Theory*. Mathematical Notes, vol. 18 (Princeton University Press, NJ, 1976)
97. Ya. Sinai, *Topics in Ergodic Theory*. Princeton Mathematical Series, vol. 44 (Princeton University Press, NJ, 1994)
98. S. Smale, Diffeomorphisms with many periodic points. In *Differential and Combinatorial Topology* (A Symposium in Honor of Marston Morse) (Princeton University Press, NJ, 1965), pp. 63–80
99. S. Smale, Differentiable dynamical systems. Bull. Am. Math. Soc. **73**, 747–817 (1967)
100. W. Szlenk, *An Introduction to the Theory of Smooth Dynamical Systems* (Wiley, NY, 1984)
101. F. Takens, Limit capacity and Hausdorff dimension of dynamically defined Cantor sets. In *Dynamical Systems* (Valparaiso, 1986), ed. by R. Bamón, R. Labarca, J. Palis. Lecture Notes in Mathematics, vol. 1331 (Springer, Berlin, 1988), pp. 196–212
102. P. Walters, A variational principle for the pressure of continuous transformations. Am. J. Math. **97**, 937–971 (1976)
103. P. Walters, *An Introduction to Ergodic Theory*. Graduate Texts in Mathematics, vol. 79 (Springer, Berlin, 1982)
104. H. Weyl, Über der Gleichverteilung von Zahlen mod Eins. Math. Ann. **77**, 313–352 (1916)
105. L.-S. Young, Dimension, entropy and Lyapunov exponents. Ergod. Theor. Dyn. Syst. **2**, 109–124 (1982)

Index

L. Barreira, *Ergodic Theory, Hyperbolic Dynamics and Dimension Theory*, Universitext,
DOI 10.1007/978-3-642-28090-0, © Springer-Verlag Berlin Heidelberg 2012